W0225697

New Advances in Distributed Computer Systems

NATO ADVANCED STUDY INSTITUTES SERIES

Proceedings of the Advanced Study Institute Programme, which aims at the dissemination of advanced knowledge and the formation of contacts among scientists from different countries

The series is published by an international board of publishers in conjunction with NATO Scientific Affairs Division

A	Life Sciences	Plenum Publishing Corporation
B	Physics	London and New York
C	Mathematical and Physical Sciences	D. Reidel Publishing Company Dordrecht, Boston and London
D	Behavioural and Social Sciences	Sijthoff & Noordhoff International Publishers
E	Applied Sciences	Alphen aan den Rijn and Germantown U.S.A.

Series C – Mathematical and Physical Sciences

Volume 80 – New Advances in Distributed Computer Systems

New Advances in Distributed Computer Systems

Proceedings of the NATO Advanced Study Institute held at Bonas, France, June 15-26, 1981

edited by

KENNETH G. BEAUCHAMP
University of Lancaster, England

Springer-Science+Business Media, B.V.

Library of Congress Cataloging in Publication Data

NATO Advanced Study Institute (1981 : Bonas, France)
New advances in distributed computer systems.

(NATO Advanced Study Institute series. Series C, Mathematical and physical sciences ; v. 80)
"Published in cooperation with NATO Scientific Affairs Division."
Includes index.
1. Electronic data processing–Distributed processing–Congresses. 2. Computer networks–Congresses. I. Beauchamp, K. G. II. North Atlantic Treaty Organization. Scientific Affairs Division. III. Title. IV. Series.
QA76.9.D5N37 1981 001.64 81-19895

ISBN 978-94-009-7762-4 ISBN 978-94-009-7760-0 (eBook)
DOI 10.1007/978-94-009-7760-0 AACR2

Supported by
U.S. Army European Research Office (USARDSG - U.K.)

The views, opinions and/or findings in these proceedings are those of the authors and should not be construed as an official Department of the U.S. Army position, policy or decision unless so designated by other documentation.

Originally published by D. Reidel Publishing Company in 1982
Softcover reprint of the hardcover 1st edition 1982

TABLE OF CONTENTS

Part 5. SECURITY AND ENCRYPTION

Part 6. PARALLEL COMPUTING ARCHITECTURE

FOREWORD

This volume contains the papers presented at the NATO Advanced Study Institute of New Advances in Distributed Computer Systems held between 15th and 26th June, 1981 at the Château de Bonas, France.

The aim of the meeting was to promote an interchange of ideas between experts in the interlinked fields of communications and computers in order to determine the essential areas for future development. Its programme was arranged to explore a number of current topics including the public data-communication networks set up by the PTTs or corresponding bodies in various countries, large-scale non-public systems such as ARPANET and its latest developments, international systems such as the airlines' SITA network, the recent and very important developments in local area networks and relevant developments by universities and other higher educational bodies. The recent moves towards formalisation and the laying down of a theoretical basis to guide future developments and standards were discussed with particular reference to the International Standards Organisation "7-layer model for Open System Interconnection" and the development of formal mathematical methods for specifying and analysing communication systems and their protocols. Consideration was also given to the theoretical techniques, and their practical realisation, now becoming available to ensure privacy and security of information transmitted over digital communication systems. Finally the penetration of the concepts of distributed processing into the domain of computer architecture, giving such possibilities as array processors and other non-von Neumann architectures formed the subject of several of the sessions.

K. G. Beauchamp (ed.), New Advances in Distributed Computer Systems, ix–x.

The scope and depth of the papers presented in this volume are an indication of the success of the Institute in meeting this aim and it is hoped that they will provide a valuable contribution to the literature in distributed computer systems.

This Institute was sponsored and financed by the Scientific Affairs Division of the North Atlantic Treaty Organisation. Additional funds were provided by the European Research Office of the U.S. Army.

The Editor would like to acknowledge this support together with the help of his co-director Dr. Howlett, Professor and Madam Simon of ASCEB and many others who assisted in the organisation of the meeting.

Finally, thanks are due to the authors of the many papers who have provided this extremely valuable compilation.

Lancaster, August 1981. K.G. Beauchamp

INTRODUCTORY SURVEY

Jack Howlett

Consultant ICL, Putney, London, U.K.

1. Three years ago, at the end of August 1978, Dr Beauchamp and I were co-Directors of a NATO Advanced Study Institute, also here at Bonas, with the title of "Interlinking of Computer Networks". Quite a number of those who took part in that meeting are here to-day and we have the pleasant feeling of meeting old friends again. The proceedings of that meeting form Volume C-32 in the NATO ASI series, and there are copies in the library here; the library, incidentally, has complete runs of the Series B (Physics) and Series C (Mathematical and Physical Sciences) Proceedings, and a selection from the more biologically-oriented Series A.

The program of the 1978 meeting covered a good deal more than the title would suggest and included accounts of the then status of some of the world's major information projects such as ARPANET, TRANSPAC and the Canadian INFOSWITCH and DATAPAC. This meeting is a successor to that; a very great deal has happened in the intervening three years and we chose the new title because of the great broadening of the field. We regret the use of the words "computer" and "computing" in these titles because they still seem to carry with them the ideas of arithmetic and scientific computing, whereas we are concerned with the much more fundamental and all-pervading concept of general information processing. But the words seem to be here to stay.

2. The phrase "convergence of computing and communications" has become a cliché but it expresses a real truth and a process of the greatest importance; and underlies the whole of this ASI. The papers and discussions in the programme fall into three broad classes:

K. G. Beauchamp (ed.), New Advances in Distributed Computer Systems, 1–8.

A. Concerned primarily with communications and dealing with networks of various types and purposes which have already been established or are being developed: their organisation, the standards to which they adhere and the services which they provide.

B. Concerned with systems combining communications and computing equipment and dealing with the interlinking of physically distributed information stores (for example, databases) and processing resources (for example, mini- or micro-computers).

C. Concerned primarily with computers and dealing with the departures from the classical von Neumann architecture made possible by recent technological advances, for example the achievement of truly parallel processing. The concept of "distributed processing" is relevant here because the practical realisation of these new architectures involves the distribution of processing power, possibly in very small units, throughout the whole system.

We do not have papers on hardware or physical technologies as such, for example on micro-electronics, LSI/VISI, optical fibre transmission or communication satellites. But considerations of these enter into many of the papers and we must never forget that developments in these fields have made all the other advances possible, practically and economically, and are still going on at breathtaking rates. My own feeling is that we are in almost a different world from that of the 1978 meeting, even though that was only three years ago.

This short introductory survey is intended as a setting of the scene for the ASI; what I shall now do is to take this broad division of the field and give what seem to me to be the important developments in the three classes: a personal view.

3. Advances in Communications

3.1 The need for public switched data-transmission networks is now completely accepted and in most advanced countries the PIT's or the equivalent bodies either have already implemented systems or are in the process of doing so; most of these use packet switching. In France, for example, TRANSPAC has proved a great success and is being expanded; in Britain the Post Office (now British Telecom), using the experience gained with the Experimental Packet Switched Service (EPSS), has developed and is now bringing into service its Packet Switched Service (PSS).

3.2 There is universal agreement also that the achievement of Open System Interconnection (OSI) is the goal to be striven for; the ultimate, corresponding to the international telephone service now available, is that the user of any piece of data-handling equipment in any part of the world shall be able, given permission, to communicate with any other by means of simple and standard procedures. Along with this has gone the agreement first that this is a long-term objective and second that if it is to be achieved in anything approaching an orderly and economical manner there must be agreement on a framework within which communications systems are designed - usually expressed as an "architectural" basis for design.

Important progress has been made in the formulation and understanding of the principle of a "layered architecture" for communications systems. In this the protocol for communication between two terminals, such as a user entering data at a simple terminal and having it processed by a program running on a distant computer, is sub-divided into a set of "layers", stacked on top of one another. Each layer of protocol is concerned with a precisely-defined part of the total process of the communication and with that only, and any coupling is only between protocols in adjacent layers. The International Standards Organisation (ISO) has proposed a division into seven specific functions which has become known as the ISO 7-layer Model. Intensive and searching discussion is going on over the details of this model, but what I feel is of very great importance is that there is a great measure of acceptance of the model as a whole as a universal standard, and that good progress is being made on agreement on details.

3.3 The importance of standards in data communication has been realised for a long time and a great deal of work has gone into identifying the areas in which standards should be applied and what these standards should be. The CCITT has made important recommendations for standards in public packet-switched digital data networks (the X-series of recommendations) and the past three years have seen acceptance of two in particular which are of very great importance:

X 25 specifying the interface between the terminal (DTE) and the network (DCE); this is already widely implemented in actual equipment on the market

X 75 specifying the interface between two public packet-switched networks

In every technical field, standards help everyone; they make life easier for the users and widen the market for the suppliers. The great danger in a new field, especially one so complex and so fast-developing as information processing and communication, is of attempting to enforce standards too early, before the real issues have become properly understood. It does seem that we have built up enough understanding in this field now to be able to formulate standards with confidence and to be reasonably certain that they will not put its development into a strait-jacket.

3.4 Ultimately we depend on manufacturers to supply the equipment we need and there is no point in specifying formal models and standards unless manufacturers are able and willing to embody these in their products. A most encouraging feature of the past few years has been the announcement by all the main manufacturers of intentions to do just this. All manufacturers are already, or soon will be, offering X-25 interface with their data terminals and there have been a number of proprietary architectures, including Honeywell's DSA and ICL's IPA, based on the ISO 7-layer model. We shall be hearing during the ASI of the relation of IBM's SNA to this.

A very effective way of getting a standard accepted is for a powerful purchasing body - the U.S.A. Department of Defense, for example - to include it in its relevant contractual conditions. I understand that the EEC is taking a strong interest in standards in the information processing field and is considering bringing them, at appropriate moments, into its contract terms.

3.5 The idea of the Integrated Services Data Network (ISDN) is gaining ground, meaning the provision of a variety of non-voice services over a public telecommunication system. Elaborate services over private networks using lines leased from public carriers have of course been in operation for some time, perhaps the best known being the very extensive SWIFT - Society for Worldwide Inter-bank Financial Telecommunications - network. Of the new public services, Viewdata in its various forms such as Prestel in Britain is now well established; others which are being developed include computing and data-processing services using equipment linked to the network and operated by the communications authority, and electronic mail.

3.6 On a very different line, but most important in my view, is the recent questioning of the role, powers and privileges of the national communications authorities such as the Post

Office in Britain and the PTTs in continental Europe. These have been either actually or effectively Government Departments and have held powerful monopolies over the whole field of their operations, including important areas of equipment supply; and have enjoyed strong legal protection. Things are different in North America but even there, where the providers of telecommunication services are competing private companies, strong constraints are imposed by the governments. Whatever view one takes of the desirability or otherwise of the monopolies held up to now by the PO and the PTTs, it is reasonable to question this in the circumstances of to-day, when technology has made such immense changes and advances and when the demands for information transmission and processing are orders of magnitude greater than they were in the early days of the public telephone systems. In Britain, the Post Office has already been split into two parts, British Telecom dealing with telecommunications and the rest with ordinary mail and the many other services such as banking which the Post Office has supplied; and legislation is being drafted which will almost certainly remove some of the monopolistic powers. These questions are very serious indeed, because it has become clear that a modern state is highly dependent on its public telecommunications services. It is therefore most important that they are studied seriously and as objectively as possible and decisions made not just on ideological grounds.

4. Advances in the Communications - Computers Combination.

4.1 The integration of data-processing and other non-voice service into public telecommunications systems is an example, and has been dealt with in para. 3.5 above.

4.2 A striking development of the last few years is that of the local Area Network (LAN). It is now a thoroughly practical and economic possibility to provide a very fast (10 Mbit/sec and more) digital transmission system for something on the scale of a laboratory or office building and to use this to link small, fast, sophisticated and cheap "personal" computers, storage units, visual display units, printers and other digitally-driven equipment. Two names for systems of this type, the Cambridge Ring and the Ethernet, representing respectively ring and straight line topology, have become well known and are discussed at this ASI. In Britain the Science & Engineering Research Council is supplying packaged ring systems to a number of universities and will take much interest in the experience of the users. The falling cost of hardware and the potentialities for greatly simplifying the software are certain to have important effects on suppliers and users, and therefore on the general computing market.

Questions of standards will have to be considered here, to allow for the need, which will certainly arise, for independent LANs to communicate with one another through public networks, and to access services provided on public networks themselves.

4.3 In Britain again the Science Research Council (now Science & Engineering Research Council and one of the 5 Research Councils financed by the government and charged with the responsibility for supporting and stimulating research in broadly-specified fields) initiated a Distributed Computing Systems Programme in 1977. To quote from an SRC document, "The primary objectives are to seek an understanding of the principles of Distributed Computing Systems and to establish the engineering techniques necessary to implement such systems effectively. In particular, this requires an understanding of the implications of parallelism in information processing systems and storage, and devising means for taking advantage of this capability". The programme has been most successful and at present has stimulated and is supporting nearly 50 separate research projects in British universities. Many of these relate to multi-microprocessor projects in which the processing and communications elements are intimately linked.

4.4 A sound theoretical basis is necessary in any technical activity, to enable equipment to be designed to meet specified requirements and to make meaningful analyses of observed performance. The mathematical-statistical treatment of telephone traffic has been developed over the years to a high level of sophistication and the techniques can be taken over to study many problems in digital communication systems. But the formal specification and analysis of protocols and interface conditions, and of parallel distributed systems, require a different type of mathematics and advances in this direction have only recently been made. The SEBC Distributed Computing Systems programme, already referred to, has stimulated much research in this field in the U.K.

4.5 There is now a sharp awareness in all countries that the very large amount of information held in computer-based systems and available for transmission over telecommunication networks is often valuable or sensitive or both and that there should be strong safeguards against unauthorised access. Some very powerful cryptographic processes have been developed during the past few years, aimed at protecting the privacy and security of digitally-stored and transmitted information, about which we shall hear during this ASI.

5. Advances in Computer Architecture

5.1 Modern technology has made possible the production of physically small, fast, simple and cheap processors in very large numbers and of fast, compact and cheap stores in units of from 4K to 64K bits, also in very large numbers. This has made it realistic to consider radical departures from the classical von Neumann architecture of a computer in which there is a clear and physical distinction between the processor and the store. One can now envisage a computer in which there are possibly very many processing and storage units with almost any form of interlinking. The key problem would be how to control such an assembly so as actually to bring its potentially great processing power to bear on any particular problem.

5.2 It is accepted that ultimately the achievement of very high processing power must involve parallelism in some form. The simplest realisation of this is the Single Instruction, Multiple Datastream (SIMD) architecture, in which at each beat the same instruction is fed to every one of the processors forming the system and each operates with this on its own item(s) of data. The pioneering machine of this type is the ILLIAC-4, designed and built in the University at Urbana in the late 1960's and installed in the Ames Air Force Base in California. This was built before integrated circuitry was in production and therefore was constrained by the discrete-component technology of its time. It has 64 processors, each quite a powerful machine in itself, and presented considerable problems of control and data routing. The machine with which I am most familiar is the ICL Distributed Array Processor (DAP) which, taking advantage of the possibilities of modern technology, has 4096 processors - which can be regarded as an array of 64 x 64 - each very simple, each with its own store of 4K bits and each connected to its four nearest neighbours. Several have been built and experience now gained shows that such an architecture gives a very powerful and flexible machine, well within the scope of to-day's technology.

Other architectures which are now being studied include the "data-flow" form; here again there are many simple processing elements but they have more autonomy than in the SIMD form and communication between them is more variable. The SERC Distributed Computing Systems programme is supporting studies in this field.

I find it most significant that a recent Japanese official report, the "Interim Report on Research and Development on 5th Generation Computers", which gives a full and exceedingly interesting discussion of the design criteria for new advanced computers, emphasises the need to move away from the classical von Neumann architecture.

5.3 The introduction of the electronic computer has had a profound effect on the techniques of numerical computation, quite apart from making it almost trivial to do calculations which are far beyond the powers of hand computation. It has led to the development of quite new methods and algorithms - and, to be fair, to the revival of some methods such as the Runge-Kutta and its variants for the integration of ordinary differential equations, which had been known for a long time but were ill-suited to hand computation. It has led also to much deeper studies in numerical analysis, such as convergence and stability of numerical processes and propagation of truncation and round-off errors in an extended calculation. It has already become clear that with a new tool such as an array processor it is profitable to look afresh at numerical problems and not simply to take over algorithms developed and refined for serial processors This can be intrinsically interesting from a purely mathematical point of view and it is my personal belief that these new developments in computer architecture will lead to new and interesting developments in numerical analysis.

DATA COMMUNICATION IN ISRAEL -- PRESENT AND FUTURE

David Biran

Senior Member, IEEE; Chief Scientist of the Ministry of Communication, ISRAEL.

ABSTRACT

Data Communication, which is the field that combines all the means of forwarding all kinds of data, will be, in the near future, one of the most important fields in the world and Israel.

This paper describes the situation in the Data Communication field in Israel, up to 1981, and the possible developments up to year 2000.

THE IMPACT OF DATA COMMUNICATION AND SCOPE

Data Communication is the link between the technologies of computing and man. Using Data Communication means that distance does not have any more influence on the capabilities and performances of man. The world has become a small place, where information passes immediately from one to another, and there are almost no borders to information.

Data Communication will be the channel through which Information Technology will affect virtually every household and occupation. It will change patterns of employment, create new jobs and new business possibilities.

The scope of this paper deals with all kinds of Data Communications which include:

* Terminal-Terminal
* Terminal-Computer

K. G. Beauchamp (ed.), New Advances in Distributed Computer Systems, 9–24.

* Computer-Computer

The present meaning of those terms, for Israel, are:

* Terminal-Terminal -- by Telegraph or Telex.

* Terminal-Computer -- by leased individual lines or by using multidrops, multiplexers or concentrators, without switching.

* Computer-Computer -- at present, computer-computer communication in Israel is almost non-existant.

In future, all these communication types will be gradually tested and established, using modern packet and circuit technologies.

COMPUTERS AND TERMINALS IN ISRAEL 1980

According to the 1980 survey [1] of the Central Bureau of Statistics, at the end of 1979, 1,088 computers and 4,929 terminals were installed in Israel in 682 establishments and institutions, covering most branches of the economy.

Computers

The annual growth of the number of various computers is given in Table A.

The size of computers can be characterized by the monthly rental and maintenance costs for a single shift.

It should be noted that increase or decrease in the number of computers by size sometimes results from changes in the size of an existing computer and not from an actual change in the number of computers.

Classification of the establishments possessing computers by firms, and number of computers in each firm, shows that in 75 percent of the firms there is one computer only, mostly (74 percent) a mini-computer. In 16 percent of the firms there are two computers in each, and in 10 percent of the firms there are more than two computers.

A firm is one economic unit, under one ownership (private establishment, partnership, stock company, etc.), that may include one establishment (then the definition of the firm is identical with that of the establishment) or several establishments. Every governmental ministry, with all its units, was considered for the purpose of this survey as one firm.

TABLE A. <u>NUMBER OF COMPUTERS AND ESTABLISHMENTS WITH COMPUTERS</u>

END OF YEAR	1966	1967	1968	1969	1970	1971	1972	1973	1974	1975	1976	1977	1978	1979
COMPUTERS - TOTAL[b]	48	66	85	139	162	206	275	321	378	439	490	610[a]	712	1,088
INCREASE OVER PRECEEDING YEAR	19	18	19	52	25	44	69	46	57	61	51	120	102[a]	376
ESTABLISHMENTS WITH COMPUTERS - TOTAL	35	50	64	96	118	147	187	215	246	281	314	407	478	682
INCREASE OVER PRECEEDING YEAR	14	15	14	32	22	29	40	28	31	29	33	93	71	204

(a) In 1978, part of the top desk calculators were not included in the number of mini-computers, due to failure to receive data from one of the firms that market this type of computers in Israel.

(b) The numbers do not include microprocessors.

TABLE B. COMPUTERS, BY SIZE

Size of Computer (by rental) and Maintenance in December		1970	1971	1972	1973	1974	1975	1976	1977	1978	1979
TOTAL		162	206	275	321	378	439	490	610	712	1,088
Giant	($75,000 and over)					4	5	6	9	10	13
Large	($30,000-$74,999)	34	44	52	52	10	17	19	21	24	23
Medium	($7,000-$29,999)					49	56	59	57	65	69
Small	($2,000-$6,999)	44	52	59	67	73	82	86	105	112	198
Mini	(less than $2,000)	84	110	164	202	242	279	320	418	501 [a]	785

TABLE C. COMPUTERS AND TERMINALS
BY TYPE OF TERMINAL, SIZE OF COMPUTER, ECONOMIC BRANCH, SECTOR AND DISTRICT

XII 1979

SIZE OF COMPUTERS ECONOMIC BRANCH SECTOR AND DISTRICT	Computers to which Terminals are Connected	TERMINALS TOTAL	Printing Communication Terminals	Display Terminals	R.J.E.	Other Types
TOTAL	381	4,929	929	3,637	287	76
Size of Computer						
Mini	186	1,321	234	1,000	55	32
Small	129	1,002	233	688	59	22
Medium	36	972	194	708	68	2
Large	20	726	160	508	38	20
Giant	10	908	108	733	67	-
Economic Branch						
Agriculture, Agricultural Services and Industry	91	562	132	404	6	20
Electricity, water and construction	4	72	5	45	2	20
Commerce, Restaurants and Hotels	39	214	50	145	15	4
Transport, Storage and Communication	22	643	105	534	4	-
Financial Institutions and Business Services	95	1,172	189	834	135	14
Public and Community Services	125	2,253	444	1,666	125	18
thereof:- Personal Services	5	13	4	9	-	-
Sector						
Private	224	2,259	486	1,559	177	37
Government	49	1,462	266	1,131	61	4
Histadrut	70	632	108	465	45	14
National	6	85	7	78	-	-
Local Authorities	30	404	32	367	4	1
Public Mixed	2	87	30	37	-	20
District						
Jerusalem	49	942	204	675	55	8
Southern	33	406	43	357	5	1
Tel-Aviv	185	2,618	375	2,009	187	47

TABLE D. TELEX - EXCHANGE CAPACITY, MACHINES CONNECTED AND TOTAL "PROPENSITY TO OWN A MACHINE"

31st March 1961 - 31st March 1980

	Exchanges	Exch.	"P R O P E N S I T Y"			Percent of Capacity	M A C H I N E S				Outstanding Applications For Machines		Annual Percent Increase	
31.3	(1)	Capacity	Total	Nos. Connected	Outstanding Applications for lines	Nos. Connected	Total "Propensity"	Total	Percent Roman Letters	**Density per 100,000**	Hebrew Letters	Roman Letters	Propensity	Machines
1961 (2)	5	446	..	263	..	60	..	280	..	12.9	..	..	..	27
1962	5	446	..	330	..	74	..	350	..	15.5	..	..	..	25
1975	12	3,380	2,320	2,195	125	65	69	2,170	69.1	63.1	59	86	12	12
1978	10	4,130	3,100	2,835	265	69	75	2,770	69.0	75.4	77	159	9	10
1979	10	4,480	3,415	3,115	300	69	76	3,020	70.9	80.4	69	198	10	9
1980	10	4,610	3,700	3,350	350	73	80	3,385	72.7	87.9	85	375	3	12

(1) The 4 exchanges within the Tel-Aviv city boundaries are counted as one unit.

(2) The inland Telex service was opened in August 1965.

Main Memory

The main memory capacity (work capacity within which the computer performs the processing) in 1,088 electronic computers, installed at the end of 1979, amounted to approximately 157,100 KB units, each KB unit being equal to one thousand bytes. The main memory's average capacity per computer was 144 KB.

Terminals

In December 1979 there were about 4,929 terminals connected to 381 computers, as compared with 2,500 terminals connected to 167 computers at the end of 1978. (Data do not include terminals connected to computers of the defence system, the defence industry, the aircraft industry and its subsidiary companies.)

The distribution of the terminals, according to the computer size, economic branches and districts, is described in Table C.

Employees

In December 1979 a total of some 8,000 persons were employed in the computer field.

TELEX

The telex network in Israel was developed in 1956 as an inland service and reached, in 1980, a total capacity of 4,610 lines with 3,385 machines connected. Tables D and E depict the situation.

In 1981-2 Israel intends to install a new exchange with about 1,000 lines capacity. This additional capacity is intended for changing the oldest exchanges, which are approximately twenty years old, and also for advancing increased capacity.

DATA COMMUNICATION LEASED LINES

The computer power and distribution in Israel requires a large quantity of lines between the terminals and the various computers. The picture and growth of the data leased lines (in numbers) is depicted in Tables F and G.

The policy of the Ministry of Communication is to give the data lines priority in installation. Since 1980, a tariff, sensitive to speed, was established.

TABLE E. INLAND AND INTERNATIONAL TELEX CALLS, 1960/61-1979/80

Year	Inland Calls (1000 Meter Pulses)	International Calls (1)							Annual Percent Increase		
		Thousand Calls			Avge. Length of Calls		Average Number of Calls Per Machine		Inland Calls (M.P.)	International Calls	
		Total	Out-going	In-coming	Out-going	In-coming	Inland (M.P.)	Inter-national (mins.)		Calls	Minutes
60/61	2,513	18.5	8.7	9.8	6.5	6.1	10,000	..	4	..	..
61/62	2,765	36.5	18.0	18.5	7.2	6.2	8,800	..	10	97	113
76/77	(18,600)	2,353.2	1,165.2	1,188.0	2.5	2.9	7,590	3,790	(-7)	14	12
77/78	(19,250)	2,737.2	1,387.2	1,350.0	2.6	2.9	7,285	4,032	4	16	15
78/79	23,000	(3,295.1)	1,701.1	(1,594.0)	2.5	(2.9)	8,470	(4,375)	(27)	(20)	(21)
79/80	24,000	..	2,011		2.4		7,500	..	4		

(1) The International Telex Service was opened in December 1959

TABLE F. INSTALLATION RATE OF DATA LINES IN ISRAEL

Bit-Rate Year	200	600/1200	2400	4800	7200	9600	Total
1969	17	1	2	....	....	....	20
1973	29	4	5	....	....	....	38
1974	66	24	30	32	1	11	164
1979	186	216	126	221	2	77	828
1980	248	294	220	308	2	153	1225
1981	294	522	565	426	11	228	2046

TABLE G. DATA LINES SITUATION IN ISRAEL - 1981

Bit-Rate Parameters	200	600/1200	2400	4800	2200	9600	T O T A L
District Lines	294	522	565	426	12	228	1739
Inter District	31	55	76	104	4	37	307

DISTRIBUTIONS BY INSTITUTIONS

Banks	1	726	136	104	..	18	485
Government	42	54	113	153	..	100	462
Universities	57	25	23	42		10	157
Others	194	217	250	127	..	100	942

Fig. 1. Telecommunications Services and Technologies Forecast in Israel

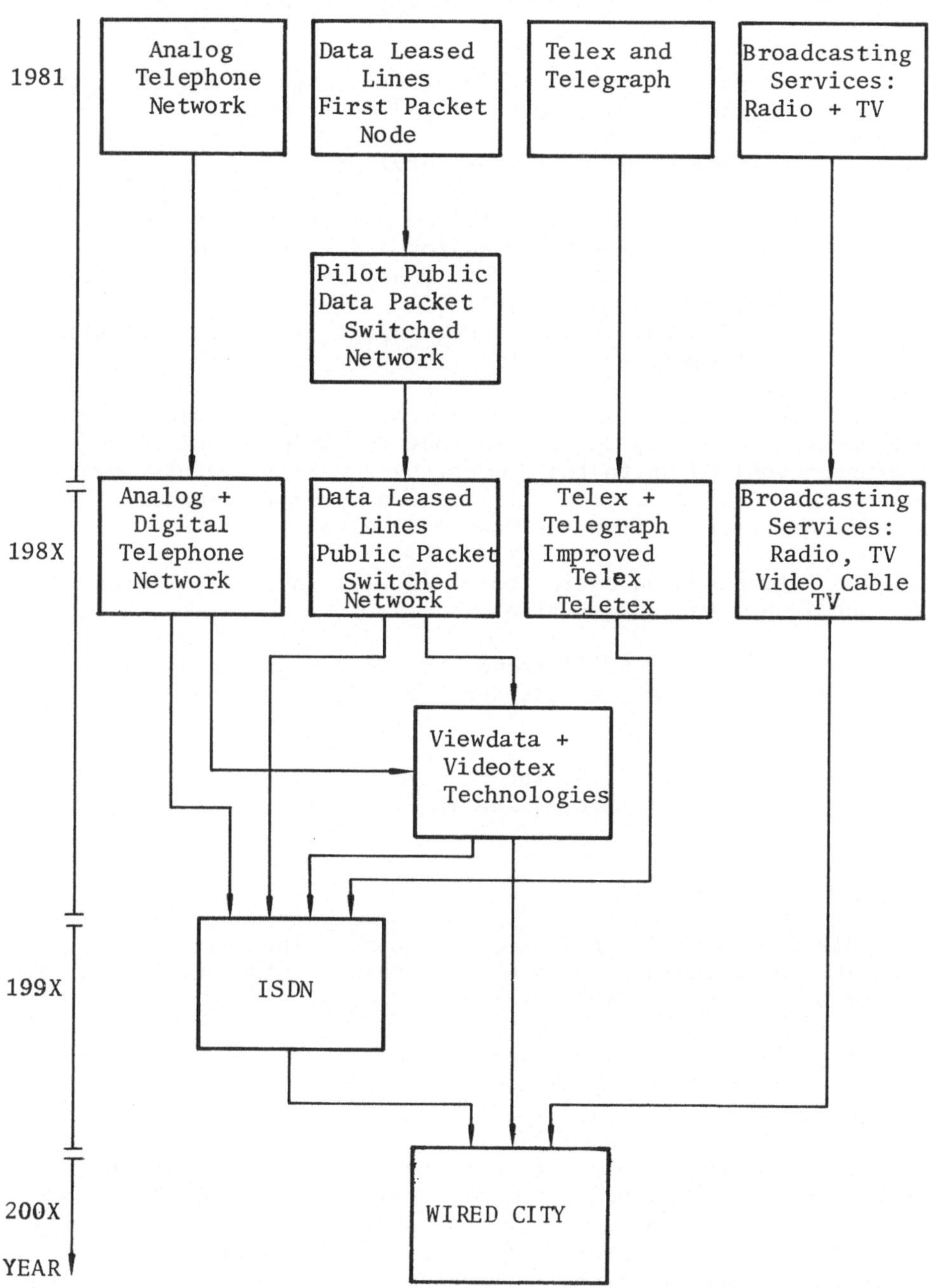

FUTURE COMMUNICATION SERVICES

Israel has started a long range planning policy with the intention to:

* Modernize the telecommunication system and start the installation of a totally digital network.

* Improve the quality of services.

* Supply all kinds of telecommunication services that are available today and/or are in the process of development in the most developed countries. Among those services, a selection will be made in order to decide upon the most convenient, in accordance with the need.

The need for an advanced communication system pushed the Government into making a decision that the Telecommunication Services--supplied up to the present by the Ministry of Communication as a civil service--will become a Government-owned company and will be separate from the Postal Services.

The future communication services will encompass: Voice, Data and in a later stage--Video.

According to the CCITT trends for an Integrated Service Digital Network (ISDN), a considerable effort will be invested in this area.

The development of the future communication services and technologies forecast is depicted in Figure 1.

FUTURE DATA COMMUNICATION

Data Communication will grow rapidly in the next decades. According to Eurodata,[4] the number of Data Terminal Equipment (DTE) will grow from 1979 to 1987 by 6.3.

We believe that the average rate of growth of DTEs in Israel will be approximately in the same range, thereby allowing for an expected 30,000 DTEs by 1987.

Following this logic, it seems reasonable to assume that there will be in Israel, by the year 2000, at least 50,000 DTEs. During the next decade a computer network will be developed and installed, which will provide for the use of all kinds of Data and Computer Communications (Figure 2).

Fig. 2. Future Data and Computer Communications in Israel

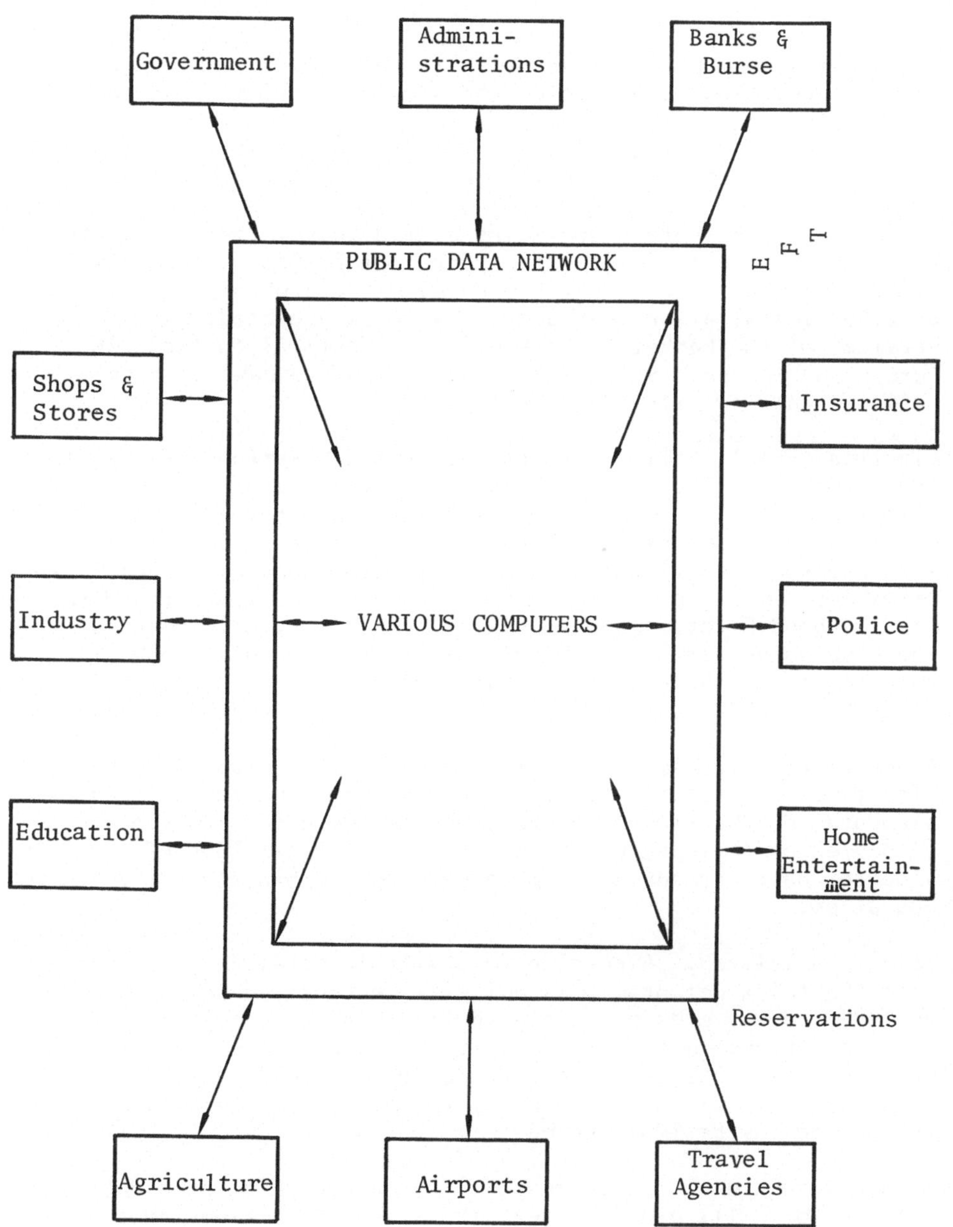

Israel has started this phase of implementation by installing a packet-switching Data Network. The first node will be installed in September 1981 and it seems that, after some trials, further installations of nodes will follow in 1982 in the major cities.

The Data Communication scene in Israel will gradually grow in importance. New technologies and services will be developed, bought, tried and implemented.

The main systems to be dealt with will be the following:

* Packet-Switching Data Network. As stated above, the implementation phase has already started. Once the system will become commercialized, it will quickly develop and probably an average of 2,500 installations per annum are to be expected. The services of the Packet-Network will be identical to those in Europe and in the U.S.A. The trend will be to have an international gateway with each country.

* Electronic-Mail. Mail treatment has become a serious problem in every country. These years we see developments of various Electronic-Mail systems. The intention is to develop an Electronic-Mail system, in both Hebrew and Latin languages, which will assist in solving the mailing problems, and will facilitate the office-work. The office of the future will have, as a backbone, a good Electronic-Mail system. An effort will be given to the development of a system which will treat both alphanumerics and graphics. The Electronic-Mail system will be utilized through the PSTN and the Public Packet Data Switching facilities.

* Alphanumerics Transmission using the home TV. A development effort will be established to have a system like the British Viewdata, or the French Teletel, on the one hand, and the Viodeotex or Antiope, on the other. The first trials of Videotex have already started in 1980 by the Ministry of Education.

* Telex and Teletex. According to available data, it is possible that the telex network in Israel will increase up to 7,000 lines by 1987 and may even be doubled by the year 2000. The telex machines, foreseen for the future, are improved versions of the existing ones, based on microprocessors embedded systems. In parallel, the telex introduced by various countries will be tested and may be implemented.

* Leased Data Lines. It seems that the introduction of a Public Data Network will almost freeze the leased Data Lines requirements. These lines will be used in future, for special needs only, for instance--when heavy traffic is expected between two installations of a customer, or for private networks of special establishments.

* Electronic Funds Transfers. The EFT network is worldwide developed. The Israeli banks have a major role in the development of the country. A big effort is being made to have a good EFT system established for use by the banks. The trend is to get the banks to make use of the Public Data Network. During 1981, some of the banks are also joining the Swift Network, and some tendency exists, as well, to implement private networks.

* Specific Services. Research and development will be done during the coming years in projects such as: Remote-Metering, Speech-Recognition and Synthesis, Telewriting, etc. It appears that there will be a good market for such systems both in Israel and in other parts of the world.

* Data Bases. As described above, Israel has quite a good computer power, relative to its size. A considerable effort will be invested in developing new computer applications and services, which will be available for internal and international usage, using the Data Communication facilities.

SUMMARY

The next two decades will be the decades of developing Telematics. Every country that wishes to enjoy a good standard of living must regard Telematics as one of the most important infrastructures.

Israel has started a new policy in Telecommunications and Telematics which will, we hope, enable the country to prosper and take part in the common efforts to improve the living standards throughout the world.

Each new system will pass a series of evaluations and pilot tests. The digitalizing of communication will pass--on a large scale, in transmission--switching and subscribers' equipment, and the passing from analog to digital will be attained in gradual steps.

REFERENCES

1 State of Israel Central Bureau of Statistics, Survey of Electronic Computers and Terminals 1980.

2 State of Israel Ministry of Communications, Statistical Yearbook, (1978/79).

3 David Biran, "Data Communications in Israel," Israel Journal of the Association of Engineers and Architects, (July - August, 1980).

4 Marino Benedetti, "Eurodata 79: The Growth of Data Communications in Western Europe," Telecommunication Journal, vol. 48, (1/1981).

4 "The Report on Electronic Mail", The Yankee Group, First Quarter (1979).

DATANET 1 OF THE NETHERLANDS' PTT

Ir. J.M. van den Burg
PTT-Telecommunications
Market Manager Business Telecommunications

SUMMARY

Datanet 1 is the first stage of the development of a public packet-switched datanetwork in the Netherlands. Its objective is to keep up with the evolving data communications needs. The network should be seen as an extension of the existing telecommunications infrastructure and also as an adaption of this infrastructure to the appearant new data communications requirements.
This paper describes the network as future users may perceive it. A way of assessing its usefulness is discussed and an approach for cost-benefit assessment is presented.

INTRODUCTION

During the last two decades the evolution of computing equipment, together with the evolving requirements for computer based information systems have generated a quickly growing demand for datacommunication links over the telephone network.
An indication of the speed at which this demand has been growing may be the following. Since 1970 the number of leased lines in the Netherlands has doubled every 2 to 3 years and the usage of these lines has increased significantly, both in terms of speed of data transfer and duration of use. In 1976 it became appearant that the future datacommunications requirements could only be met by adding a specially developed datanetwork to the existing telex- and telephonenetworks.

K. G. Beauchamp (ed.), New Advances in Distributed Computer Systems, 25–32.

USER GROUP

In that year a number of large companies and gouvernmental organisations were planning the development of widespread private datanetworks. During a meeting between PTT and these organisations it was agreed that such networks might work quite well at a justifiable cost level. It was also agreed that in many cases such networks could offer only suboptimal solutions for the foreseen datacommunications problems. The main reasons being:

- private networks - by definition - can only be used for data communication within a particular organisation and not for data communication with suppliers, clients, bankers, etcetera,
- it seemed to be cheaper to share resources, such as lines, switching- and control equipment, and also to share the staff needed for designing, implementing, operating and maintaining the network,
- once installed, the network would be available to any organisation, virtually without delay and without the risks normally involved in private development projects,
- the development of private networks would be too costly for many small and mediumsize organisations anyway.

For these reasons the Dutch gouvernment decided that PTT should install a public datanetwork and that the project planning should be prepared in such a way, that the first stage - Datanet 1 - would meet the immediate requirements of the organisations involved in the initial discussions.
In order to achieve this, these organisations and PTT formed the so called "User group Datanet 1". During regular meetings of this group many aspects of the network were discussed. Functional requirements and technical specifications were considered in relation to the structure and the level of the tariffs. The lay-out was designed according to the expected location of the eventual connections. Future procedures for connection and maintenance were presented; regulations were formulated in close cooperation between PTT and user members. Ways of comparing costs and benefits were also discussed.
The user group had 9 members, both from private companies and gouvernmental institutions. Over 130 interested organisations received copies of all documents produced.

DEVELOPMENT PROCESS

As stated earlier the first stage of the datanet was meant for the members of the user group. According to the original

plans that stage should have been operational before January 1980.
In order to complete the network and to make it publicly available, it was then to be extended by adding new facilities and enlarging its capacity.
Recently the first stage became operational, albeit with some delay. This delay, however, dit not affect the moment at which the network became publicly available. The first extension was already ordered and from september 1981 on, anyone who wants to use the network can do so.

FUTURE PLANS

For some time only connections according to the CCITT recommendation X.25 can be used, but in the course of 1983 the Packet Assembly and Disassembly facility (PAD) will be added to the network. This will enable subscribers to use some other protocols. In this respect one may think of telex, TTY-compatible and BSC protocols. It may be noted here that flexibility was considered to be one of the most important requirements of the PAD. It should be possible to add new functions in the future at reasonable cost and within a reasonable timeframe. This is to ensure that future extensions (e.g. viewdata, teletex, dialled-up X.25 connections using the public switched telephone network) can be made relatively cheaply, quickly and easily.
Connections between the Netherlands' network and networks in other countries are also foreseen from 1983 on. These shall be realised according to CCITT recommendation X.75.

COMPOSITION OF THE NETWORK

The main elements of Datanet 1 are three interconnected Packet Switching Exchange (PSEs). These PSEs will be situated in Amsterdam, The Hague and Arnhem. Fifty-seven Packet Data Satellites (PDSs) will be spread around the country and each will be connected to one PSE. The network is based upon the existing telecommunications infrastructure in the Netherlands. It should be seen as an extension of this infrastructure and also as an adaption to the new datacommunications requirements. All exchanges are located in existing PTT buildings and are interlinked using existing transmission equipment.

The functions of Datanet 1 will be controlled by a Network Operations and Management Centre (NOMC) which will be situated in Bussum. PTT staff will man the NOMC and will be able to test centrally all parts of the network, inclusing modems at user sites.

The functions provided by the NOMC can be classified as follows.

1. Network supervision: this consists of the logging of received event reports and the updating of the network status model by using periodically received activity and status reports.
2. Operational management of the network: concerned activities include switching node modules or lines into or out of service or maintenance mode, updating of parameters of tables in nodes, transferring files from the PSE to the NOMC and (remotely) controlling the loading and dumping of a node software package from a PSE to a PDS or vice versa.
3. Network debugging: realising a partial dump of node software, production of snapshots, tele-activation of artificial-traffic generators and collection and processing of the results.
4. Maintenance test activity: tele-activation of specific maintenance tests of units in a PDS or PSE, tests on subscriber-lines and the processing and presentation of test reports.
5. Statistics processing: collection and production of specific reports and logging the results.
6. Billing-data handling: collection and storage of billing information and the formatting and copying of billing data for subsequent processing.

PERFORMANCE

As far as transmission speed is concerned a choice can be made between 2400, 4800, 9600 and 48000 bit/sec. In most cases the datanet transport delay time (last bit in, first bit out) should not be more than 0,3 sec; in less than 1% of the cases a transport delay of over 0,6 sec is allowed. As regards the call set-up time, one of the design objectives is that the network can handle several thousands of calls per call.

In the initial stage the network will offer 1800 X.25 connections. The throughput of PSEs and PDSs is several Mbits per sec. Although a considerable reserve in capacity is available, extension of the network will be required as demand grows. The design of the network allows for any extension that may be needed in the future.

Future extensions will be planned to meet with the requirements of the Netherlands' data communications market.

X.25 recommendations leave some room for interpretation. One decision of the PTT was to use 'adaptive routing of data traffic'. This means that every packet finds its own route through the network. This makes flow control relatively complex, but will enable the network to handle heavy traffic flows.

Duplication of all main elements and connections gives the network a high redundancy. This produces a highly reliable service with breakdowns having only a minor impact on performance.

NETWORK FACILITIES

Permanent virtual circuits and virtual calls will be available from the beginning. Closed User Group and some other facilities will also be provided since they are considered essential by the first users.

Reverse change acceptance, packet retransmission and more facilities will be available with the second stage. They are considered as non-essential, but useful addtional facilities.

The third stage might provide multiple circuits to the same address and direct call facilities, but these are still under consideration.

USER ASPECTS

Using this network data can be transported virtually without any delay and in many cases at relatively low cost. Since most important parts of the network are at least duplicated, a high reliability may be expected. Data transfer to and from third parties will become possible. Permanent (virtual) circuits may be applied, but users can also build up and terminate incidental (virtual) connections whenever desired. Because packet switching techniques are applied, many terminals can be connected at their own speed while the host computer is connected with only a few, but very high speed connections.

Given these characteristics it does not seem so strange that quite a number of organisations are planning to use this network, not only for their internal communications, but also for the exchange of data with their external relations. At this moment feasibility studies for the latter application are being performed in various branches of trade and industry in the Netherlands.

WHEN TO USE DATANET 1

It does not appear possible to provide future users of data communication facilities with general quidelines which will always lead to the best conclusions.
However, the following few indicators may be of help:

1. In many cases it may not be wise to replace existing data communication links. Current system design may well be constrained by the limitations of existing links, and therefore it can hardly be expected that Datanet 1 will add much value to the existing system. This may result in an increased cost without increased benefit.
2. In many cases, Datanet 1 could be adventageously in designing a new system. System design will not be as constrained as would be the case with alternative data communications tools. One problem may be that not all designers are yet familiar with these new more tolerant constraints. Design of future systems may therefore continue to be hampered by limitations of the past, although these may no longer be applicable.
3. When the need arises to equip star-shaped networks with back-up facilities, Datanet 1 may prove easier and cheaper than the construction of a second star using leased lines.
4. Mesh networks based on Datanet 1 tend to be cheaper than alternative tools.
5. Datanet 1 can be attractive when data transfer to and from third parties (eg, clients and suppliers) is considered.

COST-BENEFIT COMPARISON

There are many alternatives available for datacommunication. The distance between the spot where the data is and the place where it is needed can be bridged using
- postal services,
- cars (private ones and taxi's),
- the public switched telephone network,
- the telex network,
- datanetworks(private ones and the public datanet),
- leased lines,
- etcetera,

and all these tools are being used in practice.

Before making a choice it is necessary to compare the technical characteristics of the possible alternative datacommunication tools. It is also useful to prepare an estimate of the monthly charges for these alternatives. However, to call the combination of these two activities a cost-benefit analyses is a bad habit of many computer and datacommunication specialists. Technical characteristics are not the same as benefits. In fact, datacommunication tools as such do not produce any benefits at all.

Datacommunication tools can only be of value for an organisation by contributing to the objectives of that particular organisation. Like other tools, they can only do so by being an integral part of an operational unit within that organisation. The main task of such a unit may be:

- production of goods,
- production of information,
- transfer of money,
- planning and control of an operational process,
- accounting,
- physical distribution of end products,
- etcetera.

In practice such operational units will be composed of concrete and abstract elements such as machinery, officeroom, office equipment, staff, computers, computer programmes, procedures. The cost-benefit ratio will be optimized by

- a careful selection of the elements themselves,
- a careful tuning of these elements to each other.

This has a few quite obvious, but very important implications:

1. A cost-benefit comparison between alternative tools for datacommunication can only be made in a direct way if all other elements in the operational unit can remain unchanged,which will almost never be the case.

2. The full benefits of Datanet 1 can - in most cases - only be realised by redesigning the entire operational unit by which it is to be used.

3. These benefits and the related costs can only be assessed by looking at the entire operational unit
   - as it is now and
   - as it should be in the future.

   Example

   The well known functions of accounting systems have always been:

   to systematically collect, record, process and supply

data and information for the purpose of managing and operating an organisation and in order to account for such management and operation.

The ways in which these functions were performed have changed drastically over the last 100 years (figures 1 through 4).

A comparison of the cost and benefits of the alternative systems can be made by comparing:

- input of the system (formats, quantity, deadline for delivery etc.),
- output of the system (presentation, quality, quantity, speed at which it becomes available etc.),
- management control required (general instructions, ad hoc requests),
- management reports (account, performance reports),
- cost (wages and salaries, depreciation of equipment, heating/cooling, maintenance, development, dataprocessing, datacommunication, telephone bill etc.).

Remark

When discussing cost and benefits of possible alternative systems it is important to pay some attention to

- the assumptions being made (what risk do we run if they prove to be untrue),
- the prerequisites (what are the conditions under which the assumptions and conclusions are valid),
- the flexibility of the systems
  - . in what timeframe,
  - . at what cost
  - . under which conditions
  - . by what timeframe,
  - . with what risks,

  can the system be adapted to future changes in the requirements).

INTRODUCTION OF RADIO-BASED AIR/GROUND DATA LINKS INTO THE SITA NETWORK

G.P. GIRAUDBIT

Société Internationale de Télécommunications Aéronautiques (SITA)

Following a general introduction of the activities of the Société Internationale de Télécommunications Aéronautiques (SITA) and its world-wide fixed telecommunications network, this paper describes the technical approach chosen for the integration of digital aircraft-to-ground radio communication links within the SITA network environment. A short description of the services envisaged and an overview of the functional and system architecture options together with the initial system deployment plan are given in this paper.

A - INTRODUCTION TO SITA

The Société Internationale de Télécommunications Aéronautiques, more usually known simply as "SITA", is a non-profit making cooperative organization created by the airline community to meet its needs for telecommunication services. SITA was founded in 1949 by eleven airlines : Air France, three British companies which form the present-day British Airways, KLM, SABENA, three Scandinavian companies which today are merged in SAS, Swissair and TWA. They pooled their existing telecommunications resources, which in those days consisted of radio-telegraph circuits and networks and transmitted messages by Morse code or teleprinters.

From very modest beginnings SITA has today grown into a truly international enterprise, serving 241 member airlines in 154 countries of the world, and operates the largest private telecommunications network in existence.

The SITA network permits the worldwide exchange of commercial,

K. G. Beauchamp (ed.), New Advances in Distributed Computer Systems, 33–44.

technical and administrative information for air transport and associated activities. By providing the airlines with a specialized, dedicated network, SITA ensures the rapid transmission of information relating particularly to aircraft movements and flight security. In this context, SITA holds a prominent position as a public service. Without the availability of the SITA network, member airlines would be greatly handicapped in their ability to communicate satisfactorily.

Currently, there are over 18,000 teleprinter and computer terminals in some 12,000 airline offices located in 800 cities connected to the SITA network. Included are approximately 7,000 terminals for 45 different electronic seat reservation systems, utilized by 103 member airlines.

In addition to its telecommunications network, SITA operates a large reservation system in Atlanta, U.S.A. which is shared by 30 airlines.

In 1980 SITA's total turnover reached 114 million US dollars.

SITA's worldwide operations are performed by a staff of 1750 most of them natives of the countries in which they work.

SITA's operation is truly international, with its 241 member airlines being represented at the General Assembly and on the Board of Directors, and by the various nationalities at the Headquarters and in Regional Management.

Since its creation in 1949, SITA has developed into a unique tool for worldwide air transportation which, although is based on the leasing of transmission circuits from the P & T administrations and "common carrier" organizations - is a "private" entity dedicated and restricted to the specific requirements of its users.

Services provided by SITA consist of two broad categories :

- TELECOMMUNICATIONS SERVICES, the mainstay of SITA activities since its creation, and

- INFORMATION HANDLING AND SUPPLEMENTARY SERVICES, which SITA has offered since 1973/4.

1 - <u>SITA Worldwide Telecommunications Network</u>

The SITA worldwide network today consists essentially of switching centres and teletype and data circuits, the circuits being meshed so as to provide a "fail-safe" alternative in case of outages on a particular route. The 159 switching centres which are maintained and operated by SITA, perform a message and data switching function under the store-and-forward transmission principle.

The backbone of the network is the packet-switched, meshed "High Level Network" currently composed of ten large switching nodes in Amsterdam, Beirut, Frankfurt, Hong Kong, London, Madrid, New York, Paris, Rome and Tokyo, linked by medium speed circuits operated in most cases at 9 600 bits per second (bps). It controls a second level of computerized nodes consisting of "Medium Level" centres, located in Bangkok, Manila and Sydney, and of 48 "Satellite Processors" (programmed data concentrators) scattered over the world and serving as interface devices for the connection of data terminals and teleprinters. A third level of the network is formed by Time Division Multiplexers serving as circuit concentrators, and by other types of data concentrators and manual telegraph switching centres.

In 1980 the SITA network involved over 500 medium speed and 5 000 low speed circuits all leased from the Post and Telecommunications (P & T) authorities, or recognized private operating agencies (common carriers).

The SITA network provides at present two types of communications services corresponding to two different types of traffic. Type "A" traffic refers to the conversational exchange of messages in real-time as used for instance in the remote interrogation of airlines seat reservation systems. For these relatively short (approximately 80 characters) enquiry/response type communications between a Visual Display Unit (VDU) terminal and a central computer, SITA offers a response time of 3 seconds. Type "B" traffic refers to conventional telegraphic messages of about 200 characters whose functional information relates to flight operations and safety, aircraft movements, administrative matters such as lost luggage, flight services and status, and commercial activities such as sales and reservation. For this type of non-conversational traffic SITA provides almost 100% security against message loss or mutilation.

In 1980, SITA transmitted approximately 3.1 billion type "A" and 432 million type "B" messages.

The foreseen sustained increase in conversational type "A" traffic demand, together with the need to introduce new telecommunications services, have led SITA to study and define a new network architecture, the Advanced Network, at present under development and implementation.

Intended to meet the airlines' future requirements in terms of evolving functional characteristics of communications services and of expanding traffic volumes, the Advanced Network is recognised as being the largest civilian network yet undertaken in view of the major development effort involved and of the need to achieve a smooth transition from the current to the new architecture without disrupting the 24-hour-a-day service.

The definition of the Advanced Network architecture is the result of an in-depth analysis of the functions required for the provision of the present and future communications services. This analysis led to the identification of four major classes of functions, to which are associated four specific families of systems, ie :

- User Interface System (UIS) For the support of user interface connection, concentration of data and translation of this data into internal network format.
- Data Transport Network (DTN) For the transport of data in transparent mode between two points.
- Message Storage and Handling System (MSS) For the storage and processing of messages requiring high protection (eg: Type "B" traffic).
- Network Control System (NCS) For the provision of control facilities required to operate the network.

The major advantage of the Advanced Network architecture is that it allows the independent evolution and optimum distribution and sizing of the various network resources by the removal of the present network architectural constraints.

Its modern open-ended architecture will permit to take full advantage of present and future technological advances in the fields of network engineering and of equipment hardware.

The Advanced Network development and implementation schedule is organised in two major phases :

- Phase I (1981 - 1982) : phased implementation of the Network Control System (NCS) and of the first nine nodes of the Data Transport Network (DTN); please refer to Figure 1 shown hereafter.

SITA ADVANCED NETWORK PHASE I

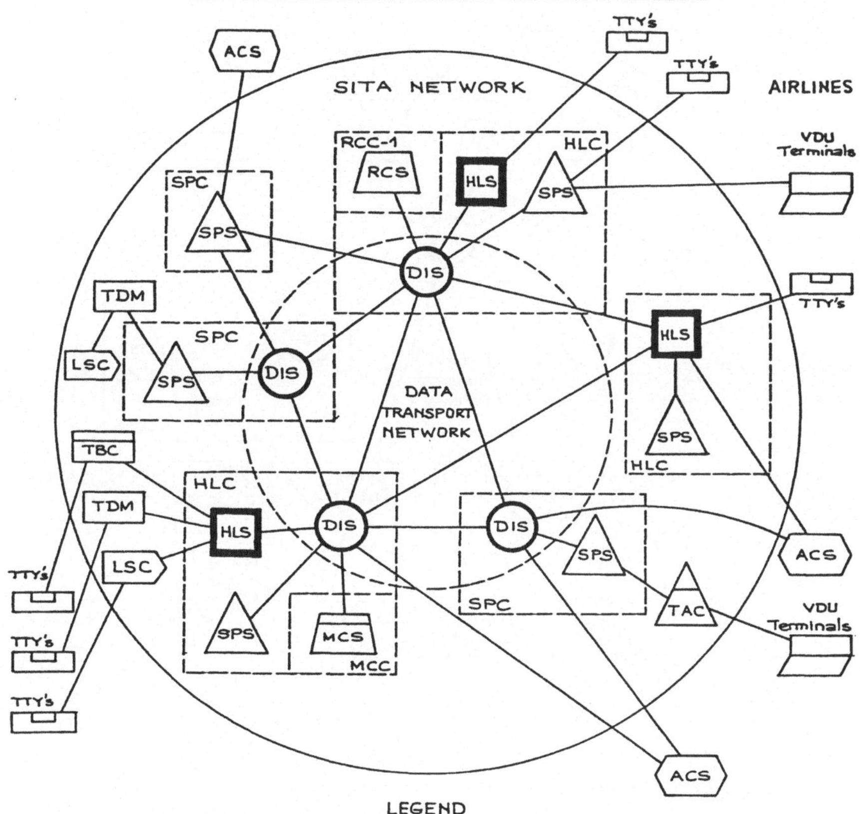

LEGEND

ACS: Application Computer System
DIS: Data Switching & Interface System
HLC: High Level Centre
HLS: High Level System
LSC: Low Speed Concentrator
MCC: Main Control Centre
MCS: Main Control System
RCC: Regional Control Centre
RCS: Regional Control System
SPC: Satellite Processor Centre
SPS: Satellite Processor System
TAC: Type A Concentrator
TBC: Type B Concentrator
TDM: Time Division Multiplexer

Figure 1

- Phase II (1984 - 1985) : phased implementation of initial UIS and MSS systems; please refer to Figure 2 shown hereafter

SITA ADVANCED NETWORK PHASE II

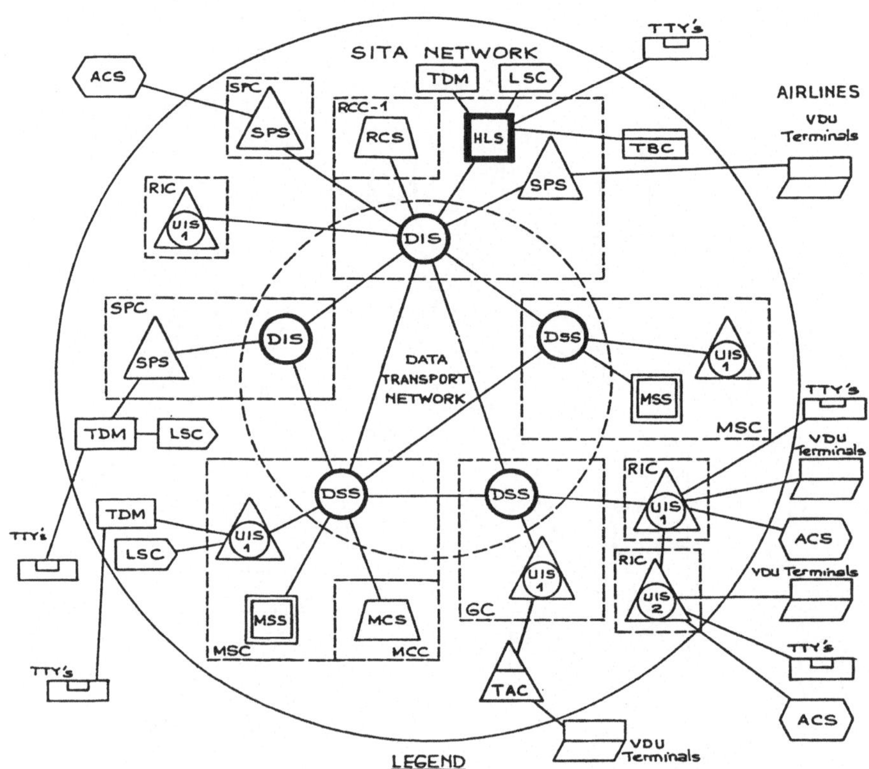

LEGEND

ACS: Application Computer System
DIS: Data Switching & Interface System
DSS: Data Switching System
GC: Gateway Centre
HLC: High Level Centre
HLS: High Level System
LSC: Low Speed Concentrator
MCC: Main Control Centre
MCS: Main Control System
MSC: Message Storage & Handling Centre
MSS: Message Storage Handling System
RCC: Regional Control Centre
RIC: Remote Interface Centre
SPC: Satellite Processor Centre
SPS: Satellite Processor System
TAC: Type A Concentrator
TBC: Type B Concentrator
TDM: Time Division Multiplexer
UIS: User Interface System

Figure 2

2 - Other Telecommunications Services

With the Advanced Network, SITA will introduce new and flexible telecommunications facilities such as :

Multi-Service Terminals. This facility will enable terminals connected to the network via medium or low speed lines to access, on a one-at-a-time basis, different airlines'

Application Computer Systems in conversational mode. The same terminals will also be able to operate alternatively in conventional mode, ie: to generate and receive type "B" messages.

Advanced Data Communications Service. This service will provide a wide range of computer-to-computer data transfer capabilities such as bulk data transmission in transparent mode, individual message exchanges, etc . . .

B - INTRODUCTION OF RADIO-BASED AIR/GROUND DATA LINKS INTO THE SITA NETWORK

In 1980, SITA undertook the general definition study of a digital air/ground communication service, hereafter referred to as AIRCOM. This preliminary work has now been completed, and efforts are at present directed towards the development and implementation of a pilot installation aimed at demonstrating the technical and operational feasibility of the AIRCOM service over the SITA Network.

1 - Service Overview

The basic function of the AIRCOM Service is to allow the exchange of data in digital form between an aircraft (whether in flight or on the ground) and the airline ground based flight operation personnel and facilities. (See Figure 4 - Aircraft/ Ground Communications).

At present, aircraft crews and ground flight operation personnel communicate by means of a combination of radio voice equipment and ground based data communications network(s); in order to illustrate a typical voice contact, let us consider the transmission from an aircraft of its Estimated Time of Arrival (ETA) to the destination station. After having selected a company allocated radio frequency, the crew will establish the communication with a ground radio operator, to whom they will spell a message consisting, most of the time, of various codes and numbers. The ground radio operator will, in turn, translate the oral message into a conventional telegraphic message which will then be forwarded to several destination addresses via the SITA and/or the Airline's network(s).

This overall radio voice operation requires human intervention at several levels, and as and as a result suffers from multiple limitations, among which we note :

- High overhead of redundant procedures required for reliable communication.
- Impossibility to transmit automatically acquired data, such as flight and engine monitoring data.

- No real time data exchange capability.
- Inefficient utilization of frequency spectrum.

The development of AIRCOM will introduce a 2400 bps air/ground data link capability for automatic, accurate, real-time data exchanges between an aircraft and a ground based communication network. This link will take advantage of the error detection and recovery features provided by modern communications protocols. It is intended to complement the existing air/ground voice communication and will provide a communication capability between the advanced avionics of current and future aircrafts and the powerful ground based airline data processing facilities.

AIRCOM data exchanges can be divided into down link and uplink messages.

For downlink (aircraft-to-ground) messages, the following information can be transmitted :

- Times : out of the gate, off the ground, on the ground, in the gate.
- Estimated time of arrival.
- Fuel status.
- Crew on board.
- Engine data from AIDS (Aircraft Integrated Data Systems) records.
- Passenger service data.
- Maintenance data.
- Weather collected data.

For uplink (ground-to-aircraft) messages the following information is envisaged :

- Time reference.
- Weather update.
- Flight operation data (flight plan, weight and balance).
- Data destined to flight management computer.

The availability of the AIRCOM service to exchange the above data will open the way to the progressive automation of key air-line operation applications such as flight operations (flight movement supervision and flight management applications), aircraft maintenance and engineering, etc . . . AIRCOM will, in addition, provide the capability for new, on board, passenger services.

2 - AIRCOM System Architecture

From a technical standpoint, the AIRCOM service will be provided by a network of dedicated VHF ground stations (referred to as remote ground stations) supporting the aircraft-to-ground data links and connected to the nearest SITA Network Interface System (ie: Satellite Processor or any other type "A" access facility), (see Figure 3).

All AIRCOM traffic received by the remote ground stations will be routed as type "A" data block(s) via the SITA Network to a SITA "AIRCOM Service Processor" central computer system in charge of AIRCOM service supervision and AIRCOM message processing (addressing and routing, format and code conversions, end-to-end communications control etc.). From the AIRCOM Service Processor messages are then sent to the airline final ground destination facility (terminal or computer system) via the SITA network. (see Figure 5).

This system architecture makes extensive use of the existing SITA Network infrastructure and facilities, thus minimizing system deployment investments and circuit costs. (See Figure 6).

In order to provide a world wide service coverage which takes into account the development already in progress in the United States with ARINC's ACARS system, AIRCOM will ensure compatibility with ACARS and make use of the same on board avionics equipment (see Figure 7).

3 - Deployment Plan

SITA plans a phased deployment of AIRCOM. At the end of 1981 a pilot AIRCOM unit including two remote ground stations and a basic AIRCOM Service Processor will be available for evaluation of the service by the airlines (KLM is the first airline considering participation in the experiment). The operational deployment will start by the installation of 50 VHF ground stations offering the coverage of major international airports and air routes in Western Europe and Mediterranean Basin. (See Figure 8). In a second step the coverage of this geographical area will be extended to offer contiguous en route coverage above 20 000 feet.

From this initial phase and according to airline needs the extension of the service coverage to other parts of the world, such as Asia, Africa and the Middle East, is planned by the addition of ground stations connected to existing SITA type "A" access facilities.

SITA AIRCOM SERVICE (The ellipse A marks the flight operation centre of a given airline)

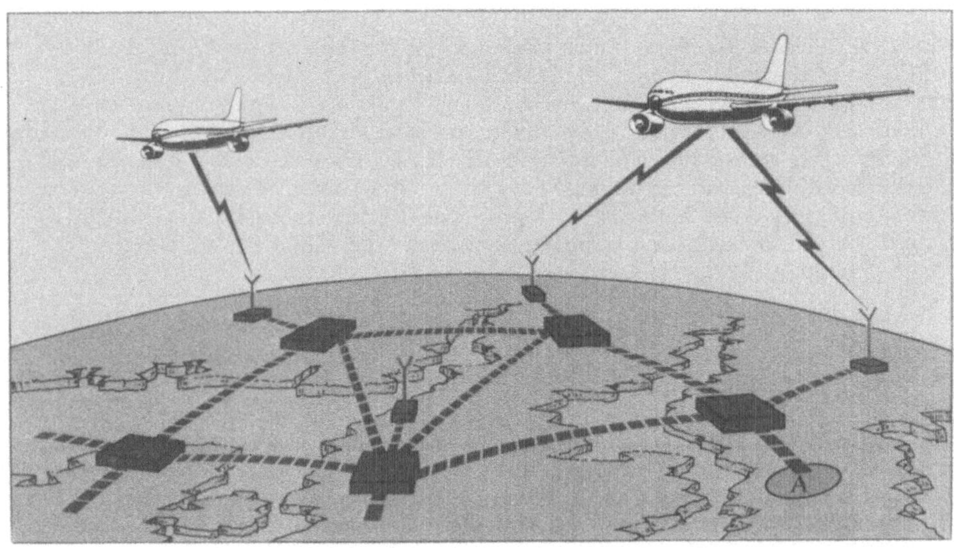

Figure 3

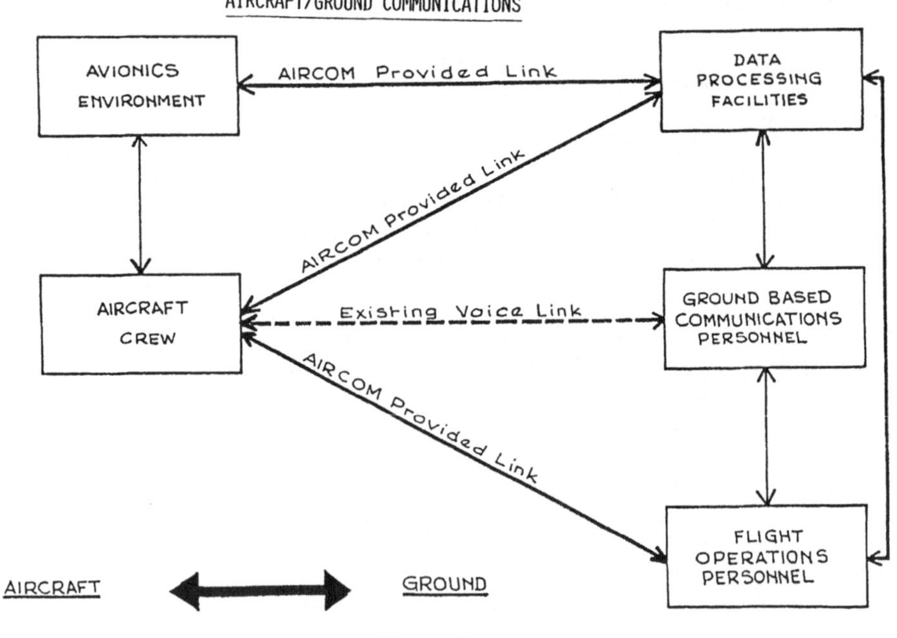

Figure 4

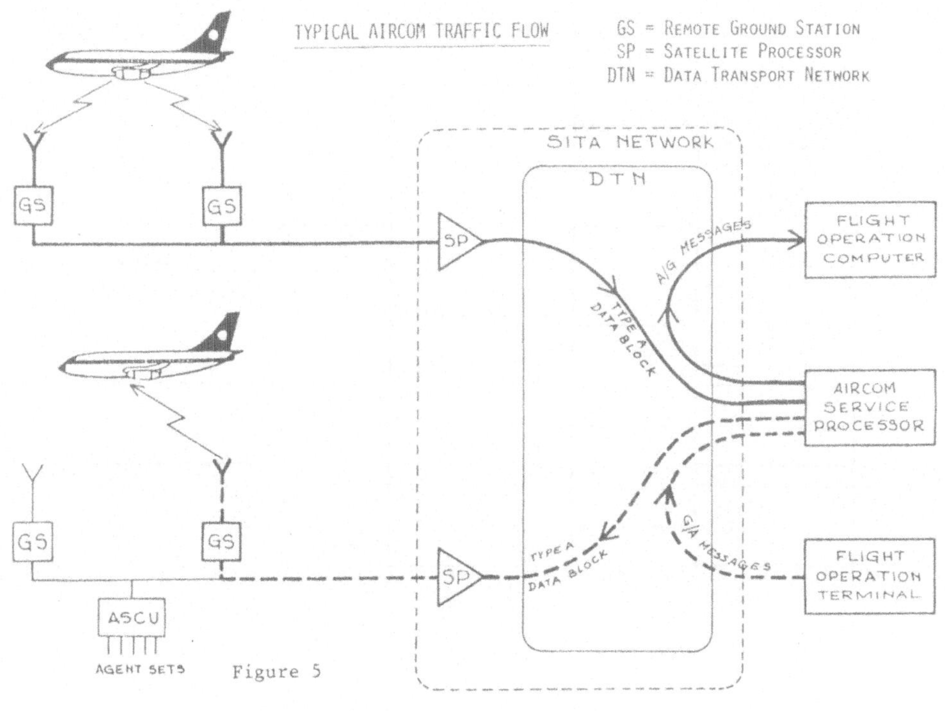
TYPICAL AIRCOM TRAFFIC FLOW
GS = REMOTE GROUND STATION
SP = SATELLITE PROCESSOR
DTN = DATA TRANSPORT NETWORK
SITA NETWORK
DTN
GS
GS
SP
TYPE A DATA BLOCK
A/G MESSAGES
FLIGHT OPERATION COMPUTER
AIRCOM SERVICE PROCESSOR
GS
GS
SP
TYPE A DATA BLOCK
G/A MESSAGES
FLIGHT OPERATION TERMINAL
ASCU
AGENT SETS

Figure 5

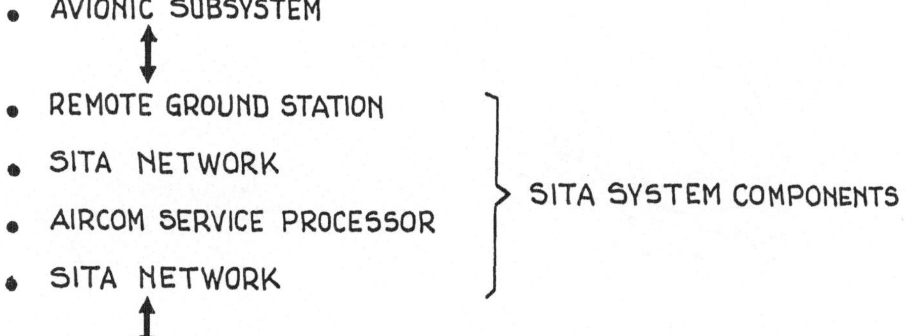
AIRCOM SYSTEM COMPONENTS
AVIONIC SUBSYSTEM
REMOTE GROUND STATION
SITA NETWORK
AIRCOM SERVICE PROCESSOR
SITA NETWORK
SITA SYSTEM COMPONENTS
USER GROUND FACILITIES:
FLIGHT OPERATION APPLICATION COMPUTER
FLIGHT OPERATION TERMINAL

Figure 6

Figure 7 AIRCOM AVIONICS BLOCK DIAGRAM (ARINC 724)

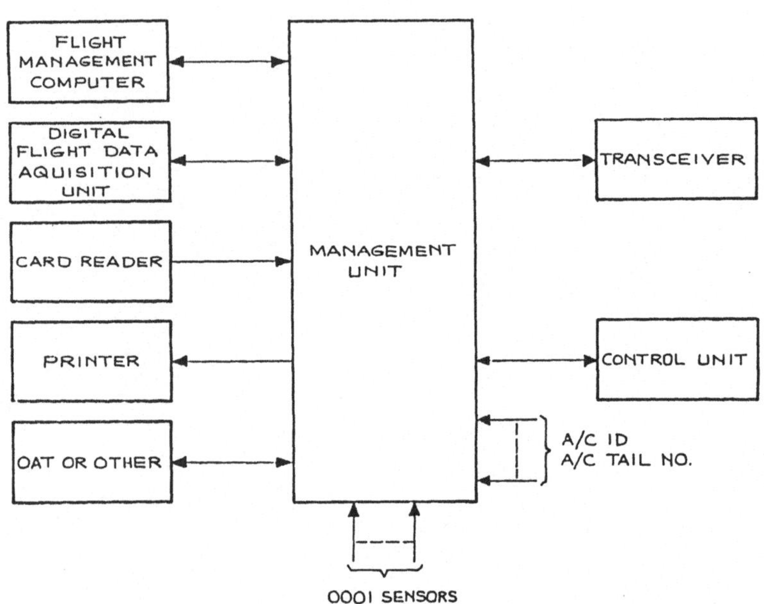

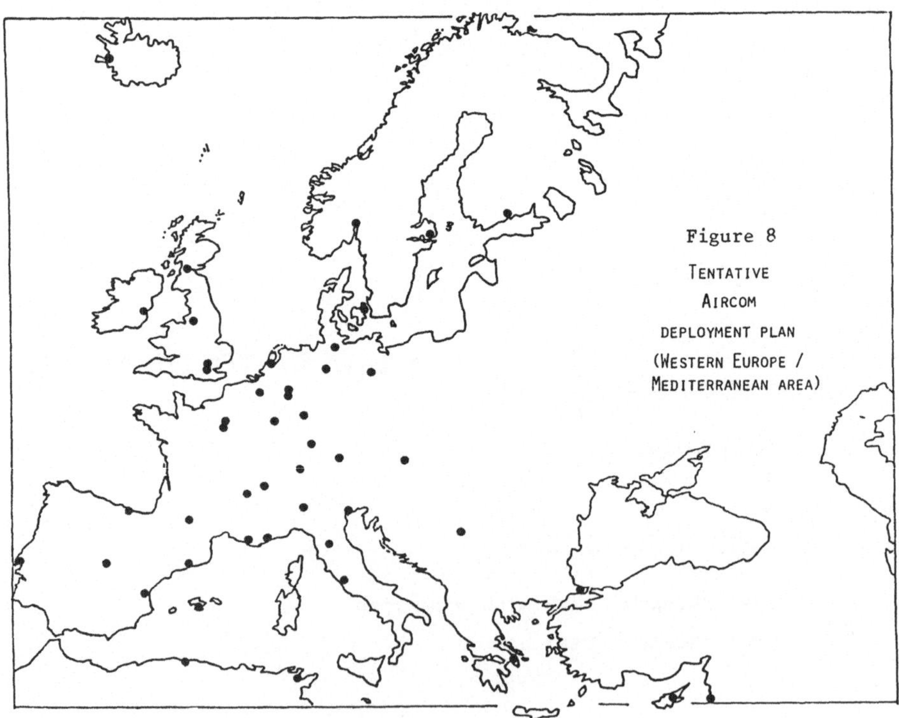

Figure 8
TENTATIVE AIRCOM DEPLOYMENT PLAN (WESTERN EUROPE / MEDITERRANEAN AREA)

PRESENTATION OF THE NADIR PILOT-PROJECT

HUITEMA Christian

Centre National d'Etudes des Télécommunications (CNET)

Pilot-project NADIR was launched in 1980 by the french government. It is sponsored by the Agence de l'Informatique (Ministry of Industry) and PTT. It is intended to study new types of applications made feasible by high speed satellite communications systems, such as the French "TELECOM1". Some exemples of these applications are remote processing, distributed data base, voice/image/data message switching. In its very first phase, NADIR is presently developping basic tools such as transferring large data files and interconnecting local networks using high speed (up to 2 Mbit/s) satellite links. "TELECOM1" will be launched in mid-83. In the meantime, experiments will be conducted by using either existing satellites (e. g. OTS) or the ANIS satellite simulator, specially developped by the project. Results of the first experiments are expected to be available during spring 1982.

1 - GENERAL DESCRIPTION

The pilote-project NADIR was launched in 1980 by the French government. It is sponsored by the Agence de l'Informatique (Ministry of Industry) and the Direction Générale des Télécommunications (Ministry of PTT). This project is intended to study new distributed data processing applications using a satellite broadcast network, such as the french satellite system "TELECOM1". Satellite broadcast networks will allow new distributed data processing applications by taking advantage of high transmission bandwidth capabilities (very large file transfer, voice, image and data processing), as well as broadcast capabilities (simultaneous query of a set of data bases, file updating and teleconferencing).

K. G. Beauchamp (ed.), New Advances in Distributed Computer Systems, 45–50.

The purpose of the NADIR project is to experiment computer-to-computer communications via the french satellite "TELECOM1". It mainly consists of the implementation of a set of computer applications, taking advantage of the capabilities of the satellite transmissions. The availability of a satellite communications system should facilitate the development of regular teleprocessing applications. On the other hand, it should also offer new possibilities due to the low cost of high data transfer rate (up to 2 Mb/s on TELECOM1) and broadcasting capabilities naturally offered by satellite systems.

A sample of such possible new applications is given below :

- Remote service access
  The remote users will be offered services comparable to existing local services in terms of available transmission rate.

- Computer load sharing
  Computer load can be shared between remote computers in the same way as it is done today for locally interconnected computers.

- Distributed computer system
  The processing could be distributed among several specialized computers independently of their geographical position.

- Failure recovery
  The load of a computer center could be transferred to another one in case of computer failure, by permanent file updating of the backup computer. This is made feasible by the availability of high transmission rates.

- Broadcasting of software
  Satellite systems could be used for distributed computer systems maintenance : a single maintenance center could use broadcasting capabilities to distribute new software releases towards various remote computers in a very inexpensive way.

- Broadcasting of data
  In some companies, a large amount of data need to be broadcasted and periodically updated from a central site. This could be performed in a very efficient way by using the broadcasting facility of a satellite network.

- Electronic mail
  Satellite broadcasting could be used for electronic mail distribution. In addition, the availability of high transmission rate should allow to develop new electronic mail systems including text, voice and image messages.

- Distributed stock management
  Simultaneous queries of a set of stock files can be done by using the broadcasting capabilities while these files are locally updated.

- Distributed account management
  Customer account management of a banking system can be done locally in each branch. Broadcast satellite systems could be used for communications between headquarters and the branches.

2 - THE TELECOM1 NETWORK

TELECOM1 will be launched in 1983 by the European launcher "ARIANE". The TELECOM1 system will include :

- The satellite : five transponders, each of them transmitting 25 Mbit/s using 12-14 GHz transmission
- The ground-stations, including an antenna, a TDMA equipment, and a local switch.
- A command station.

The command station will share the satellite transmission capacity between the stations. Customer devices will be connected to the local switches. PABX, Visioconference rooms, X21 or V35 DTE's could be connected to the switches.

From an X21 user point of view, TELECOM1 will be seen as a Circuit Switched Data Network. "Transparents" data circuits will be provided, with a propagation delay of roughly 300 ms, and a bit error rate (BER) lower than 10 E-6 during 99 % of any period of time. Optionnally, encryption and/or a 4/5 forward error correction (FEC) will be available. If FEC is used, the BER should be lower than 10E-10.
Various data rates can be selected, from 2400 bit/s to 2 Mbit/s, including 4.8, 9.6, 64, 128, 256, 512 and 1024 Kbit/s.Circuits can be established either on a call per call basis (X21), or using a "reservation" scheme.

Circuits may be of three different types :

- point to point (simplex or duplex)
- broadcast (with or without return circuits) : each information bit sent by a "master" member of the communication is received by all the "slaves" members
- conference : each information bit sent by any member of the communication is received by all the other members ... as long as only one member is trying to send information at a given time.

3 - BASIC TOOLS STUDIED BY NADIR PROJECT

In order to support new feasible applications using satellite systems, three major sets of basic tools should be developed :

- Utilities for data transport in a point-to-point or multipoint way,
- Utilities for local area network interconnection,
- Utilities adapted to specific applications such as :
  . electronic mail
  . data base management and query

These three sets of tools will be studied according to the characteristics of satellite broadcasting systems, considering the constraints of such systems (i. e. the propagation delay - 300 ms for geostationnary satellites -), and their availability on high transmission rate and broadcasting.

3.1. Transport utilities

The transport utilities concern first file-to-file transfers in both point-to-point and multipoint links.
The high transmission rate offered, joined to the large propagation delay, broadcasting capabilities and variable bit error rate, leads to study new transport protocols.
These new protocols should optimize memory buffers and transmission efficiency by using appropriate error recovery algorithms.
For a point-to-point data transfer, error recovery can be done by repeating the only erroneous frames, in two possible ways :
during the transmission of the file itself, or in separate phases, after the file has been completely once transmitted.
For a multipoint data transfer, the problems come from the fact that each individual receiver must send acknowledgments to the transmitter. Recovery of all erroneous frames may decrease the transmission efficiency.
On the second hand, the transport utilities include also messages transfer in a point-to-point or multipoint way. The main purpose for the messages propagation delay, in order to have a good response time. To avoid repeating erroneous messages, it is possible to use self correcting codes that will decrease the apparent bit error rate, at the cost of a loss of transmission bandwidth.

3.2. Utilities for Local Area Network Interconnection

One possibility of efficiently use the link capacity by message transfer consists in loading the link by multiplexing several kinds of traffic corresponding to various users.
A very interesting case is when the various users are connected to several local area networks (LAN). Those LAN's can be interconnected using either point-to-point either multipoint satellite links (e. g. "conference"). Messages transfer protocols will be used. Addressing and flow control problems have to be solved.

3.3. Utilities for electronic mail

There are several kinds of electronic mail utilities :

- computer-aid teleconferencing
- image and voice packets transmission
- voice packets transmission on a multipoint link.

The basic tools for these transmission utilities will be studied independently and dedicated to each kind of utility. Voice and image coding technics could be used to build voice and image messages that can be processed in the same way as text messages, for a point-to-point or broadcasting transmission.

This kind of messages may accomodate a certain amount of residual errors during the transmission. Nevertheless, voice messages need real-time transmission, therefore new protocols have to be studied by considering these transfer characteristics.

3.4. Distributed data bases utilities

Three main topics will be considered :

- Broadcasting query

The broadcasting facilities of a satellite system allows the user to ignore the real geographical data location. A broadcasting query may have several replies coming from various data locations, for which new protocols will provide controls to collect every reply and manage the queries.

- Broadcast updating

High data transmission rate for a broadcasting satellite system allows easier data bases updating which can be also more frequent.

- Distributed data base management

The two basic tools listed above will lead to a new data base management using satellite system capabilities. The possibility to easily transfer a complete data base may change the actual point of view about technics developed for distributed data base.

4 - INTERNAL TOOLS FOR NADIR PROJECT DEVELOPMENT

Because the launching of "TELECOM1" french satellite is planned for mid-83, the protocols tests and experiments will be conducted first on a simulator.
This satellite simulator has been defined corresponding to the external "TELECOM1" characteristics, and will be available by the

end of 1981. Using the satellite simulator will provide more facilities in testing and evaluating protocols.
Software utilities are also designed in order to evaluate protocols efficiency according to different parameters (data transfer rate, bit error rate, etc...)

5 - NADIR PROJECT FIRST EXPERIMENTS

The NADIR project is jointly developed by the INRIA (Institut National de Recherche en Informatique et Automatique), and the CNET (Centre National d'Etudes des Télécommunications) and the project team includes members from various informatics companies.

The project is actually studying new protocols adapted to point-to-point and multipoint files and messages transfer at a high data transmission rate on satellite link.

AUTOFLOOD - A FLEXIBLE TEST SYSTEM FOR PACKET SWITCHED NETWORKS

C. G. Miller and M. J. Norton

British Telecom
Systems Evolution and Standards Department

INTRODUCTION

This paper describes the "AUTOFLOOD" test system developed by British Telecom (BT) for the commissioning and acceptance testing of Packet Switched Data Networks. The paper describes the history leading up to the development of AUTOFLOOD and the design and development of the hardware and software of the tester, and describes the operation of the tester in some detail.

EXPERIMENTAL PACKET SWITCHING SERVICE TESTING AND EXPERIENCE

When British Telecom ordered equipment and software for the Experimental Packet Switching Service (EPSS), it was realised that a requirement existed for some means of proving to BT's satisfaction that the network software, hardware and firmware were correct. This led to BT developing independent test software to run on machines similar to those used as switching nodes, which was capable of testing the network. Another tester was built and programmed which ran on completely different hardware and was designed primarily for testing customer interfaces.

In addition to developing tester capability, BT had the responsibility for developing software for online billing and statistics collection. Finally, there was software to be written for the off-line processing of the billing and statistics information. All these tasks gave BT considerable experience using Ferranti Argus 700 machines for real time

K. G. Beauchamp (ed.), New Advances in Distributed Computer Systems, 51–67.

systems, and with the problems concerning testing of complex packet switching systems.

Another important spin-off resulted from the close involvement with the contractor during the software development phase; it was possible to identify areas of weakness in the standard Ferranti operating system, and a secure operating system was developed by BT as part of the on-line billing collection system. A further improved version of the same operating system was used as the heart of the AUTOFLOOD tester system.

TESTER SYSTEM REQUIREMENTS

The primary requirement of a tester is that it should be reliable - it is essential that if faults are found by a tester, there should be no doubt that it is the system under test that is misoperating, because the personnel carrying out the testing must have confidence in their tests. It is also vital that the originator of the software under test should be convinced of the validity of the tester's results. This is partly to avoid disagreement as to who is responsible for the misoperation, and partly to convince the contractor that the tests the purchaser is carrying out are comprehensive, thereby discouraging the use of quick and ill-thought-out "fixes" to faults.

The general aims of the tester are twofold - firstly it must be capable of proving that the network responds correctly to any packet or character string stimulus, and secondly that the network performs correctly under heavy traffic load, and at the specified speed. It should also be capable of providing good monitoring information, and check for correct responses while only generating output as a result of an incorrect response, since the user must not be presented with an unmanageable amount of information.

For functional testing, the tester must be able to send out any possible packet (or character string in the case of the asynchronous lines) and compare the response received from the network with that expected. While it is perfectly possible to use a tester into which the user enters all the test data in hexadecimal, it is extremely tedious to write long tests in this way, since the majority of the data can be generated automatically. It was decided at the outset that the test input data should be in the form of a series of statements in a language specifically designed for testing "X" series packet switching systems. Further, that wherever possible, the system should be intelligent enough to generate default values for fields that did not need to be specifically controlled as part of the test. An example is the automatic maintenance of sequence numbers. For normal tests there is no need to enter the sequence numbers used for every packet, but if required it is easy to override their automatic generation, thereby allowing full flexibility whilst retaining maximum ease of use. Ergonomically, this is advantageous, since it concentrates attention on what is being tested, rather than being distracted by every detail of the packet. The user can see immediately what part of the packet has been altered.

The test philosophy is that for every stimulus, the network should give a predictable response. By and large, this is true, but there are certain exceptions which give the test-writer problems. There are difficulties, for example, where the network has to generate a random value as part of a packet. The X75 call identifier is a case in point: when setting up a call over an inter-node or inter-network line, the STE generates a unique value for the call identifier as part of the network utilities field. Special measures have to be taken to deal with variable fields, and it is essential that they do not have any effect on the validity of the tests. Practical solutions have been found to these problems, and the general philosophy is adhered to as far as possible.

For traffic or load testing, the tester will not usually be carrying out complex functional tests as part of the traffic load, although it is possible to use it in this way. Normally, for straightforward measurement of network throughput, calls would be set up, then data packets passed from end to end as fast as possible. Another common test is to set up and clear calls as fast as possible, and in many ways this is a more severe test.

The throughput capability desired for the tester is of the order of four hundred frames per second maximum. This value was suggested as being adequate to provide a significant load on the switch nodes to be tested. TESTER HARDWARE REQUIREMENTS

Experience with the EPSS test apparatus led to several design requirements for AUTOFLOOD. For the EPSS testing, all the test data was held in the memory of the tester. Apart from the fact that this was expensive on memory, it also meant that tests had to be prepared off-line and loaded from cassette. For this reason, it was desirable that AUTOFLOOD should have some form of mass on-line storage. The obvious choice was to use flexible discs (floppies), since the media are easily portable, and relatively robust.

In addition, the EPSS test gear had a limited throughput when carrying out load or traffic testing. For this reason, it was decided that AUTOFLOOD should incorporate the newer and faster Argus 700 G processor, rather than the Argus 700 E used for EPSS.

Finally, since it was felt that monitoring facilities were rather limited on the EPSS tester, a fast Visual Display Unit was specified so that mimic diagrams and statistics could be displayed and continuously updated.

The use of a fast Direct Memory Access (DMA) multiplexer channel driving the High level Data link Control (HDLC) protocol interface cards was retained since this enabled some of the lowest level tasks to be delegated from the main processor to the DMA channel, making the HDLC lines interrupt driven at frame level. The use of a software multiplexer channel using programmed I/O for the asynchronous lines was considered adequate for the intended application.

The final hardware build for the first AUTOFLOOD was therefore:

1) Eight HDLC ports (2.4k bit/s to 48k bit/s)

2) Eight asynchronous ports (50 bit/s to 1200 bit/s)

3) Argus 700G processor

4) 96k Bytes of memory (48k 16 bit words)

5) Twin single sided double density floppy discs

6) Fast cursor addressable Visual Display Unit (VDU) with private memory

7) Fast multiplexer DMA channel to drive the HDLC cards

Since the tester had to be capable of being transported to the USA for the UK National Packet Switched Service factory tests, it was decided that Ferranti's normal rack-mounted equipment would not be practical. For this reason, the equipment was housed in two short racks which could form pedestals for a desk. Splitting the equipment in half in this manner meant that each part could now be lifted by two men without the use of special lifting equipment. It also posed some interesting problems with providing high-speed bus interconnections between the two units, and finding a connector scheme that made it impossible to interconnect the units wrongly.

When assembled, a table top is placed over the two pedestals, with the VDU and floppy disc drives on top. A later version of AUTOFLOOD will have the floppy disc drives incorporated into one of the pedestals, allowing rather more free space on the working surface. A normal asynchronous terminal used as a console device and the keyboard for the VDU are also accommodated on the table top.

TESTER SOFTWARE REQUIREMENTS

The broad requirements for the software are that it should be reliable and easy to use. It should also be written using structured techniques to facilitate any future modifications.

The main feature responsible for the reliability of the software has already been mentioned - the high security operating system. The use of this operating system forces the programmer writing the user processes into a certain discipline. The problem of ease of use has been addressed by the design of the test language, and the software in the tester which uses it.

The principal software modules are identified as:

1) The Operating System (Executive)
2) Command decode and Editor
3) File handling
4) Disc device control program
5) General Character Handler
6) Software multiplexer program
7) VDU screen formatter
8) HDLC card driver and Level 2 handler
9) Test syntax analyser
10) Test sequencer/Level 3 call control
11) Test language interpreter
12) Debug utility
13) Failure print utility

Some of these modules are part of the memory-resident system, while others are held on disc and brought into memory when required. This saves memory for modules that will never be required simultaneously, and makes for ease of updating should a module require modification - a very useful facility when de-bugging the system.

All modules were designed and written specifically for AUTOFLOOD, with the exception of the General Character Handler, Disc File Handling, Disc device control, and Software Multiplexer - these only required slight modification from software already in use on other systems.

The Test Syntax analyser, Test sequencer and Test language

interpreter were all designed around the test language devised for AUTOFLOOD. The language was based on what was then the most definitive document on X25 - the Grey Book. In practice, the changes that X25 has seen from the Grey Book have been incorporated without any real difficulty. Certain X75 facilities have required more effort to incorporate.

There now follows a short description of the purpose of each module:

The Operating System:

The function of the operating system is to schedule processes for running when necessary, to handle inter-process messages, to administrate the use and passage of Dynamic Working Storage, to perform initial processing and vectoring of interrupts, and to perform system validity checks. In addition, the maintenance of system clock and calender and of certain VDU fields is handled by the operating system. The operating system forms the subject of a separate section of this paper.

Command Decode and Editor:

Command decode accepts commands from both the user keyboards and converts them into the required actions; typically this will involve initiating a series of messages between various processes. Some actions are carried out within command decode itself, e.g. setting the system time, but most will require the activation of other processes.

For convenience, the editor is made part of command decode, mainly because various functions (such as accessing disc files, manipulating character strings) are useful to have readily available to other processes. The editor was designed as a simple interactive editor useful for making small changes to disc files. It is by no means an attempt to provide the more comprehensive facilities usually found in a program or text editor.

Other functions which reside in command decode include the facility to copy a disc, load disc files, delete files, obtain a directory listing, and list files to the console.

File handling:

The function of the file handling program is to maintain the file name to disc mapping table (directory) and to process requests for file manipulation from other processes. An important feature of the filing system is that a flexible file structure is used, with the physical disc being allocated to files in a paged manner so files can expand or contract without being relocated or leaving unusable parts of the disc embedded between files. The majority of requests for file handler activity come directly or indirectly from command decode/editor. The file handler uses the disc device control program to actually interface to the hardware disc controller, which is a DMA device.

So that the speed of editing and syntax analysis are not unduly limited by the speed of disc access, a fairly large amount of memory has been reserved for disc buffering, and this buffering is maintained by the file handler, so that the user processes are unaware of it. A number of separate buffers are used, so that if several files are being read from or written to, a buffer can be used for each one.

Disc Device Control Program:

The disc device control program merely translates requests for transfers to and from disc into the appropriate commands for the DMA channel which controls the disc. It also monitors the performance of the discs and drives by analysing any excessive use of repeated read or write attempts; these being carried out automatically by the hardware. The DMA channel interrupts the processor when a series of transfers is completed, and the interrupt is vectored through the operating system to the disc device control program.

General Character Handler:

The general character handler is responsible for setting up transfers to and from the asynchronous devices in much the same way as the disc device control program, except that in this case, the commands are executed by the software multiplexer channel rather than a piece of hardware. The responsibility covers the console device and the asynchronous test lines. It is also capable of driving a line printer, paper tape reader and punch if required, although this feature is not used in the AUTOFLOOD.

Software Multiplexer Channel:

In much the same way as the disc controller channel and HDLC line channel operate, the software multiplexer is continually scanning for commands for transmitting or receiving characters on the asynchronous lines. The multiplexer is run by the operating system at regular time intervals; in fact it simulates the action af a DMA channel which can, if speed is of the essence, be substituted for the software version.

VDU Screen Formatter:

The formatter carries out maintenance of the VDU screen layout, and controls most screen updating tasks, including periodic updating of the various monitor formats provided on the VDU. It handles messages for the VDU in a similar way to that in which the general character program handles messages for asynchronous ports. The actual I/O to and from the VDU is handled by the operating system. The operating system uses spare processor time to send characters to the VDU screen, so unless the tester is heavily loaded, the speed of updating is very fast. Even if the machine is under heavy load, a mechanism exists whereby the updating speed cannot drop below thirty characters per second.

Link Handler:

The link handler performs the dual function of maintaining Level 2 protocols and controlling the transmission and reception of frames on the HDLC cards via the DMA multiplexer channel. It is capable of completely automatic action at Level 2 using either LAP or LAPB protocols, but has interfaces with the Test Sequencer which can override the automatic action, and force fields to particular values if necessary. Like the disc controller, the DMA channel interrupts at the end of a transfer, and the interrupt is vectored through the operating system.

Normally, on receiving a frame, the link handler will see if it is to be handled automatically - e.g. a Level 2 supervisory frame - and either deal with it, or pass it up to the sequencer.

Test Syntax Analyser:

When it is desired to run a test, the syntax analyser parses the test file and generates a number of files and memory resident data structures which are used by the test sequencer to control the test. The syntax analyser carries out extensive checking that the text defining the test is valid, and control will only be passed to the test sequencer if this is the case. The syntax analyser carries out any macro expansion that may be written into the test so as to reduce the amount of processing required at run time. If an error is encountered by the analyser, the line number that is at fault will be flagged, together with a suitable message indicating the nature of the error.

Test Sequencer/Level 3 Call Control

Having processed a test file into the intermediate files and data structures, control is passed to the sequencer. It is the sequencer that forms up packets and passes them to the link handler for transmission. Similarly, when the link handler receives a packet, it passes it to the sequencer for checking. The sequencer is multi-threaded, and is capable of supporting up to sixty four virtual calls. Automatic maintenance of virtual circuits is provided, so that normally generation and checking of sequence numbers would be left to the sequencer. The interface between the syntax analyser and the sequencer provides for overriding sequence numbers (and any other desired field for that matter) if desired. Other functions include execution of requests made in the test file for writing a message to the console log and/or disc file, connecting and disconnecting lines at Level 2 (via the link handler), and waiting for a certain length of time before executing the next event.

Test language Interpreter:

Although the test language is moderately readable to those used to it, to others it can appear rather cryptic. In addition, a high degree of macro expansion is available through the test language, and this can be confusing if trying to understand what a test is supposed to do. For this reason, the interpreter has been provided to take the test file after the syntax analyser has performed syntax checking, and convert it into a more verbose and readable English form.

Debug Utility:

The debug utility is provided mainly for programmers developing the software. Facilities include examining and changing the contents of memory, examining (and changing) certain operating system data structures, and inserting, trapping, and removing breakpoints. So that the disc resident programs can be supported, means of accessing the disc in a carefully controlled way have been provided. It is also possible to simulate a system failure as if a fault has been detected in any chosen process.

Failure Print Routine:

It is to the Failure Routine that control is passed if a software fault or irrecoverable hardware fault occurs. The routine prints out information about the last process to run including its data structures, stacks, dynamic work space owned, and similar information about the current state of the operating system. The object is to provide the engineer with sufficient information to identify the fault, whether it be hardware or software. For more about the failure print routine, see the section on software development.

THE OPERATING SYSTEM

The Ferranti Argus range of processors feature hardware write-protect of the memory in blocks of 4k words under software control. Although the standard operating system makes use of this to the extent that a program will normally have access only to its data area and program area, this means a program can still corrupt itself, even if other programs are still intact. The secure operating system applies protection to the program itself, and to save large rounding wastage, it is therefore necessary to separate the program from its write access data. This forces the programmer to use certain methods of reserving the required amount of memory for write access variables. Similarly, the memory stacks used for program return linkage and temporary data storage must be reserved, and all programs have this write access memory allocated by the operating system in a contiguous block at run time. Operating in this way does form a small restriction on the way the programs are written, but the restrictions are not significant, and well known solutions exist for them. Effectively, all the programs are now in read-only-memory, so even if an error is detected by the operating system, it is necessary only to re-initialise the read/write area of memory and re-start the system, without re-loading the entire memory. As an extra check, when the system starts up, a checksum is formed of the part of the memory that is read-only, and if an error occurs, the sum is re-calculated and compared with the stored value, restarting only being allowed if the checksum agrees.

The stacks are checked by the operating system every time the program is rescheduled to ensure that overflow or underflow has not occurred; if it has, the system is restarted after printing out information relating to the failure so that the programmer has enough information to isolate the bug. Indeed, if the machine generates a software failure for any reason, it will print out information which is usually sufficient to isolate the cause of failure, be it faulty hardware or a bug in the software.

In common with most real-time software systems, buffers of various sizes (called "Dynamic Working Storage" or DWS) are available for processes to use for message passing and data storage. All requests for allocation and de-allocation of DWS are handled by the operating system. A table is maintained containing information as to which process is using each page of DWS at any one time, so that if a process "invents" a page or works on a page which has already been de-allocated, the operating system will print out failure information and restart the machine. In addition, any DWS pointer passed through the operating system is checked for validity before passing it on or allocating/deallocating it. Inter-process messages are treated with the same caution, and again, if an error is found, the system will generate failure information and restart.

Naturally, all this checking constitutes a certain overhead for the machine, and a facility exists whereby the checking can be inhibited. This feature is normally only used when it is essential to squeeze the last ounce (gram?) of performance from the machine - typically when carrying out load testing. The increase in throughput is typically around thirty percent. For most purposes, however, it is usual to leave the checking switched on, just in case something untoward should happen - it does not interfere with normal operation. In fact it is really two operating systems in one: a debug system for program development, and a production system designed for high efficiency.

TESTER DEVELOPMENT

The use of the secure operating system contributed very considerably both to the short time in which the software was developed, and to its subsequent integrity. Faults which were detected only after the AUTOFLOOD was in full-time use were not very large in number, and could in general be corrected without need for long debugging sessions, since the information contained in the failure print-out was enough to give the designers very strong pointers as to where the fault lay. The failure print routine lists out the current contents of the stacks, the local variables and any buffers currently owned by the process that was last scheduled, and similar information for both the interrupt handling and the "organiser" part of the executive. In addition, certain executive data structures relating to the last process are printed, and the current values of all machine registers for all modes (process, interrupt, and organiser) are printed. In the case of certain processes, e.g. the link handler, information relating to the last stream within the last process is printed - in the case of the link handler, this would pertain to the last used port.

The software was developed jointly by the Authors, and was completed in approximately nine months. The majority of the software was developed before the hardware was delivered, thanks to the use of another machine with a similar hardware build. In fact, the development machine was based on an Argus 700 E with a 5MByte cartridge disc, but the 700E and 700G processors are software compatible, and the cartridge disc and flexible disc systems were sufficiently similar for the same software to be used with a few system parameter changes.

The operating system, command decode, debug system, file handling process, and failure print routine were developed first, this giving a suitable environment in which to develop the link handler, syntax analyser and test sequencer processes. After those, the VDU formatter, and test interpreter processes were developed and enhancements to command decode were carried out. Since then some minor changes have been made to most of the modules, and extra features have been added to command decode.

USING AUTOFLOOD

As has already been mentioned, one of the aims of AUTOFLOOD is that it should be easy to use. A brief description of the user commands will now follow.

The commands available from either of the two user positions are:

ABANDON	ADDCALL	CALENDER	CLOCK
CONNECT	DEBUG	DELETE	DIRECTORY
DISCONNECT	DISPLAY	FREE	HOLD
LIST	LOAD	MODIFY	PRINT
RECOVER	RUN	STOPCALL	TEXT
TIMESTAMP	TRAFFIC	TRY	VIEW
ZARCHIVE	ZBINARY	ZKILL	ZLOAD
ZMASK	ZMOVE		

Taking the more basic commands first, CALENDER and CLOCK are used for setting the system calender and clocks, while DIRECTORY is a means of obtaining a list of files on the disc and their sizes. It also calculates the free disc remaining. The list can be directed to either the VDU screen or the console from either keyboard.

DEBUG is a means of invoking the debug utility from the keyboards, and is usually only used by system designers.

Commands for handling files are LOAD which requests text from the keyboard, DELETE which deletes a file after a verification question, LIST and VIEW which produce a full or partial listing of a file to the console or VDU respectively, and MODIFY which invokes editing mode - again either peripheral can be used.

The commands which control starting and stopping of tests

are TRY which will cause the test file to be analysed and run once through, TRAFFIC which is similar but which will only start up traffic call macros defined in the test file, and RUN which starts up traffic call macros and causes repeated attempts to run the functional parts of the file.

The CONNECT and DISCONNECT commands are used to put a particular packet line in and out of service at level 2. This can also be done under program control as part of a test. The HOLD and FREE commands are used to force AUTOFLOOD to send out a level 2 receive not ready and receive ready frame respectively.

Once a test has been started, the ABANDON, ADDCALL and STOPCALL commands can be used to control the aborting of the test or merely stopping and starting specific virtual calls. TIMESTAMP can be used to make the tester append a time field on the end of all data packets leaving the tester, and it will compare the time when the packet arrives back at the tester with the contents of the time field in order to establish cross-office times. The TEXT command is used to invoke the test interpreter so that an English version of the test is produced.

The DISPLAY command will initiate the monitor formats on the VDU, and has parameters relating to which format is required and what the update period is to be. The PRINT command is often used in conjunction with the monitor displays on order to obtain a hard copy of the information currently displayed on the screen.

The remaining commands will not be used in day to day running, but are more of the nature of utilities. RECOVER forces a system restart, and can be used if a test has gone so badly awry that it is quicker to restart the machine. Before recovery is allowed, the software checks that the memory resident programs are intact. ZKILL is the means by which the operating system policing is turned off, and would usually only be used if maximum tester throughput is required. ZBINARY, ZLOAD, and ZMOVE are all concerned with moving files - ZBINARY loads system files to the disc from paper tape or the second disc, ZLOAD does the same for text files (e.g. tests), and ZMOVE is used to move a text file from the system disc to the second disc. ZARCHIVE is used to make a copy of the system disc if security back up is required. Finally, ZMASK is used to control the level of comment produced during test runs.

Testing Method.

To carry out a test, a suitable series of statements must

be written in the test language and loaded onto the AUTOFLOOD disc. This can either be done by typing it in directly using the LOAD command, or it can be prepared on another machine and loaded from the second disc. The TRY command is then used to invoke the syntax analyser which will parse the file and inform the user of any syntax errors. The MODIFY process can then be used to correct the file, and then the test can be TRYed again. AUTOFLOOD will start sending out the packets as requested, and comparing the response with that expected. If a discrepancy occurs, the test will stop and information will be printed out indicating the step in the test at which the error occurred, and the difference between the expected and observed events. The user must then examine the discrepancy and decide if it is the test that is wrong, or if the network is out of specification (or both). The test can be run between line limits, so that if a large test has been prepared, providing it is written in such a way that it can logically be started at some point into the test, it is possible to try out each part without making modifications between attempts.

In this way, the tests eventually become "debugged", and a number of discrepancies will have been found to exist between the expected and actual operation of the network. Extra work will then be needed to decide if the discrepancy is a fault or a misinterpretation of the specification by either party. Another possible reason is that the network parameters have not been set up as was envisaged by the user of the tester.

OPERATIONAL EXPERIENCE

AUTOFLOOD has been used on IPSS, Euronet, and PSS(UK). In each case some AUTOFLOOD problems were found, but in all cases it was possible to fix the problem or provide a satisfactory way to work round it, and testing was never delayed by AUTOFLOOD. The main experience was with PSS, so it is this that will be described here.

It was decided at an early stage of the PSS project that it would be essential to carry out as much testing as possible at the contractors site, so that the fault discover/rectify/test cycle could be kept as short as possible. Accordingly, the contractor agreed to provide accommodation and power for AUTOFLOOD and a small test network so that software validation could be carried out. AUTOFLOOD was shipped to the USA for the duration of the tests. This posed interesting customs problems quite apart from the difficulty of packing the machine sufficiently strongly for air freighting. All these were solved, however, and the machine arrived in USA in February 1980, on a one year temporary import carnet in a very large, solid, wooden crate. A number of interface cables were shipped with AUTOFLOOD, and prior to that a number of BT modems had been shipped and commissioned so that the compatibility of the equipments would be tested. Some of the test circuits were therefore connected via pairs of modems, and others via modem eliminators. Agreement had been reached with the contractor as to a suitable topology for the network under test, and for a suitable mix of port profiles.

Initially, when an attempt was made to run the tests that had been prepared beforehand, it was found that virtually every test produced a discrepancy. Part of the reason for this was that the people writing the tests were fairly new to X25, and some invalid assumptions had been made. The problems were, however, largely due to the rather elastic X25 recommendations available at the time. The tests were modified in some cases, and in others a fault report was raised. In some cases, it was necessary to use an independent passive line monitor to demonstrate the fault to the contractor's satisfaction before he would accept the fault as being in the network rather than in AUTOFLOOD - perhaps not surprising, since the AUTOFLOOD software was if anything newer than the network software. After a while, however, confidence was reached that the results of AUTOFLOOD tests were correct!

As time progressed, the tests evolved and eventually, a large suite of tests accumulated so that new releases of network software could be checked by re-running the tests in the full knowledge that a new discrepancy would be due to a mistake in the network software (with the proviso that fixes to some faults may produce different but valid results in which case the test has to be modified). Since this suite of tests is now reasonably comprehensive, when a new release of network software is given to BT for testing, fairly thorough X25, X28, X29 and X3 tests can be made within minutes of loading onto the nodes (or even test nodes), and a verdict as to whether the new software is usable can be reached quickly.

As faults are fixed, the tests to prove that the fault is cleared can be incorporated into the test suite, forming a valuable check that if a fault is cleared, it does not re-appear in the next issue - it is by no means unknown for a programmer to slip up and incorporate an old module into a new system!

During the period of testing in the USA, a small number of hardware faults occurred with AUTOFLOOD. Initially, an HDLC interface card failed, but it proved possible to repair it in the field, later the Argus 700G CPU failed, but a spare had already been procured in the UK against such an eventuality and it was possible to take it out within a day or two. Apart from one or two other faults which were found to be due to loose connectors and similar easily correctable problems, no failures occured - this in spite of an air conditioning failure which resulted in the test area temperature on one occasion rising to over 140 degrees Fahrenheit! In the last case, the only casualty was the floppy disc currently in use... it melted.

In addition to checking asynchronous character terminal and packet protocols, AUTOFLOOD was used to provide sample calls with precisely known characteristics for checking the accounting part of the software.

A very extensive area of trouble was network performance under high traffic load. This too would have been difficult (if not impossible) to measure without the facility of AUTOFLOOD, since no test gear existed at the contractor's site which was capable of generating the traffic throughput of AUTOFLOOD. The traffic testing showed that initially the network was very prone to failure under traffic load, and that the network did not meet the specified throughput.

Once the hardware and software had reached a stage at which it was possible to operate a network in reasonable agreement with CCITT X25 and with a reliability adequate for further testing with customers, the software was shipped to the UK and loaded onto the hardware which had been installed during the period of software acceptance testing. The project then entered a phase known as Network Acceptance Testing. AUTOFLOOD was also returned to the UK, and its use continued for some months finding out more information about known faults, locating new ones, and carrying out special investigations into customer related problems. AUTOFLOOD is still in use mainly for testing new releases of software, and for providing background traffic on what is as yet a fairly lightly loaded network.

SUMMARY AND CONCLUSIONS

The BT AUTOFLOOD proved an extremely powerful and flexible tool during the software acceptance test phase of PSS. A large number of faults were found during that time, and the six week period originally set aside for acceptance had to be extended considerably. It is quite certain that without a machine such as AUTOFLOOD, it would have been impossible to locate all the faults that were found at such an early stage.

Interest has been shown by several other organisations in using AUTOFLOOD to test their own implementations, or those they are buying. It remains to be seen how applicable our test tables are to other implementations. It may well be that they will require a certain amount of modification, but it is not anticipated that any changes will be needed to the AUTOFLOOD operating software apart from some changes already specified to enhance its X75 capability which, although desirable, are not essential for X75 testing.

The reason AUTOFLOOD was developed was that no tester offering the required facilities could be found "off the shelf". AUTOFLOOD is capable of driving packet level protocols and asynchronous protocols simultaneously, it can operate up to eight packet ports simultaneously, can send or receive any frame or packet, and can support a throughput of about four hundred frames per second. The monitoring facilities enable easy measurement to be made of network throughput and cross network delay times, and to see at a glance the performance of a particular line into the AUTOFLOOD.

It has been found that these facilities have all been useful in enabling comprehensive testing of a network as complex as a packet switching network, and the fact that they are combined in one piece of test gear was highly expedient bearing in mind the location of the contractor.

The views expressed here are the author's personal views and do not represent British Telecom policy statements. Acknowledgement is made to the Senior Director of Technology Executive for permission to make use of the information contained in this paper.

References:

1. British Telecom Packet Switching Service Technical user guide:

2. CCITT Recommendations X25, X28, X29, and X3.:

3. Packet Switching System Tester - Test Sequence Control Language: C. G. Miller, BT document issued July 20th 1980

4. Packet Switching System Tester - Command Structure: M. J. Norton, BT document issued November 11th 1979

5. Packet Switching System Tester - System Build: M. J. Norton, BT document issued March 1980

6. Experience in Software Test Techniques For Packet Switching Exchanges: M. J. Norton, Proceedings of the Third international conference on Software Engineering for Telecommunications Switching Systems

INTEGRATED NON-VOICE SERVICES IN THE ELECTRONIC OFFICE

P Drake & M J Norton

British Telecom
Product Development Unit &
Systems Evolution and Standards Department

This paper outlines some of the problems and trends in computer communications within and between offices. Four central themes are introduced. The first discusses information handling requirements and user problems related to the wide variety of available terminals and information services. The theme finishes with the ideal solution, an integrated non-voice service network. The second shows the evolution from centralised to distributed architectures and its effect on networks and non-voice services. The third describes how existing non-voice services eg telex,teletex, videotex etc can be integrated into levels 5-7 of the open system interconnection (OSI) model. The fourth then describes existing wide area (WAN) and local area networks (LAN) in relation to OSI and the electronic office.

From this two types of electronic office networks are discussed. In case 1 an Ethernet local area network operating under a distributed operating system provides resource sharing and dynamic multiplexing at each workstation/server. In case 2 a digital PABX provides a terminal access network to a number of servers. Terminals strictly converse with one server at a time. These servers may be interconnected but this is independent of the PABX.

The paper then shows a possible interconnection of voice / non-voice ISDN LANs and WANs.

K. G. Beauchamp (ed.), New Advances in Distributed Computer Systems, 69–86.

1. INTRODUCTION

British Telecom in particular, and the PTTs in general, are committed to provide integrated voice/non-voice communication services (ref1). In the future, the PTTs are intending to provide such services using the proposed CCITT defined Integrated Service Digital Network (ISDN) (ref 2).

Simultaneously, the computer industry is evolving towards distributed processing and multi-vendor attachments as hardware costs fall and software costs rise. This is leading the industry into providing common interface standards and a greater emphasis on standard modules communicating through standard protocols.

The development of the OSI model (ref 3) is the logical outcome of this convergence between the computer and communication industries. The model provides a guideline to build an integrated non-voice service. The first realisation of this service is likely to be in the electronic office which exhibits most of the problems associated with a global service on a smaller scale.

In order to understand the problems of the electronic office it is necessary to perform a top down analysis, starting with the user's information handling requirements and finishing with the network solutions. The result of this is that three distinct trends can be ascertained. The first, is the evolution of computer systems from centralised to distributed as represented by local area networks (LANs). The second, is the evolution of the mail service from electronic message services into an electronic mail system. The third, is the evolution from analogue to digital circuit switching networks which will act as access circuits to centralised and distributed computers. Each of these three trends complement each other.

2. INFORMATION HANDLING REQUIREMENTS

Future electronic offices require that office personnel operate in a new way and in a new style organisation. The success depends on the man-machine interface and on the non-voice services providing effective information (ref 4). As there is no clear definition of what is effective information, these systems must be highly adaptable. Such adaptability will only come if the system is based on modular units connected by standard interfaces (see section 2.3).

Many offices have information handling problems which are difficult to quantify. Data only becomes information to the business user when it is transformed into a form useful for presentation and decision making. Information quality is more important than information quantity and identifying decision makers and their information requirements is the first step in problem solving.

2.1 User Types

Fig 1 illustrates a typical division of office personnel into four main groups (ref 5).

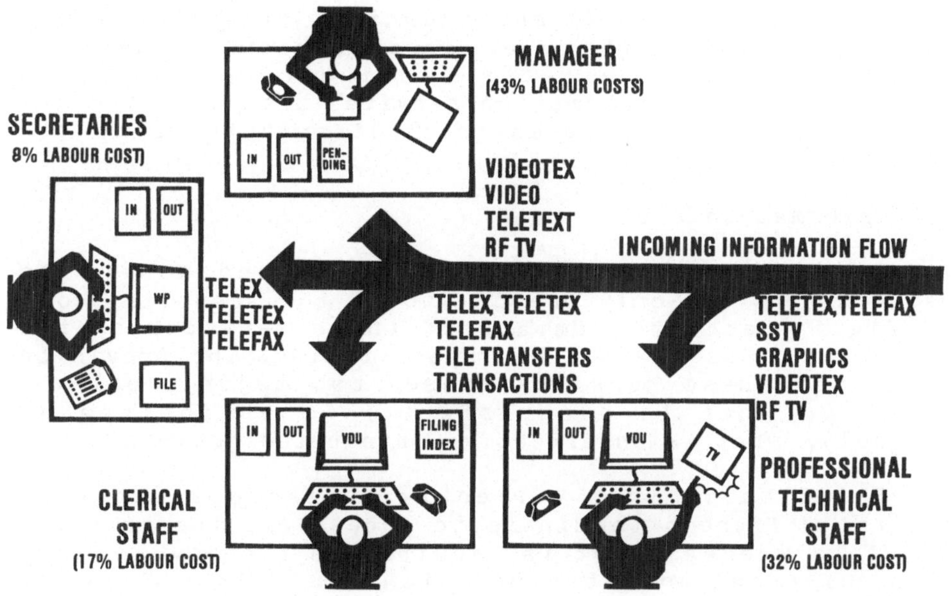

Fig 1: Information Flow in the Electronic Office

Management (43% costs) including all adminstration from 1st line to senior positions require filtered information presented in a natural environment eg colour, graphics and high resolution. This information should be easily accessed with minimum training. This group require decision support systems, videotex services (private and public) and electronic mail facilities.

Secretaries (8% costs) are defined as aids to management and act as mail offices to filter out the important mail, maintain limited filing systems, provide letter typing facilities. They also act as an information desk on the whereabouts of office personnel. This group mainly use the electronic mail services eg telex, teletex, fax etc.

Clerical (17%) staff are responsible for all those office procedures and information structures which are sufficiently well defined and routine that comprehensive filing systems and operational summaries can be maintained. This group mainly use paper systems analogous to database services, both locally and remotely.

Professional (32%) and technical staff cover all the specialists within an organisation eg drawing office staff, architects, engineers etc. Their requirements for information services are unlimited and they will be the prime users of distributed computer systems.

2.2 Information Flow

Currently, most office systems and non-voice networks provide only one or two of the services shown in fig 1. Hence the impact of these services on the organisation as a whole has been limited. The provision of all of these services causes two major problems. Firstly, what is a satisfactory user interface? Secondly, what is an effective integrated network?

Fig 2 shows that currently a user must acquire several different terminals to access all the provided services whereas what is required is the development of a universal workstation which can be interfaced to all the services. However such a workstation should be modular and allow the support of only one service (for cost reasons) and yet be upward compatible with a workstation supporting all services. This leads to the concept of a class of non-voice services which are upward compatible.

The problem is compounded by the requirement that one of these workstations may require access to a number of different common services at the same time.

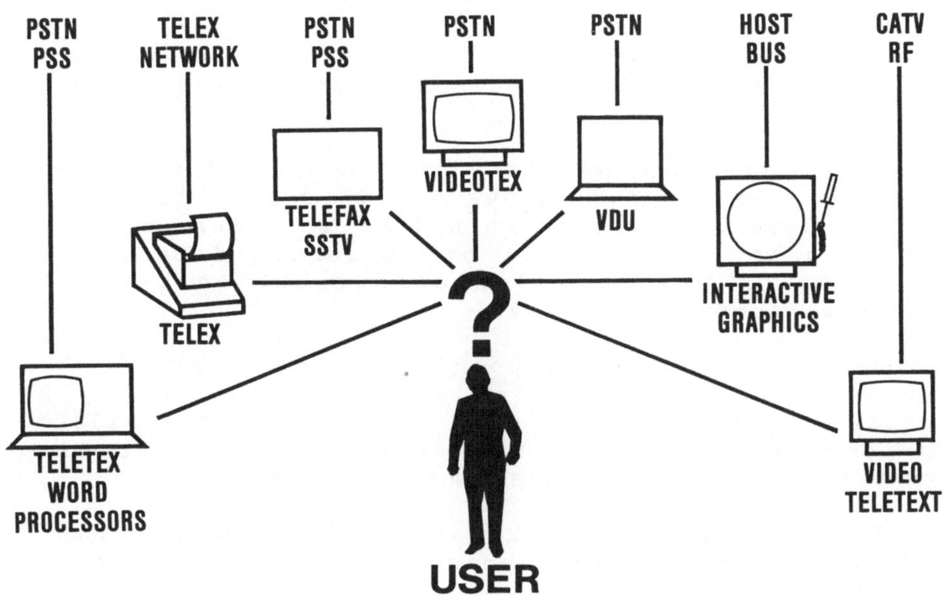

Fig 2: The User's Dilemma

The problem is further compounded by the requirement for forward and backward compatibility between terminals and common services. This concept is of central importance. Forward compatibility means that information structures must be designed in such a manner that future workstations will be able to access old data. Backward compatibility (which is more difficult) means that existing old terminals will be able to display and manipulate new information designed for advanced workstations (ref 6).

Most existing non-voice services do not allow this forward / backward compatibility. The reason for this is that each service was designed to meet the needs and technology of the moment. Hence telex, telefax, viewdata, broadcast TV, interactive graphic terminals are closely associated with the type of information they display. Future services must separate terminal characteristics from information structures. This is the reason for defining the terms user and resource services (see section 2.3) and the classification of presentation protocols in level 6 of the OSI model (see section 4).

2.3 The Ideal Integrated Non-Voice Service

The user's ideal solution to the above problems is based on two concepts called user service and resource service (ref 7).

User service supports a user via a friendly man-machine interface and allows him to enter, manipulate and retrieve information in a form most meaningful to him. A user service will be implemented on a user workstation which will also support an external standard network interface.

Resource service supports a common service eg database, printer, special processor, which may be required by several users. A resource service will be implemented on a resource server which will also support an external standard network interface.

Fig 3 shows two workstations each supporting a different user service and capable of communicating with each other and with all the resource servers.

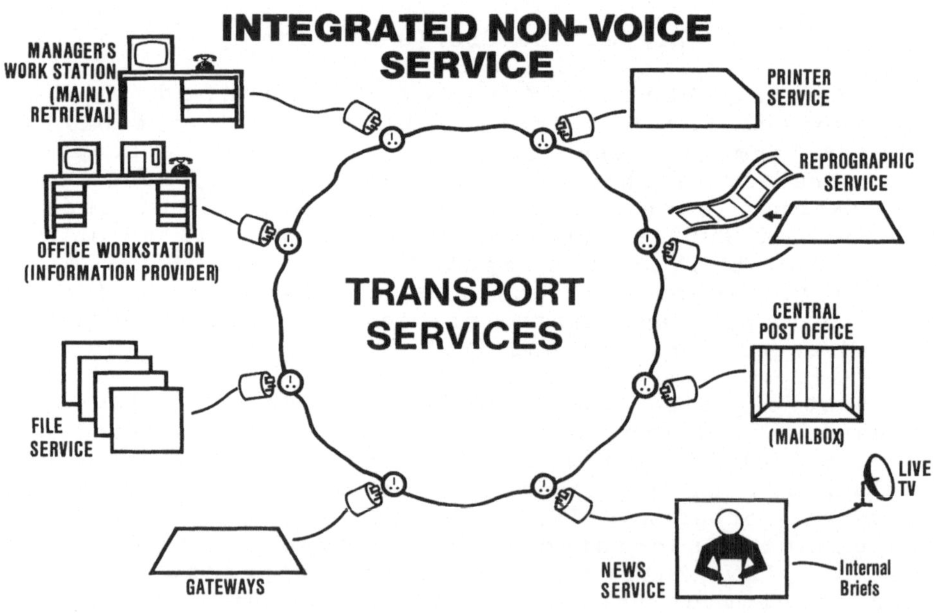

Fig 3: Integrated Non-Voice Service

The two different user services represent support to two main groups of users, namely users who are information providers eg secretaries, clerical and professional staff in the electronic office, and users who are information retrievers eg management. This division into groups is meant only to reflect the main interest and both groups can be information providers and receivers. The workstations may also be customised by downloadable software from a "software warehouse" server.

The number and type of servers will depend on the cost and user requirements. Basic servers include printer, file manager, central post office, reprographics unit, news, and network management units.

Workstations via their transport interface could also support value added facilities eg electronic mail, teleconferencing and multi-call capabilities. The support of these facilities requires that part of the workstation would be permanently connected to the transport system and that this part would be "open" to unsolicited mail, messages and enquires. For those workstations which cannot support this, a value added server may be provided akin to the post office, where a workstation will collect it's mail at it's convenience.

Because workstations and servers are connected by a standard transport system and independent of each other new workstations and servers can be added and removed to meet the requirements of the particular office. The result of adding or removing a workstation is a personal one for the particular user. The result of adding or removing a server is to change the facilities and quality of service offered to the users.

3. OFFICE SYSTEMS EVOLUTION

3.1 User Service

In the centralised approach non-intelligent or semi-intelligent terminals were augmented by user support software in a central computer. The association between the user's actions at the terminal and the centralised user support software was very close.

Hence the access network for remote terminals required high reliability, transparency, realtime response and two way working eg point to point or circuit switched (real or virtual) connection. Many of the interactive communication protocols eg X28/X29/X3, videotex etc reflect this two way interaction and close association of remote terminal and central user software. The virtual circuit concept of X25 is primarily to maintain this close association.

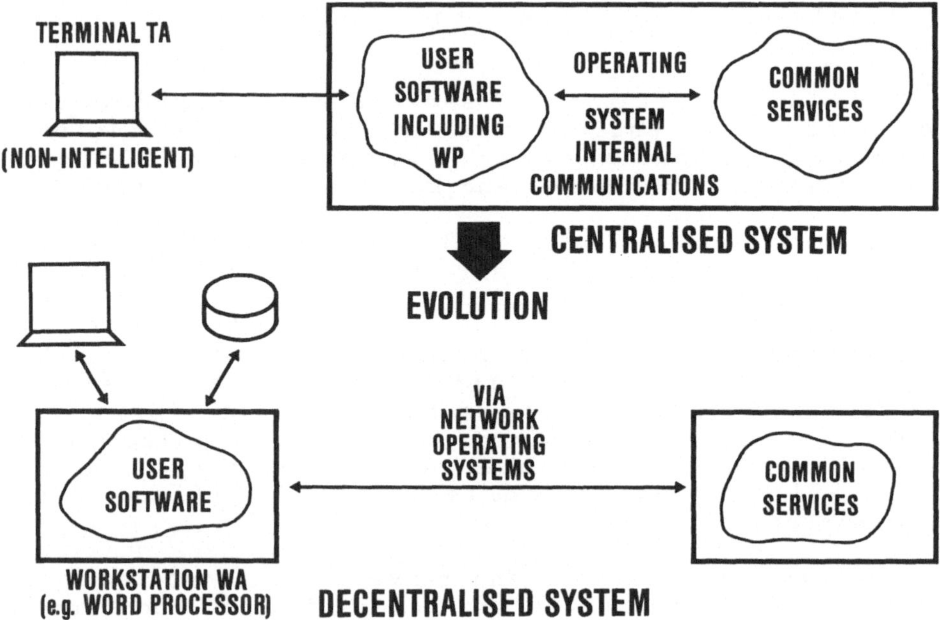

Fig 4: Evolution of User Support Software

The trend to distributed systems and the implementation of user support software in the user workstation (see fig 4) alters the need for this static association to a central system. In the distributed system user workstations require multi-associations with servers and each other and thus require dynamic multiplexing capabilities.

3.2 Resource Services

The resource services are also evolving from centralised to distributed architectures. This involves servers requiring multi-association with each other based on the following three dimensions: processing, database structure and control organisation (ref 8).

3.3 Distributed System Trends

From the above, three trends are seen in electronic

office systems. The first is where the user can access multi-facilities without being aware of the physical structure of the system. The system may be a central computer or a distributed network. In both cases the user sees the same facilities although with different qualities of service. The user is able to interactively "pull" information from the system. Characteristics of the underlying network are high bandwidth, resource sharing, complex process synchronisation and multi-access. The primary office user would be clerical staff for database adminstration, and professional / technical staff to access specialised servers, databases etc.

The second trend is where a user is aware of a local and remote environment ie the user requires to send information to a distant user or system. The user does not expect global interactive access to all servers when he is in this mode. He expects to send and collect information akin to the postal service. This trend which started with telex, and then message switching services fits under the generic name electronic mail. There is no global distributed operating system masking the physical structure from the user and providing a global interactive service. Characteristics of the underlying network are medium bandwidth, no resource sharing, simple session control, low network activity and multi-access. The primary office users are management and secretaries.

The third trend is where remote workstations interface to centralised or distributed systems via digital circuit switch networks. The remote workstation provides all the facilities of the first and second trends with the one exception, that multi-access is difficult from a circuit switched channel. Hence the workstation in general converses with only one server or user at a time. Such a configuration would require third party servers to act as mailboxes, enquiry points for unsolicited information. The third trend is likely to be seen in offices which have existing digital PABXs and require access to remote centralised or distributed systems.

In practice an organisation which has a large number of electronic offices will provide centralised and distributed systems which will exhibit each of the three trends.

4. THE OSI MODEL AND ASSOCIATION SERVICES

The OSI model has been defined (ref 3) in order to provide the modularity required in distributed systems. This is achieved by making the peer to peer protocols on one level independent of the service at lower levels through the use of a standard service access method. The definition of this access method does not make any assumption on the internal structure of the lower level service and hence levels become independent (ref 9). This allows each level to evolve and change without affecting higher or lower levels. Thus new hardware, software, networks and information services can be incorporated in a system without disturbing the existing capabilities.

Using this concept and the 7 layer model this section looks at existing non-OSI services and trys to integrate them into the OSI structure. This is done firstly for the association services (levels 5-7) and in the next section for the transport/network services (level 1-4).

The session level 5 provides three dialogue services: one way interaction corresponding to electronic mail; two way simultaneous (TWS) interaction corresponding to distributed operating systems control; and two way alternative (TWA) interactions corresponding to interactive user access protocols.

The presentation level 6 reflects the wide window of data descriptions from bit encoding to document oriented protocols. An integrated service must provide the forward and backward compatibility described early.

The application level 7 has already been discussed in the previous chapters. A number of application groups exist including electronic mail, distributed file management, distributed processing, teleconferencing, process control, information provision, and information retrieval.

Looking at X28/X29/X3 and the teletex S61/S62 protocols (ref 10) it can be seen that there is a trend towards upward compatibility with each other. This is being reflected in the fax standards where the new group 4 protocols are attempting to be defined on the base of the mixed mode teletex standards. Similiarly videotex standards are evolving to include text, mosaic graphics and geometric graphics within a defined standard.

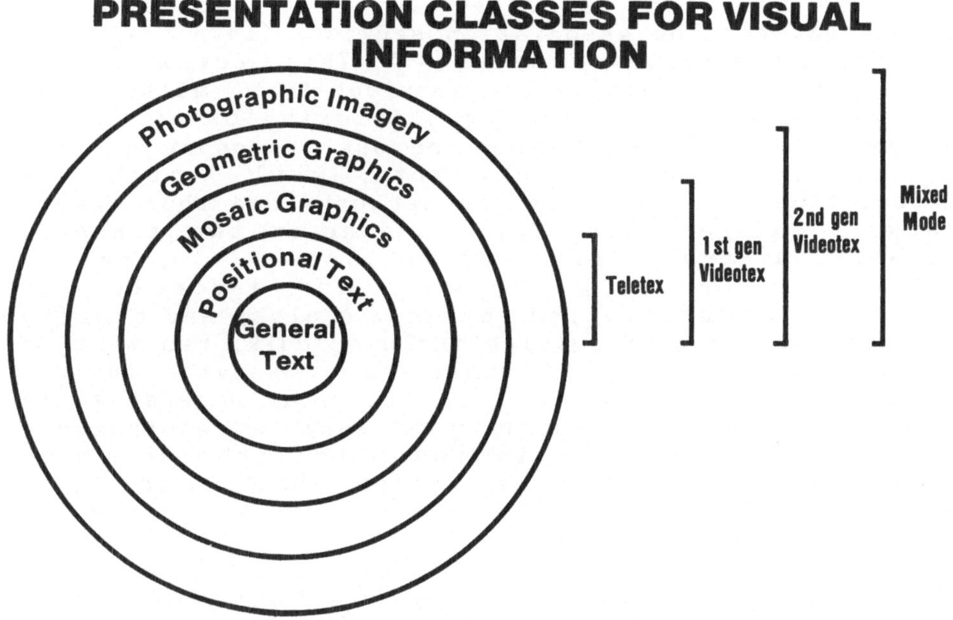

Fig 5: Proposed Presentation Classes

Fig 5 shows the logical progression of the presentation classes for visual information, based on forward and backward compatibility. The development of a class of presentation protocols should allow a simple text only terminal to access complex visual images and display only the text portion of these images (backward compatibility). Similiar, complex image displays should also be capable of accessing text only database servers (forward compatibility). Another feature of this compatability is that high resolution images should be displayable on lower resolution displays and vice versa. Thus it is seen that presentation protocols should not only consist of the application information but should also described the display media for which they were intended. This description will be called presentation control and with it information can be described in it's ideal form. Terminals and workstations can then display the information in the manner most suitable for the terminal capabilities.

4. TRANSPORT PROTOCOLS AND NETWORK TECHNOLOGIES

4.1 Transport Service

The transport service (level 4) provides the bridge between the association services (level 5-7) and the network services (levels 1-3). This service allows either different association services to be multiplexed on one network ie upward multiplexing or different network services to converge on one transport service ie downward multiplexing or both. Level 4 masks the association services from the networks and thus ensure network independency. This is done in two manners: firstly, by providing local to global / logical to physical addressing schemes; and secondly, by offering the association services a class and quality of service. Class of service includes bulk, realtime or dialogue data transport; qualities of service include error detection, recovery, flow control. Depending on the cost versus class/quality of service requirements the transport service is there to optimise which network service a particular session should used. It is also there to provide alternative network services in case of network failures.

CCITT are currently discussing a number of transport negotiation classes including a simple class 0 (similiar to teletex S70), a basic class1, a flow control supported class 2, an error recovery class 3 and error detection and recovery class 4. Which of these classes are required for a session will depend on the cost/network capabilities trade-off. For example, a point to point network will have less problems than a mesh network and hence a lower level transport negotiation class will probably be sufficient for a particular quality of service.

4.2 Network Services

Networks services can be divided into local area networks (LANs), wide area networks (WANs) and gateways connecting all combinations of the two types.

Local area networks are a combination of distributed processing and backend storage networks (ref 11). Implementation of this combination has taken different forms depending on where the network designers decide to place the network (see section 3.3). The various forms are digital PABX/office controller, packet switched star, ring, loop and multi-access bus. The differences between the types are dependent on the level of coupling between the workstations and servers.

Rings, loops and buses are characterised by their ability to provide the correct media for closely coupled workstations and servers ie they are conducive to operating system environments. LANs could also be used for the loosely coupled electronic mail requirements.

Packet switched star networks can also provide many of the above capabilities but with reduced performance and a higher protocol overhead. However, where such a LAN requires to communicate with other similiar LANs via a packet switched WAN then this solution as some practical advantages especially in minimising gateway developments.

Digital PABXs provide only access circuits to centralised or distributed servers and hence are not suitable as a media for a local network operating system (see section 3.3).

Wide area networks available to interlink office systems and LANs are as follows: the analogue public switched telephone network (PSTN); the telex network; the X25 packet switched network, the private wire point to point network; the message switched network; the cable television network; and finally, the broadcast network. Each of these networks offers different facilities but non provide all. Within the UK it is planned to provide a private circuit digital data network based on X21 interfaces upto 48kbits/sec or 64 kbits/sec transparent in 1983; a circuit switched digital network for voice or data using System X and thirdly a satellite business network. Each of these, along with the packet switching service, will form the basis of the future ISDN planned for trials starting in 1983.

Of crucial importance in an integrated network is the development of gateways to connect LANs to WANs, LANs to LANs and WANs to WANs. The telephone networks have had gateways for many year at the international boundaries between national networks and packet switched networks have also developed such internetwork gateways. LAN/WAN gateway development will be difficult because of the wide differences in the characteristics of LANs and WANs and the continually changing environment of LANs which will evolve much more rapidly than WANs.

5 SOME POSSIBLE IMPLEMENTATIONS OF LAN/WAN ARCHITECTURES

In order to illustrate the various aspects of the electronic office in term of information handling, OSI protocols and network topologies, two examples will be examined. The first represents the closely coupled LAN and the second the digital PABX/server network LAN solution.

5.1 Case 1 Closely Coupled LAN - Ethernet

Ethernet is an example of a contention multi-access bus LAN which with it's high connectivity, equal access from all stations, high speed, and non-blocking throughput, provides an environment on which a local network operating system (LNOS) can be built. Similiarly, loop and ring networks have equivalent characteristics. Ethernet has only been defined for levels 1,2 (ref 12) of the OSI model. Hence data is sent and received in level 2 frames using an access protocol with "transmitframe" and "receiveframe" primitives. As all stations on the bus can listen to all data sent from each sending station then full connectivity between all stations (workstations and servers) is available.

Ethernet has not yet standardised on a level 3 protocol. This network protocol could be a packet switching virtual call protocol or a packet switching datagram protocol or even a circuit switching protocol ie continous bit stream on one channel with no packetisation. The virtual call protocol will probably be specified in accordance with X25 but minus many of the facilities fields which are unnecessary because of the high reliability of the level 2 protocol in terms of lost or mis-sequenced packets. However, for many LNOS a datagram level 3 service would be the closest to inter process communications within existing operating systems. Hence it is likely that a datagram network protocol will be specified to aid the operation of the LNOS scheduling, inter process activations etc. A circuit switched level 3 protocol will be necessary to meet the realtime requirement of voice traffic and some higher level bit stream protocols.

5.2 Case 2 Limited Access LAN - Digital PABX

For electronic offices which have or will have a digital voice PABX then the addition of 64kbits/sec circuit switching is an easy progression. Similiarly for offices which only require limited flexibility where one workstation requires to be connected to one server at a time and only switches between servers after long intervals then the digital PABX is the best solution. Its main weakness is it's limited switching speed in a dynamic multiplexing environment.

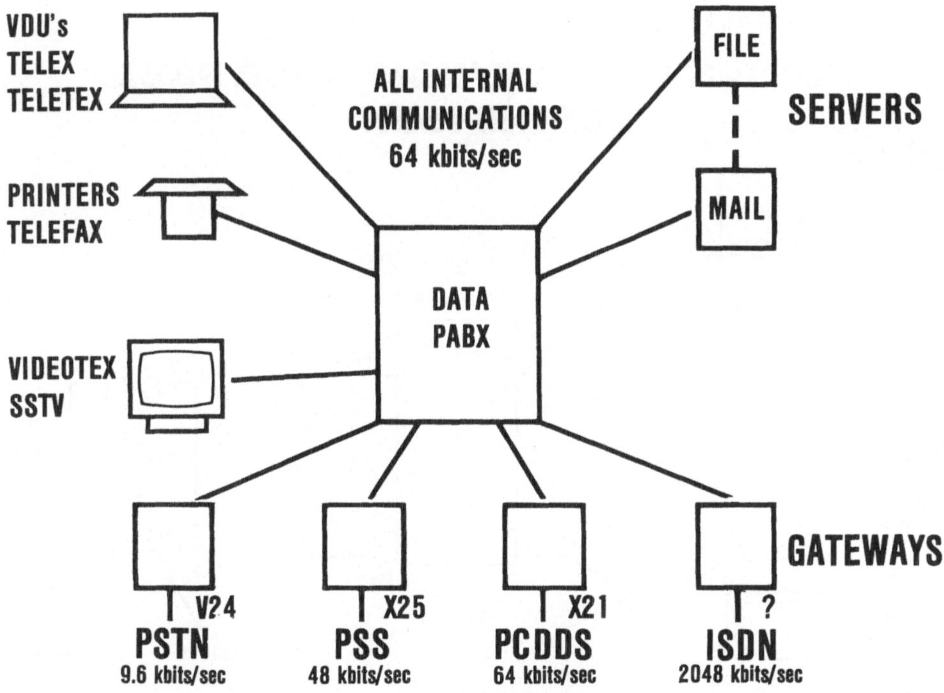

Fig 6: The Data PABX

However, these limitations are also its strengths in that it is simple to comprehend, can be added to a well developed product line and is less sensitive to misuse. Fig 6 shows how workstations would access servers and also illustrates the affect of a number of gateways to WANs.

6. FUTURE DEVELOPMENTS

The ideal integrated non-voice service requires

many years of further development. The development of presentation and session protocols is needed before services can be truly integrated. The transport protocol is the central pivot in an effective service because it masks the evolving networks from the evolving association services. One fundamental problem must be solved before an integrated network is available. This is the interworking of circuit and packet switching systems within the LAN and WAN environments. Circuit switching is based on the human requirement for a continous speech circuit. Packet switching is based on the machine requirement for multi-associations.

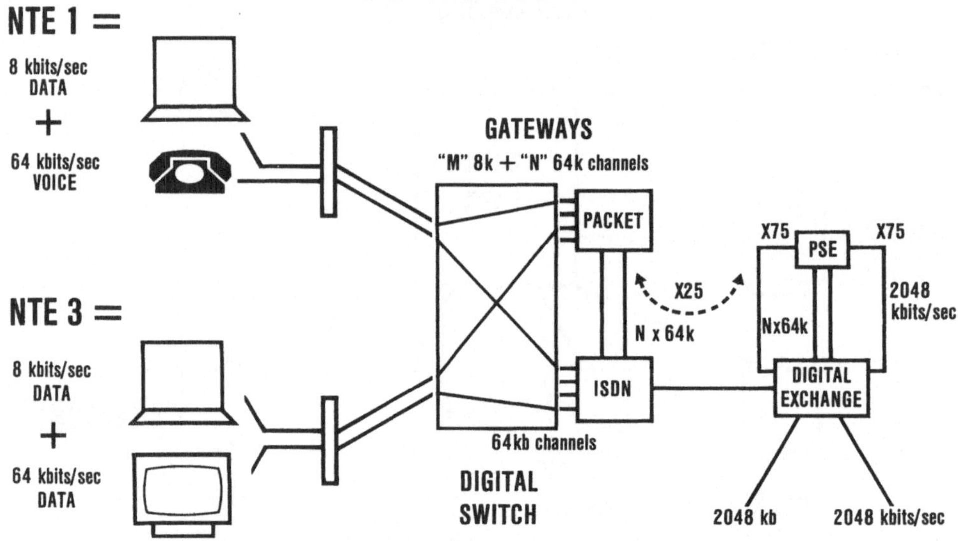

Fig 7: Possible Future Solution

One possible network implementation of this is shown in fig 7 where access from a digital PABX allows both voice and non-voice integrated traffic to flow through a future ISDN network. Voice using the circuit switching element of the PABX and ISDN. Non-voice using its LAN (ring, bus, packet node) and hence via a gateway to the packet switching exchange at the ISDN network local node.

7. ACKNOWLEDGEMENT

The views expressed here are the authors' personal views and do not represent British Telecom policy statements.

Acknowledgement is made to the Senior Director of Marketing Executive for permission to make use of the information contained in this paper.

References

1. Non-voice Network Services - Future Plans. PTF Kelly ONLINE

2. Preparations for the Evolution towards an ISDN. AG Orbell ICC Boston June 1979.

3. Reference Model of Open Systems Interconnection. ISO/TC 97 SC 16/N227.

4. Integrated Office Information Systems. M D Zisman Infotech State of the Art 1979 Office Automation.

5. Technology and the Office of the Future. B W Manley, Electronic and Power IEE Jan 1981

6. Picture Description Instructions (PDI) for the Telidon Videotex System. H G Bown, C D O'Brien, W Sawchuk, J R Storey, CRC No 699-E, Department of Communications, Canada.

7. Open System Architecture for Military Command, Control and Communications. P Drake, MEDE 80, Wiesbaden Oct 1980.

8. Distributed Computer Systems Impact on Management, Design and Analysis. G A Champine, R D Coop, R C Heniselman, North Holland Publishing Co 1980.

9. Open System Architecture. H Zimmerman, Proceedings of Eurocomp 78, Online Series.

10. Teletex and it's Protocol. GA Routhorn, PA Carruthers. IFIPTC-6. International Symposium of Computer Message Systems, 6-8 April 1981.

11. Introduction to Local Computer Networks. H A Freeman, 5 th Conference on Local Computer Networks. Minneapolis Oct 6-7 1980.

12. The Ethernet. A Local Area Network Data Link Layer and Physical Layer Specification. Version 1.0 Sept 1980. IFIP WG 6.4, Local Computer Networks, Dec 1980.

13. The Integrated Services Digital Network (ISDN) and its use for Text and Data Communications. P Bocker, P R Gerke. IFIPTC-6, 6-8 April 1981.

OPEN SYSTEM INTERCONNECTION AND THE INTEGRATED SERVICES DIGITAL NETWORK

W.A. McCrum

Department of Communications, Ottawa, Canada

Recent advances in digital technology have made it technically and economically feasible to consider application of digital techniques to the telephone network. Investigation of the economics of adding other services for support of machine-to-machine communication points to a scenario where a wide range of services are carried on an Integrated Services Digital Network (ISDN).

This paper discusses the driving forces leading to development of an ISDN and deals with the importance of Open System Interconnection (OSI) concepts to its orderly evolution. The application of OSI protocol and interface standards to the ISDN is presented as a potential solution to the problem of consolidation of diverse networks.

TELECOMMUNICATION NETWORK EVOLUTION

For many years the telephone system has met the basic communication/information exchange needs of developed Western countries such as Canada. However, the emergence of information-oriented societies and the continual declining cost of micro-electronics has resulted in communication facilities of the telephone system becoming inadequate. Consequently beginning in the late 1960's separate telecommunication networks began to appear, tailored to the more exacting performance requirements for communications between machines configured for distributed computing applications.

K. G. Beauchamp (ed.), New Advances in Distributed Computer Systems, 87–96.

These dedicated data networks, both public and private, employed digital technology and were designed to accommodate a broad array of terminal devices and systems. In fact the search for improved productivity behind the proliferation of terminals and systems was what led to such networking.

The rapid advance of technical innovations was such that the existing telephone system with its own inertia against change through sheer size and complexity was unable to respond to meet the networking needs in a timely and economic way. Even large corporations in the business of design and manufacture of high technology communications products faced this problem. To bring order into the chaos of interworking between their own terminal and system designs they introduced the concept of an integrated proprietary network architecture. IBM alone had over 30 different network access methods and 15 different line protocols when it introduced its System Network Architecture (SNA) in 1974 - other major corporations followed IBM's lead shortly thereafter.

Now that digital technology may be applied economically to the telephone system the telecommunication carriers are accelerating their investigation of the technical and economic feasibility of a single digital network supporting a wide range of services both voice and non-voice. Although separate networks for data, telephony and other services are justifiable for the time being, the desirability of service integration for the purpose of user convenience and network economy will ultimately lead to a common user network - the Integrated Services Digital Network (ISDN). As with the equipment and system suppliers, so will the communications service suppliers require an integrated network architecture to ensure an orderly evolution of their business.

NETWORK AND SERVICE INTEGRATION

Network integration and service integration are not new concepts and it is well known that the existing telephone system has supported data, voice, facsimile, and broadcast program traffic for many years. However, performance limitations in areas such as error rates, bandwidth, and phase distortion, together with the need for various terminal handling features, plus digital to analogue conversion, have accelerated the development of network alternatives. These alternatives which include digital private line, circuit and packet switching, appear to be re-converging with the application of digital technology to the voice network.

In Canada, for example, over 96 percent of all households are connected to the telephone network. This network, which provides such widespread telecommunications coverage, is an obvious vehicle to be exploited to the fullest extent for provision of additional non-voice services. From the carriers perspective this would provide the broadest possible sharing of network resources thereby avoiding costly duplication of facilities, administration, and maintenance functions. From the users perspective such an integration of services potentially provides the required degree of service accessibility and interconnectivity at lowest possible cost.

At the present time a number of developments are taking place in parallel in the area of telecommunication networks evolution.

(1) Public switched data networks are being implemented with appropriate protocol and interface standards. Examples of these networks are Datapac and Infoswitch (Canada), Transpac (France), PSS (U.K.), and Telenet (U.S.A.).
(2) Local area networks are being implemented with vigorous activity underway to establish protocol and interface standards. Examples of such networks are Ethernet, Mitrenet and NBSnet.
(3) Integrated digital transmission and switching for voice communications is being implemented with appropriate interface and signalling protocol standards under development.
(4) Additional non-voice services are being planned for the integrated digital voice network which will ultimately result in the Integrated Services Digital Network (ISDN).

It is expected that subscriber requirements and economic benefits will eventually result in the ISDN encompassing at least the data communications networks mentioned in (1) and (2) above. Progress towards this scenario, where public data networks and local area networks are overlaid on an integrated digital voice network, would be greatly facilitated if each system were evolved in conformance with a basic reference network architecture.

NETWORK ARCHITECTURE

The ultimate goal in new communications systems development is to have universal accessibility and interconnectivity like the telephone system of today. The difference is that the intelligence for many types of access will reside in terminals and not the human operator. Since the intelligence in one terminal is required to communicate with that in another, clearly a common communications structure must be used. This is the thrust behind the standards development work of the International Telegraph and Telephone Consultative Committee (CCITT) and International Organization for Standardisation (ISO) aimed at establishing a standard for Open Systems Interconnection Architecture, known internationally as the Reference Model for Open Systems Interconnection.

The Reference Model for Open Systems Interconnection (OSI) organizes in hierarchical levels all possible functions necessary for the correct exchange of information between two entities wishing to communicate by means of a communications network. In the Reference Model structure the various protocols may be standardised ensuring compatibility at various levels as required.

Within both CCITT and ISO draft Reference Models for OSI have been developed with good alignment between them. It is expected that complete alignment will be achieved. The models developed so far comprise seven layers which may be viewed as subdivided into two sets:

- o a 3-layer user-oriented set viz. the application, presentation and session layers
- o a 4-layer data transport-oriented set viz the transport, network, access link and physical layers.

A layer is formed by a group of functions providing a layer protocol. Entities operating at the nth layer communicate by means of an nth layer protocol. The nth layer provides a service to the $n+1^{th}$ layer only and uses the services offered by the $n-1^{th}$ layer only. Each layer is completely independent of the other so that modification of its functions does not impact another layer.

Figure 1 illustrates the seven-layer Reference model showing peer to peer protocol interrelations and layer interfaces.

The Physical Layer provides mechanical, electrical, and procedural functions in order to establish, maintain and release physical connections between the user and the network.

The Data Link Layer provides functional and procedural means to establish, maintain and release data link connections. These connections are made over already established physical connections.

The Network Layer provides the functional and procedural means to exchange data between the user and the network. It relieves the Transport Layer from concerns about switching, routing, and

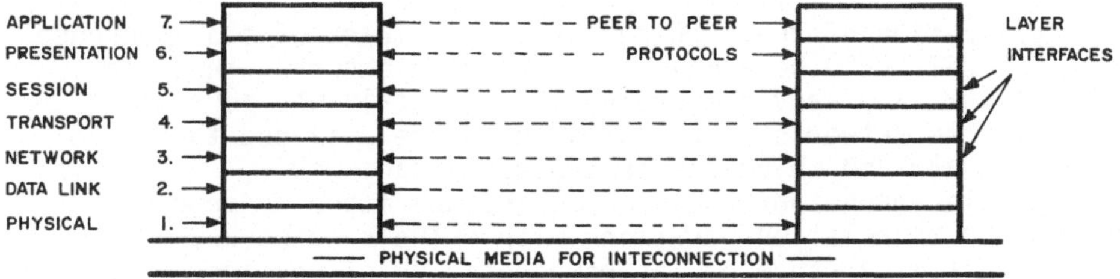

FIGURE 1
Seven layer reference model showing peer to peer protocol interrelation and layer interfaces.

network access flow control.

The Transport Layer provides for transparent transfer of data and essentially enhances services provided by the Network Layer.

The Session Layer supports the interactions between entities in the Presentation Layer by essentially providing the "tools" in logical groupings. Included in its functions are the control of data exchange and management of session dialogue.

The Presentation Layer provides the set of services which may be selected by the Application Layer to enable it to interpret the meaning of data exchanged. Thus in a sense this layer organizes the tools required by the Application Layer to execute its function so that the tools are presented in logical sequence.

The Application Layer is the highest layer in the Reference Model and the protocols of this layer provide the actual information processing. Within this layer are the ultimate source and sink for data exchanged.

As shown in Figure 1, peer to peer layer protocols provide for data exchange between systems - all layers need not be present in a given system. Interface protocols provide for data exchange across layer interfaces. The Open Systems Interconnection Architecture provides the environment in which standard protocols and interfaces may be defined.

In the case of the ISDN, present efforts internationally are oriented towards development of an architecture providing adequate flexibility to permit building a wide range of non-voice services on the emerging digital voice network, and to identify specific service types.

Application of the architectural principles of the Reference Model for OSI would be a major step towards ensuring that the ISDN networks and services evolve in an orderly way permitting attachment to networks of products which are designed to a known standard and which can communicate meaningfully. It would enable the network developers and operators to provide mapping and translation functions within their networks, permitting a wide range of interconnectivity of devices. These mapping and translation functions, being designed around the framework established by the Reference Model, would consequently have wide applicability and require a minimum of customizing.

The widespread acceptance of a Reference Model as the basic framework within which devices, systems, and networks evolve, is the first step. Development and standardization of interfaces and protocols within this framework is a vital subsequent step. In addition, since the new communications networks and systems are very much software-based and the layering and interfacing requirements are based on software partitioning, it is important that techniques for precise protocol description, testing and verification be available. It is also necessary that clear definitions exist for the functions, services and input/output descriptions of the layers if an orderly application of the OSI concept is to become widely accepted and standardized. These needs are reflected in the program of work being undertaken by ISO and CCITT.

SUBSCRIBER INTERFACE TO THE ISDN

The ultimate objective of subscriber interface development for the ISDN is to have a single standard access type employing a single interface standard. Such a standard might be modular in the sense of permitting groupings of functions selected from an overall structure to suit various service and subscriber terminal complexities in a cost effective and efficient manner.

Determination of the appropriate subsets of access types will require considerable analysis and experimental work.

The OSI Reference Model concepts provide an important basis for subscriber interface development if a universal service-independent subscriber network access is to be realized. In this regard the principle of layering is of great importance and the work already done to establish open system interconnection standards for public data network access provides some guidance for structuring a potentially universal ISDN interface.

ACCESS TYPE CHARACTERISTICS

Subscriber access types to the ISDN may be characterised by the application involved (e.g. voice, computer data transfer, facsimile), information flow (e.g. data rate, format) and the information channel type. The work within CCITT aimed at standardizing the ISDN access types has resulted in a characterization of the following information types.

Type v: signals corresponding to a conventional (analogue) telephone subscriber station.

Type f: digital voice at 64 kbit/s.

Type d: standard types of data communication according to CCITT Recommendation X.1 and including already standardized as well as possible future standardized user classes (e.g. using the X.21 and X.25 interfaces).

Type s: customer network signalling for control and monitoring of network resources.

Type w: digital information at n x 64 kbit/s representing, for example:
- i) digitally encoded broadband audio signals;
- ii) high speed non-voice services (e.g. fast facsimile) or still picture transmission;
- iii) combinations of i) and ii).

Type t: telemetry information at very low rate conveying for example:
- i) customer alarms;
- ii) signals for remote control of equipment on customer's premises;
- iii) remote meter reading.

Type d': slow speed data which may or may not include currently standardized types of data communication. It may correspond to a type of data communication service that, subject to future standardization, should be suitable for message interleaved handling with the signalling information type on the customer access.

FUNCTIONAL INTERFACES FOR LOCAL DIGITAL ACCESS

The development of reference diagrams designed to show the functional module and interfaces in local digital access to the ISDN has been progressing within CCITT Study Group XVIII. Figure 2 illustrates possible functional modules and interfaces. These, in general, correspond to different access types or access type combinations.

Type A interfaces typically correspond to external customer terminal equipment interfaces for specific services. For example, interface A2 may correspond to the Recommendation X.21

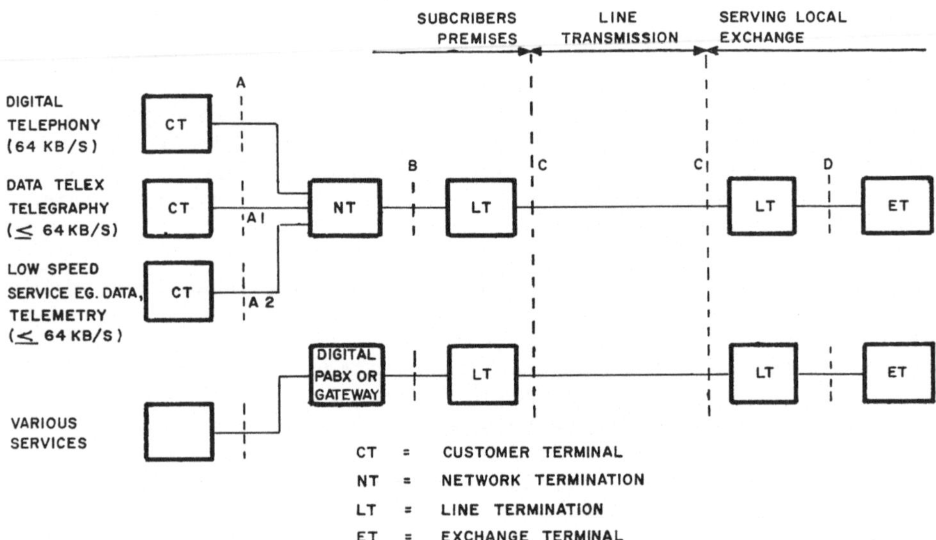

FIGURE 2
Possible functional interfaces for digital local access

or X.25 data terminal equipment interfaces which conform to the architectural principles of the Reference Model for Open Systems Interconnection.

Type B interfaces provide an access type which is independent of individual subscriber terminal type and line transmission technique.

The network termination (NT) provides an appropriate mapping between the particular A interface and subscriber access type defined at B. Interface type C is mainly oriented towards the line transmission technique while interface D relates primarily to the characteristics of different access types.

The case where access is from a local area network is also illustrated in Figure 2. Here the network termination may be a gateway, PABX or other type of concentration and translation device.

Figure 3 illustrates the scenario where a multifunctional terminal accesses the ISDN through a universal multi-service interface Ax. The range of functions supported could include those for digital telephony and multipurpose data applications. This is conceptually the most desirable interface type and is based on the assumption that it is possible and desirable to standardise a set of interface functions covering the network

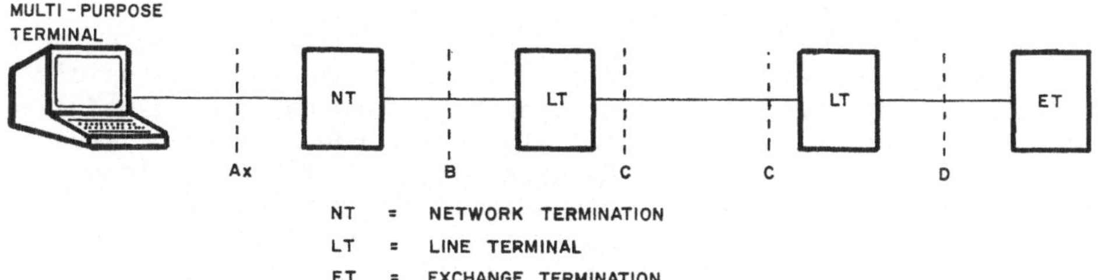

FIGURE 3
Multi – service interface

dependent requirements of a wide range of terminal types. Determination of the exact specification of this interface and the multi-functional terminal access protocol features are items currently under study by the international telecommunications community.

CONCLUSIONS

It is the thesis of this paper that the Reference Model for OSI appears to provide an adequate architecture for the ISDN supporting a wide range of access types and information flows.

The basic requirements for any communication process are the establishment, information transfer and termination of the interchange. This holds true for voice, data, facsmile and other communication processes. In the specific case where digitalization of the information has taken place the information interchange essentially becomes the interchange of sequences of digits. It seems logical therefore to expect that a common connection, transfer, and termination process, could be used for exchange of a wide range of information types.

Computer/data communications is concerned with the correct exchange of information between machines. Due to the exacting nature of this interchange a comprehensive grammar has to be used to provide for flow control, error recovery, segmenting and other requirements of the specific application. Digital voice communication has similar requirements. However, in this case, tolerance to errors is relatively high, flow control is not useful especially if the communication is taking place in real time, and contiguity of data flow is required.

Examination of these requirements shows that the communication system requirements for digital voice and those for data have considerable commonality. It therefore seems plausible that a common communication system architecture based on a modular structure would be capable of satisfying the needs of both service types. Such a structure, the Reference Model for Open Systems Interconnection, has already been established for the exchange of information between processes that are "open" to one another by virtue of their mutual use of the applicable standards. In addition a number of OSI protocols have been standardised and implemented. Thus an excellent basis has already been laid for the evolution of an ISDN capable of meeting the service and acceptable cost requirements of both the traditional telephone user and the rapidly expanding user of non-voice telecommunication services and especially computer/data communications.

REFERENCES

1. T.G Fellows, J.M. Hogg, "An Approach to the Design of an ISDN", Proceedings of NTC '80 Volume 4, Dec.1980.

2. W.A. McCrum, "Canadian Public Switched Data Networks - Infoswitch and Datapac", NATO Advanced Study Institute Series, Interlinking of Computer Networks (1978).

3. P. Hsi, T. Lissack, "Local Networks' Concensus: High Speed", Data Communications, Dec. 1980.

4. H. Zimmerman, "OSI Reference Model - The ISO Model of Architecture for Open Systems Interconnection", IEEE Transactions on Communications, Volume Com-28, No. 4, April 1980.

5. N. Accarimo, et al, "Customer Access Protocols for Integrated Services Digital Networks", Computer Communications, Vol 2 No. 6, Dec. 1979.

6. CCITT Study Group XVIII, "Extract from the Report of the Inter-Regnum Meeting of Experts on ISDN Matters" (Innsbruck 12-15 Jan. 1981).

DESIGN ISSUES FOR HIGH SPEED LOCAL NETWORK PROTOCOLS

Franklin F. Kuo

University of Hawaii at Manoa

With recent advances in optical fiber technology, it is now feasible to consider data communication systems with speeds up to 1000 Mb/s. Because of their high bandwidth, low delay, and low error rate characteristics, optical fiber communication systems seem to provide an almost ideal transmission medium for high speed local networks. This paper deals with the implications of the use of fiber optics in the design of high speed local network protocols.

1. TRANSMISSION MEDIA

With recent advances in optical fiber technology, data transmission at speeds up to 1000 Mb/s is now technically feasible. Optical fibers may be used in a variety of applications where twisted copper wire-pairs, coaxial cables, and waveguides are now used for the transmission of data; these applications range from short data links and equipment connections within a building and between buildings, to long trunk circuits between cities. This report deals with the implications of the use of fiber optics for local network applications upon local network communication protocols. Using fiber optics technology, high speed local networks with throughputs of 10 Mb/s to 100 Mb/s are possible. The discussions here will examine protocol implication of local nets with data rates in the range cited above.

Without going into the details of fiber optics technology the transmission medium can be described as having the following characteristics:

K. G. Beauchamp (ed.), New Advances in Distributed Computer Systems, 97–105.

* low transmission delays
* wide transmission bandwidth
* low error rates

These characteristics give rise to the possibility of "ideal" transmission medium for local networks--low delay, high throughput, and almost error free (error rates of 1 bit in 10^{20} are commonly cited). What do these ideal characteristics imply from the standpoint of computer protocols? In the following are discussions of some specific issues.

2. PACKET SIZE

It is well known that low delay requires short messages, short queues, and few control messages, while high throughput requires long messages, long queues, and low overhead. Are these performance trade-offs applicable to high speed local networks? According to McQuillan and Cerf (1), delay is defined as "time between transmission of the first bit and delivery of the first bit". Its components are:

* speed of light ~ distance
* transmission delay ~ message size/circuit rate
* processing delay
* queuing delay ~ system load.

On the other hand throughput is defined as "number of bits sent divided by time between transmission of the first bit and delivery of the last bit". The components of throughput are:

* effective bandwidth of processing equipment
* effective bandwidth of transmission media.

For fiber optics local networks the distances are short, the transmission delays are negligible, and the effective bandwidth of the transmission medium is very wide. However, the other components of delay and throughput cited above have nothing to do with the characteristics of the transmission medium, but are related to the inherent processing power of the devices attached to the local network. Since some of these devices can be terminals or microprocessors with little processing power, the throughput/delay trade-offs of a high-speed local network are limited by the processing power of these least capable members of the network. The implications as far as packet size is concerned are these:

a. Packet sizes should be kept short in order to match the sizes of the buffers of the terminals and/or microprocessors on the net.

b. Because bandwidth is inexpensive and readily available on a local network, there is little motivation to keep the size of the header or overhead bits down on a packet. Thus a highly inefficient packet in ARPANET consisting of a single character or word of data together with multiple words of administrative information would be regarded as acceptable in a high speed local net.

c. To keep node processing down, a standard header format with fields in fixed locations is recommended. Multiple packet types, optional fields, etc. all tend to increase communications processing at the node. Metcalfe (2), in fact, recommends more spacious fields than one would think necessary since "try as you may, one field or another will always turn out to be too small."

In summary, small packet sizes are recommended for high-speed local networks in order to match the buffer capacities of the microprocessors and terminals attached to the network. However, since bandwidth is inexpensive, larger header overhead can be tolerated.

3. FLOW CONTROL

Because of the wide bandwidth available, congestion control on the transmission medium is not necessary. Thus, link level flow control is not a major requirement for high speed local networks. However, buffer management for the various hosts attached to the network is still an important issue since different hosts generate and absorb data at varying rates. Thus, a transport-level flow control mechanism is needed. Since flow control is closely tied to buffer availability, the amount of buffer space needed for efficient operation under different circumstances is an important factor in protocol performance.

A simple dynamic flow control mechanism that does not function efficiently in conventional long-haul packet switching networks might work well in a high speed local network. This mechanism is based upon a simple start-stop command. The receiver issues start and stop commands that place the sender in a transmission-allowed or transmission-blocked state. Since in high-speed local nets, these commands suffer little delay in transit, the flow control mechanism that would require somewhat more processing power is based upon granting "credits" for transmission. The receiver grants credits for a certain amount of data to the sender so that both know exactly how much data will be exchanged. The number of credits provided by the receiver is frequently called the "window size" when credits are expressed relative to packet sequence numbers. In a high speed local network where control commands can be transmitted with very little delay, the start-stop flow control scheme seems to be preferable because of its simplicity to the "credit" flow-control scheme.

4. ERROR CONTROL

Since the transmission medium is almost error free, it means that link-level error control is unimportant. Since the link-level error control feature is found in almost every data link control protocol in use today (HDLC, SDLC, ADCCP, BISYNC, etc.), it raises the question whether to adopt one of the present data link control procedures unchanged and apply link-level error control, even when it is not needed, or to adapt or develop another version of the data link control procedures to better suit the characteristics of the transmission media.

It is clear, however, that end-to-end error control is needed in a high speed local network. This is because of the fact that the devices (computers, peripherals, and terminals) are inherently more error-prone than the fiber optics transmission medium. This means that error control at the transport protocol level is needed. As to the best way to implement the transport level error control, it appears that error detection with retransmission (ARQ) is probably preferable to forward error correction. This is because low transmission delays will produce fast acknowledgements, so the efficiency of an ARQ scheme is much greater in a high speed local network than for conventional transmission media. Moreover, for a local network many of the devices attached to it will not have much processing power, so the requirement of forward error correction might place an undue burden upon the device

5. DATAGRAM VS. VIRTUAL CIRCUIT PROTOCOL

In view of the fact that a local network might have considerably more simple hosts such as intelligent terminals and microprocessors than long haul networks, the low-level transmission protocols should be kept at a level that the simple hosts can handle. This argues strongly for a datagram protocol to be the basic transmission protocol. That datagram protocol would provide the basic responsibility of delivering a single addressed packet to one or more of its destinations. Above the datagram layer, however, should exist a virtual circuit layer for those hosts capable of supporting a transport station which can multiplex a number of virtual circuits.

6. INTERNETTING CONSIDERATIONS

In order to allow for growth and evolution, local networks should be designed to allow interconnection with other local networks and long-haul networks. For other local networks, the concept of a bridge is useful. As described by Clark et al. (3) a bridge contains

> "two network interfaces, one appropriate to each of the subnetworks it interconnects, a limited amount of packet buffer memory, and a control element, which implements an appropriate filter function to decide which messages to "pull off" and buffer until it has an opportunity to retransmit it to the other subnetwork."

It there is substantial speed disparity between the two local networks, the bridge must have either extensive buffer capacity or the ability to regulate the flow of information from the higher speed network. The long distance bridge concept which Clark et al. describes might be adapted for this purpose (3).

For interconnection with a long-haul network, a gateway is required. The simplest gateway would be for the case where the long-haul network offered a datagram interface to the local network. However, for most commercial long-haul networks the interface offered is a virtual circuit (e.g. X.25). Therefore, a virtual circuit protocol should be implemented on the local network which is as close as possible to the long-haul virtual circuit model. Since the local network probably will not require the full range of functions available on a long-haul virtual circuit model, a subset of the long-haul virtual circuit protocol could be implemented for the local network. However, according to Clark et al. (3), the compatibility of the virtual circuit protocols between long-haul and local area network does not answer the question of how the features as flow control, buffering and speed matching should be implemented. Standards should be proposed for gateways between local networks and X.25-based long-haul networks.

7. ADDRESSING AND ROUTING

For high speed local nets, routing is not an important issue. Since delay is so low, optimal routes are not significantly different from suboptimal routes. Routing is only important at a gateway or bridge, and internet packets must be sent to the correct gateway or bridge for forwarding.

In local area datagram networks, message exchanging between two cooperating entities such as ports requires that each entity knows the network address of the other. Sometimes entities are known by their names rather than their network addresses. In such cases, it is advantageous "to maintain, as a network service, a facility which will take the name of a desired entity and give back its network address" (3). In a high-speed local network with many simple hosts, the network directory service can perform very effectively since the delay between queries and responses can be quite low due to the high speed transmission media. In general, central network services become more efficient when they can be accessed through wideband, high speed transmission media. The only efficiency constraint then becomes their own processing power in dealing with queries and responses.

8. HIGH-LEVEL PROTOCOLS

High level protocols are those protocols primarily concerned with performing remote operations across a network. The low-level virtual circuit and datagram protocols discussed in previous sections are "communications protocols" whereas the high-level protocols discussed here are "resource sharing" protocols. In terms of the ISO Open System Interconnection Architecture Model (ISO/TC97/SC16), the high-level protocols discussed here are at the session or presentation level (4). We will concentrate on three high level protocols: terminal, file transfer, and remote job entry protocols, which provide basic services for the users of a local network. Terminal protocols establish mechanisms that allow efficient and flexible terminal access to networks. Terminal protocols not only allow a user to access a time-sharing service through the local network, but can also be used as a character-oriented network interprocess communication facility. File transfer protocols allow users to manipulate remote file systems and to transfer files from one host system to another. Remote job entry protocols provide users with a mechanism for submitting jobs to various batch services on a network. Many of the problems encountered in these protocols recur in more complex forms in more sophisticated protocols (e.g. network mail protocols, distributed data-base protocols) which may be built on top of them.

In this discussion we are primarily concerned with heterogeneous network protocols, i.e. those protocols which deal with networks of heterogeneous computers, terminals, peripherals, etc. The common problem that all three high level protocols share is that they require substantial network software effort to implement. Moreover, in long-haul networks each host or terminal offering the remote service is required to have a copy of the high level protocol software within its own physical memory space. Since a heterogeneous computer network can have many different varieties of computers and terminals, the task of programming each of the devices for high level protocols is expensive and time-consuming. For long-haul networks, it is not feasible to have a central facility which acts as a mediator between two entities wishing to engage in a cooperative task, such as remote job entry. This is because of the excessive time-delays encountered in having such a central facility. However, in high speed local networks, where time delays are very short, such central facilities are feasible and even advantageous. Not only do they provide a centrally supported software facility, but they also reduce the requirements that each user terminal or host retain a full copy of the high level protocol within its own memory space. The use of the central high level protocol facility for each specific protocol will be discussed in subsequent sections. However, in each case, it can be described as shared facility, somewhat like a re-entrant compiler on a time-shared computer. Note that a re-entrant compiler can be shared among a number of users without the necessity of each user having a copy of the compiler in his own working space. Similarly, a central high level protocol facility can be shared by a number of processes without each possessing a complete copy of the protocol in its own file space.

9. TERMINAL PROTOCOLS

In order to accommodate the wide variety of different kinds of terminals a virtual terminal protocol (VTP) is commonly used in which a network virtual terminal (NVT) is defined as the network standard. The terminal side of a connection maps the output of its terminal into the NVT format for transmission to the host. The host then maps the NVT format into its local form. Each host of the network then only needs to support one terminal type (the NVT). In order to allow not only terminal to process communication, but also process-process and terminal-terminal interactions, the NVT software should reside in both sides of a connection, thus leading to a <u>symmetrical</u> view. Such VTP schemes are expensive in that network resources must be dedicated at every host and terminal to support the NVTs. For a high-speed local network environment, a centralized NVT is feasible so that each side can access a single virtual data structure that performs the functions of data translation, option negotiation, echoing, and interrupt signaling. Such centralized NVTs were not previously possible in long-haul networks because of response-time limitations. However, in high speed local networks, their use can be highly advantageous, not only in conserving memory space at each node, but also because of software support for a single facility is more efficient than for multiple computer and terminal-types.

10. FILE TRANSFER PROTOCOLS

A File Transfer Protocol (FTP) defines the set of rules for the transfer of files from the file system on one host to the file system on another. In heterogeneous networks the purpose of an FTP is to establish a network virtual file system (NVFS) which allows a process in a local host to access data stored on any remote host as if the data were stored locally. To accomplish this, canonical or virtual file formats are defined in an FTP so that the source will map the file into the proper virtual format and the destination will map the received virtual form into the proper local format.

Other very important functions of the FTP are maintaining access control and directory information across a network. Access controls are required for reasons of security and privacy. Directory information is needed for efficient file management.

Recent FTP designs partition the FTP into two separate protocols: a data transfer portion and a file management portion. This partitioning allows the FTP to be partially off-loaded onto a network front-end, so that the front-end performs the data-transfer tasks and the host performs the file management tasks. In a high speed local network, a central FTP facility which performs the front-end data transfer tasks for all hosts can be quite advantageous. In addition to performing data translations from a local host format to the NVFS format, the central facility can also act as a security filter for access control and maintain current directories of important network files. The central facility will also relieve a host from some (but not all) of its data and file conversion tasks of going from its local form to the network canonical form, but more importantly will relieve the host of all network access control burdens. It should be emphasized that the efficiency of the central FTP facility is strongly dependent upon how it is implemented, and not simply dependent upon the high-throughput character of the local network. However, because the wideband transmission available, the central FTP concept should be carefully considered in future local network architectural designs.

11. REMOTE JOB ENTRY PROTOCOLS

Remote job entry (RJE) protocols are, after remote terminal protocols, the most important network requirement in the current data-processing environment. Remote Job Entry Protocols (RJEP) provide the network user with the flexibility to use a single implementation of RJE software with a variety of batch systems. The RJE protocol sits halfway between the file transfer protocol (FTP) where high throughput is a requirement, and the virtual terminal protocol (VTP) where low delay is a primary goal. As such it could draw upon facilities of each protocol without having to develop a completely unique set of functions. For economy reasons, the RJE protocol could be implemented easily if it is co-located with the central FTP and VTP facilities. We will not dwell on this point further.

12. CONCLUSIONS AND ACKNOWLEDGEMENTS

In this paper we have presented a number of issues concerning the design of protocols for high speed local networks. With increasing attention being paid to the application of fiber optics links to local networks, the paper addresses the key question of whether conventional communication protocols are appropriate for transmission media with speeds of 10 to 100 Mb/s.

This work was partially supported by the U.S. National Bureau of Standards and by the Office of Naval Research under contract N00014-78-C-0498.

13. REFERENCES

(1) McQuillan, J.M. and Cerf, V.G., *A Practical View of Computer Communications Protocols,* IEEE Computer Society, 1978, pp. 4-5.

(2) Metcalfe, R.M. and Boggs, D.R., *Ethernet: Distributed Packet Switching for Local Computer Networks,* Comm. ACM, July 1976, pp. 395-404.

(3) Clark, D.D., Pogran, K.T. and Reed, D.P., *An Introduction to Local Area Networks,* Proc. IEEE, November 1978, pp. 1497-1517.

(4) International Standards Organization (ISO), *Reference Model of Open Systems Interconnection Architecture,* Version 3, November 1978.

DIRECTORY SYSTEMS FOR COMPUTER MAIL IN INTERNETWORKING ENVIRONMENTS

Jose J. Garcia-Luna-Aceves, Franklin F. Kuo

University of Hawaii at Manoa

In this paper we present design considerations of directory systems for computer mail. Directory systems are analyzed based on a hierarchical architecture for computer mail systems and the emphasis of the paper is on large systems and system interconnection. The paper describes the organization of the directory system's databases into (logical) levels according to the nature of the information stored in such databases, and discusses the design issues associated with the management of such a distributed database. These issues include:

a. How the information is structured and distributed.
b. How to control access to the data.
c. How to process identification queries.
d. How to ensure integrity and security of information.
e. How to update the directory system.

1. INTRODUCTION

With the merging of techniques for the communication, storage and retrieval of information in office environments, there is a large market for computer-based message systems. Many systems already exist and many more are expected to be in operation in the future (11). However, today's message systems differ from each other in many ways (2). Thus, the future computer mail environment will be heterogeneous, with a large number of users, many organizations, countries and systems involved. For messages to be delivered, the system-level address of the recipients of the messages must be obtained. This aspect of computer mail system design, the provision of on-line identification services,

K. G. Beauchamp (ed.), New Advances in Distributed Computer Systems, 107–124.

requires a directory system to maintain the users' system addresses. Even though directory systems (also named identification database systems) are essential for the effective operation of computer mail systems, especially when large user communities and system interworking are involved, very few studies have been undertaken on the subject (5), (6), (7). In this paper we present design considerations of directory systems for computer mail in internetworking environments.

Section 2 specifies a general model for computer mail systems in which addressing and delivery services are transparent to the users. This model specifies the elements and operation of computer mail systems. The rest of the paper specifies the directory system that maintains the information necessary for that form of system operation. Section 3 describes the organization of the directory system, specifies its functions and delimits the services needed for its operation. Section 4 specifies the structuring of information in the directory system. Section 5 describes the issues associated with the management of the information and the distributed control of the directory system. Section 6 summarizes the main concepts introduced in this paper and points out areas in need of future research.

2. A FUNCTIONAL MODEL FOR COMPUTER MAIL SYSTEMS

2.1. Components

We model a computer mail system by partitioning the system into functional components, each dedicated to a specific set of computer mail functions. There are three different types of functional entities in the proposed model (6): MAILBOX, MAILER, and GATEWAY MAILER.

MAILBOX (MBX) is the entity responsible for the processing, storage and retrieval of user messages. A mailbox serves as the interface between its message system user and delivery services. This component consists of:

a. Message processing modules to compose, edit, retrieve and archive user messages.

b. Communications software to transfer user messages to and from the entity dedicated to message delivery (the local mailer).

c. User files where user messages are permanently stored and (optionally) a personal directory with users' system mailbox addresses is maintained.

d. The message workspace where undelivered messages and messages being composed are maintained.

MAILER (MLR) is the component process responsible for the delivery of messages to and from a specific set of mailboxes and the identification of the system mailbox addresses of the recipients of the messages. The mailer is formed by:

a. Communication software modules dedicated to the communications of the mailer with its local mailboxes and other mailers.

b. The message buffer where messages to and from local mailboxes are temporarily stored.

c. The mailer directory database which maintains addressing information and time-stamped records of message deliveries.

A mailing network is formed by the union of logically connected mailers and corresponds to a public or private message system. Thus, mailing networks (and their corresponding mailers) could be managed and owned by one or more organizations.

GATEWAY MAILER (GMR) is the entity responsible for internetwork communication. Each mailing network has associated with it a gateway mailer, which represents half of the gateway between a given mailing network and any other mailing network. The gateway mailer is formed by modules similar to those of the mailer described above.

A computer mail system consists of a set of interconnected mailing networks. Each mailing network has its own standards to process, deliver and structure user messages. Consequently, a computer mail system is a collection of heterogeneous mailing networks which communicate with each other by means of internetwork protocols via the gateway mailers.

2.2. System Operation

In a computer mail system the sender of a message must enter some meaningful information about the recipient so that the system can identify the recipient and deliver the message. In today's message systems the sender of a message has to enter the formal address (e.g., NIC IDENT code (1), home street address, mail stop) of the recipient in terms of the addressing standards of the system. This may work well for the case of a small system with a homogeneous set of addressing standards. But in an internet environment it would not be feasible for the message system users to handle system addresses of all the various mailing networks involved because of the differences in addressing formats and standards between mailing networks. On the other hand, it would be very difficult to fix general addressing standards that could be effectively used for internet message delivery and at the same time be feasible enough to be used by humans. Either the flexibility allowed to the users for address specification would be very restricted, or the delivery procedure would be very complex and the users would have to specify too many fields in the addresses. Because of this, in our model delivery and addressing are functions transparent to message system users. The sender of a message should not be concerned with the recipient's system mailbox address or how the message is delivered. To provide such services, a user oriented naming standard is needed to name the recipients of the message in a form as flexible as a postal address for example, and be independent of the addressing standards internal to each mailing network. The system must assist the sender in identifying the system mailbox address from the user-oriented description entered by the sender and then deliver the message according with the address obtained.

We define a user-level naming format called the NOLS address (6), (7), which consists of four major fields that contain information about the recipient of the message, as is shown in Table 1. When the sender of a message enters his message, he also enters a NOLS address with what he knows about the recipient's name and/or title, organization, geographical location and (perhaps) message system.

A NEXUS is an end-to-end virtual connection established between the sender's and the recipient's mailboxes (6). A NEXUS relies on internetwork connections between gateway mailers and intranetwork connections between mailers and their local gateway mailers. A NEXUS address is the specification of a NEXUS in the system. It is specified at the internetwork level as is shown in Table 1. The internetwork address is mapped into intranetwork addresses that follow standards particular to each mailing network, as is shown in Table 1.

Message delivery is carried out in two phases: The NEXUS establishment phase and the message delivery phase. During the NEXUS establishment phase, the system maps the NOLS address (user oriented standard) into a NEXUS address (system oriented standard) and establishes the end-to-end virtual connection. The message delivery phase consists of the dispatch of the message through the NEXUS.

To map NOLS addresses into NEXUS addresses (and those into intranetwork addresses), the system mailbox addresses of message system users must be maintained somewhere in the system.

Table 1. Naming and Addressing Formats

FORMAT TYPE	COMPONENTS
NOLS address (naming)	<N field> <O field> <L field> <S field> N-field -- Contains information about the recipient(s) of the message, such as his name and/or title. O-field -- Contains information about the organization(s), group(s) or system server(s). L-field -- Contains information about the geographical location of the organization(s) or group(s) referred in the O-field. S-field -- Contains information about the System which offers the computer mail services to the recipient(s).
NEXUS address (internet addressing)	([sender's system mailbox address]) , ([recipient's system mailbox address])
network mailbox address	Any format defined in the mailing network.
system mailbox address	<internet address of mailing network> <network mailbox address>

In an internet environment it would not be feasible or desirable to maintain all such addresses in the database of each mailer or even a centralized entity. The information is, by its very nature, distributed and must be organized in such a way that each organization can maintain its own information according to its own needs. Users of the system should be given a simple and integrated view of the information distributed in the system. In our model each mailer maintains only the mailbox addresses of the users served by that mailer, together with a set of mailer addressses (pointers) that the mailer associates with user locations and/or organizations. The NOLS address entered by a sender constitutes a query to the system for identifying the correct system mailbox address. As is shown in Figure 1, the procedure followed to resolve such a query is a store-and-forward process in which the NOLS address is forwarded among mailers and gateway mailers according to the network (inter-network) pointers they maintain. When a mailer (gateway mailer) receives a query, it examines the fields of the NOLS address and based on a search of its directory database, it decides whether to forward the query or to reply with a positive or negative acknowledgement. Once the sender's mailer obtains a positive acknowledgement, the NEXUS between sender's and recipient's mailboxes is created and the message can be delivered (Fig. 1). Various gateway mailers and/or mailers may have to be queried when a NOLS address lacks certain key information, such as name of organization and/or location.

3. DIRECTORY SYSTEM ORGANIZATION AND FUNCTIONS

3.1. Need for a Distributed Organization

The future environment of computer mail systems will be such that:

a. Many message systems (public or private) will exist on national and international bases, each with its own addressing standards;

b. Message system users (individuals and organizations) will belong to one or more message systems;

c. There will be many identification databases maintained by organizations containing the information that the organizations need or can afford to maintain;

d. Message systems and private companies will offer public information services, and there will be many differences among the services offered.

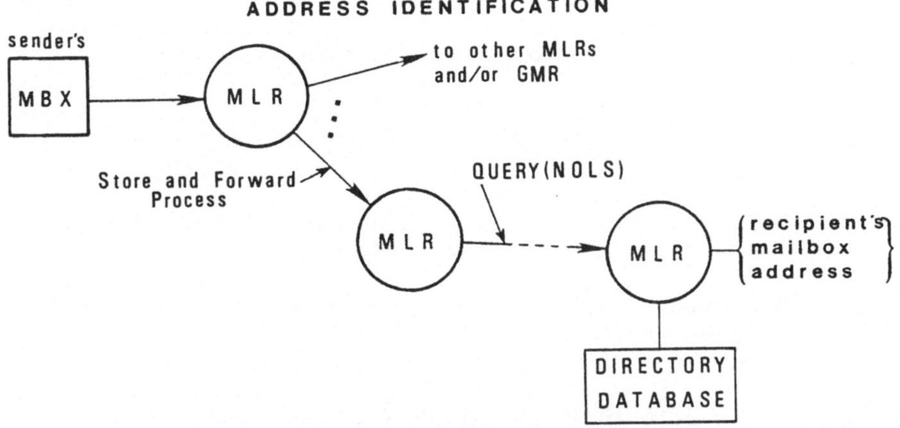

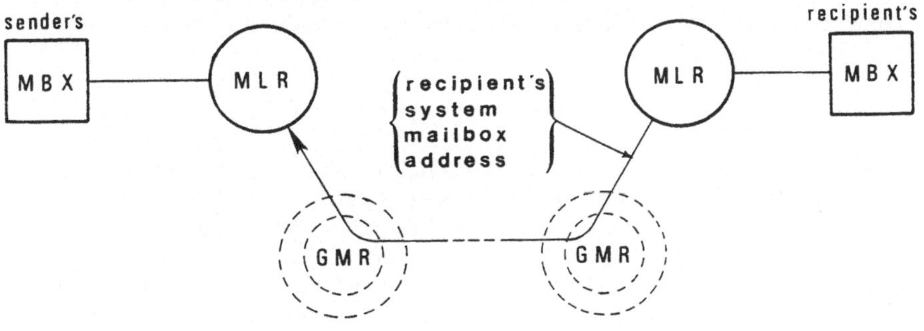

Fig. 1. System Operation

Under such circumstances, the question is not whether the directory system of a computer mail system should be distributed or not, but how to effectively organize the various components of a federation of many heterogeneous database systems to ensure that:

a. An integrated view of the information is presented to the users.

b. Each organization is free to maintain its own information according to its own needs.

c. The computer mail system can provide efficient identification services to all its message system users despite the differences among the various databases.

The directory system is a special-purpose distributed database system aimed at the maintenance of system-level addresses. According to our model, the directory system is formed by the personal directories of mailboxes, the directory databases of mailers and gateway mailers, and software modules to manage such a distributed database.

3.2 Organization of the Directory System Databases

As shown in Figure 2, we organize the databases of the directory system in four levels: the user level, the local level, the network level and the internet level. The local level of the directory system is formed by the union of <u>local directories</u>. Each mailer maintains a local directory with complete identification information about the users served by that mailer only. Such a directory specifies who the message system user is, where he is and (perhaps) what he does (5). The structure and management of the local directory of a mailer is independent of the rest of the system. A local directory corresponds to the identification database of an organization, a branch of an organization, or a regional computer mail server.

The network level of the directory system is formed by the <u>network directories</u>, each of which can be centralized in a single mailer or distributed among the various mailers of a mailing network (Fig. 2). This database is a directory of organizational directories (i.e., the local directories) that allows the mailers to find out where in the network the information about an organization (or one of its branches) is located. That is, the network directory contains mailer addresses (pointers) associated with the organizations served by the mailing network. When a NOLS address referring to a remote organization is received by a mailer, it consults the network directory to find out where to forward the query. This level of the directory system is

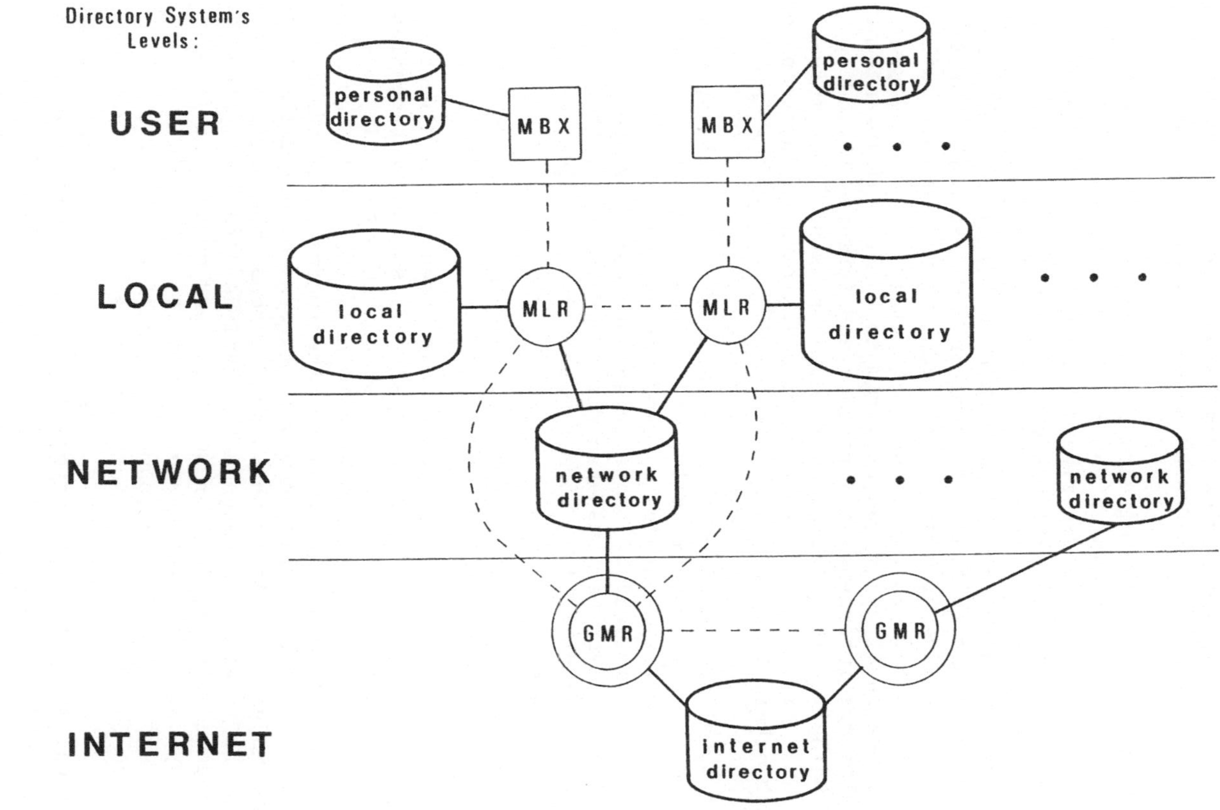

Fig. 2. Organization of Directory's Databases

concerned with the integration of distributed organizational information into a network-wide database. In the future, this "directory of directories" facility will be implemented by computer mail services, the organizations with their own networks and third parties (e.g., large information companies) (11).

The internet level of the directory system is formed by an internetwork directory, which can be either centralized in a single gateway mailer or distributed among the various gateway mailers of the computer mail system. This database plays the same role as a network directory, but at the internet level. That is, it is a directory of network directories that allows the gateway mailers to find out where in the system the information about a mailing network is located. When a gateway mailer receives a NOLS address referring to an organization or a geographical region (i.e., a country) remote to its mailing network, it consults the internetwork directory and forwards the query correspondingly. This internetwork database should be implemented as a joint effort of the parties desiring to interconnect with each other.

The user level of the directory system is formed by the personal directories that are (optionally) maintained in the mailboxes. A personal directory contains system mailbox addresses (plus some extra information) of those recipients commonly addressed by the sender. Each user manages the information contained in his personal directory in a form completely independent of the rest of the system.

3.3. Directory System Software Modules

The software necessary to manage the databases of the directory system is a distributed database management system that handles both the communication between the various computer mail processes (mailboxes, mailers and gateway mailers) and the database management operations at each site. The functions of these software packages can be partitioned into layers following the ISO's Reference Model (8). As is shown in Figure 3, the software that controls the interaction between the various database systems corresponds to the session, presentation and management (application) layers. The existence of transport services (4) is needed to support the establishment of logical connections between process in different host computers and communication networks. The management layer supports the functions particular to the management of the directory system's databases, integrating them into a single entity. These functions are:

a. To provide the message system users with a unified, global view of the information stored in the directory system's databases, and to allow them to enter system-wide queries;

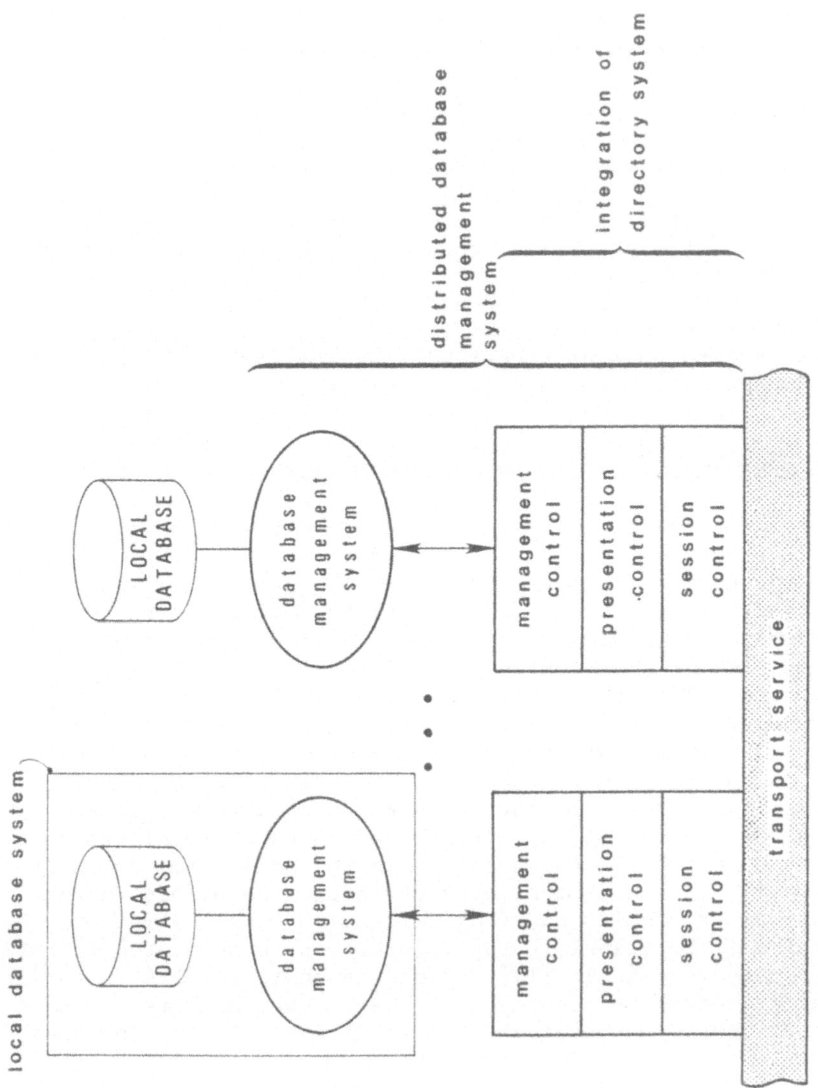

Fig. 3. Integration of Databases in the Directory System

b. To permit the exchange of information between heterogeneous databases according to a standard structure of information;

c. To ensure consistency and security of the information maintained in the directory system;

d. To resolve system-wide queries; and

e. To provide error recovery.

The session and presentation layers support the establishment of reliable end-to-end virtual connections between computer mail processes and the exchange of information in standard formats. In this paper we are only concerned with the management layer of the software of mailboxes, mailers and gateway mailers. The existence of both presentation-level services (8) and the local database management systems will be assumed.

4. STRUCTURE AND DISTRIBUTION OF INFORMATION

A system-wide data model is needed to describe the (logical) structure of the information maintained in the directory system and to structure the message system users' queries. In our model a system-wide user schema is defined that gives the users a unified view of the logical structure of the information, independent of the distribution of information and the structure of the system. The user schema is the External Schema (3) of the directory system. Based on this schema a set of rules can be defined to structure NOLS addresses. Each mailing network has a network schema that standardizes the logical structure of the information distributed in the network. Such a schema is the Conceptual Schema (3) of the network and is used by mailers to plan the processing of the queries. In the same form, an internet schema is defined at the internet level. Each local directory in a mailer has its own (logical and physical) structure. Individual mailers translate the information communicated in NOLS addresses into formats locally defined in their local directories. A mapping procedure is necessary at the gateway mailers of those mailing networks whose network schemas differ from the internet schema.

The Entity-Relationship data model (2) is a good candidate to represent the internet and network schemas of the directory system because the model is rich in semantic information about the data and can be easily translated into different data models. Since the exact structure of NOLS addresses is yet to be defined,

Table 1 only gives the type of information that could be effectively used on a system-wide basis.

In our model information is distributed by schema instances. A schema instance consists of the information whose structure is defined by a schema, and is therefore semantically complete. The semantics of the information (i.e., types of entities, relationahips and attributes) is defined by the organizations at the local level, and by the Directory System Administrator (DSA) at the network and internet levels. Therefore, at the local level the local directory of each mailer maintains semantically complete information about the local users. At the network level a network directory maintains semantically self-contained information about the organizations in the mailing network. Either one, various, or all the mailers of the network have a copy of the network directory. The same applies to the internet level, where the gateway mailers of the system have a copy of the internet directory.

As it has been pointed out in (9), distribution of information by schema instances reduces communication costs. In addition, the hierarchical distribution of information by schema instances of the proposed model reduces the complexity of the communication protocols. On the one hand all detailed information about individuals is maintained with no (network-level) redundancy. On the other hand the information that has to be maintained at network and internetwork levels refers only to organizations and mailing networks and not to individuals, and changes in such information are infrequent.

5. DIRECTORY SYSTEM CONTROL

The procedures used for the control of the directory system rely on the hierarchical organization of the system and the distribution of information by schema instances. The control of the directory system is carried out in three levels: local, network and internetwork.

5.1. Query Processing

In our model a mailer can resolve queries that refer to its local users but not queries that refer to remote users. These queries must be forwarded to remote mailers. The processing of a query depends on the distribution of the network and internet directories. Throughout this discussion we will assume the correct operation of mailers and gateway mailers and the consistency of information in the directory system.

When a sender enters a query to his mailer that refers to remote users, the sender's mailer must determine where to forward the query. If it has a copy of the network directory, the mailer determines which mailers maintain the information about users in organizations and/or geographical locations that fit the data in the sender's NOLS address. Accordingly, the sender's mailer forwards the query to all those mailers. Each such a mailer locally processes the query (based on its local directory) and as a result it either obtains the recipient's mailbox address or a list of "similar names" (containing the information about a set of users that fit the sender's description), or the queried mailer cannot associate any local user to the NOLS address. If the sender's mailer does not have a copy of the network directory, it simply forwards the query to a mailer that has a copy and such a mailer continues the processing of the query. The same query processing procedures apply at the internet level depending on the distribution of the internet directory.

Note that the exact form in which a query is processed may not be known by the sender's mailer and various mailers (gateway mailers) may have to be queried if the NOLS address lacks key information such as location. Each of the queried mailers (gateway mailers) replies to the sender's mailer (gateway mailer). Once the sender's mailer obtains the replies from all the queried processes, either a NEXUS has been established (Fig. 1) or the queries have failed to identify the system mailbox address(es) necessary for message delivery.

5.2. Directory Updating and Error Recovery

Because of the organization of the directory system and the nature of the information maintained at the network and internet levels, the updating procedures are very simple. At the local level, no database synchronization is required between mailers since each mailer independently maintains its local directory. Local locking is only required to ensure the consistency of the local information. At the network level, mutual consistency of the various copies of the network directory can be obtained with no need for synchronization among mailers. Each mailer sets aside a workspace (private storage) for every transaction where the information accessed by the (read or write) transaction is copied. Therefore, various readers and one writer can concurrently access any entity in the network directory. There are practical forms to implement this scheme (10), (15). Using local locking (10) each mailer ensures the local consistency of information. The DSA issues (time-stamped) updates to all the mailers of the mailing network when a change occurs at the network level. Each mailer processes the update in an asynchronous form with respect to the other mailers. If a mailer receives a query (NOLS address) that refers to an organization whose system address has just changed and the mailer has not yet updated such data in its copy of the network directory, the mailer will erroneously forward the query to the organization's former mailer. The same type of error would occur if a mailer crashes and its copy of the network directory is out of date when the mailer comes back to operation. To recover from network-level identification errors, a forwarding mechanism is used that works as follows:

Each command (query, update or message) contains the following control information:

- A command identifier
- The next distination of the command
- A forwarding list that identifies the mailers (gateway mailers) that have handled the command.
- A time-stamp (14)
- The identification of sender and receiver

When a query is received by a mailer it determines whether or not the NOLS address refers to an organization that has changed its system-level address. If that is the case, the query is forwarded to the appropriate mailer and an update is sent back to the mailer that issued the query to update its database. The update contains time-stamped information elements and the information stored in the mailer's database also has the time-stamp of the last update. When the mailer that issued the query gets the updating command, it checks the time-stamps of the information stored and transmitted and determines if its database is out of date, in which case it is updated.

This forwarding mechanism is loop-free because of the forwarding record included in every command. Any mailer with out-of-date information is detected by means of the use of time-stamps. Therefore, error recovery from process failures or differences caused by updates from the DSA is supported by our model. Repeated updates can also be detected by the time-stamps in information.

The same procedures described above apply to the internet level. Error recovery procedures require that the information stored in local directories be never destroyed, since it is locally maintained by mailers. If a copy of the network (internet) directory is destroyed, a new copy can be transmitted to the crashed mailer (gateway mailer) without penalty because it is a small portion of the entire database.

6. CONCLUSIONS

In this paper we have presented a general framework for the design of address directory systems for computer mail and we have proposed various techniques aimed at internet milieus and large computer mail systems. The organization of the directory system we propose is such that each organization can independently maintain its own local identification database. The individual organization databases are then integrated into a system-wide distributed database, presenting the users of the system a unified view of the information. The form in which senders describe the recipients of their messages is independent of the structure of the computer mail system and its delivery and addressing procedures. Such user-oriented descriptions (NOLS addresses) constitute queries to the system and not as physical addresses as in the postal service. The system maps those descriptions into system-level addresses (NEXUS addresses) needed for message delivery in a form transparent to the users. Various studies related to the design of computer mail protocols exist in the literature (13), but the role of directory systems for system address identification have been overlooked. International standards are needed to specify a common logical structure of identification information and thus permit the open interworking of heterogeneous identification databases.

Inside small networks, provisions could be made to overcome the necessity of a two-phase delivery procedure such as the one we propose. For example, users could be asked to enter very specific NOLS addresses and use them as the formal system-level addresses for delivery. Message delivery could be allowed to public bulletin boards maintained in mailers. In this form, messages could have a chance of delivery even if the recipient's mailbox address was not obtained.

The simplicity of the control procedures of the directory system we propose rely on the hierarchical organization of information. If redundancy is introduced at the local level (i.e., more than one local directory contains information about a given user), synchronization among the various mailers is required to deal with local-level updates and as a consequence, the control procedures become much more complex.

7. ACKNOWLEDGEMENTS

This work was partially supported by the U.S. Office of Naval Research under Contract N00014-78-C-0498 (F. F. Kuo); and by the National Council of Science and Technology under CONACYT-Mexico (J. Garcia-Luna-Aceves).

8. REFERENCES

(1) ARPANET Directory, Network Information Center, SRI International, NIC 41472, July 1977.

(2) P. P. Chen, "The Entity Relationship Model-Toward a Unified View of Data," ACM Transactions on Data Base System, Vol. 1, No. 1, March 1976.

(3) CODASYL-SDDTG, "Stored-Data Description and Translation: On a Model and Language," Information Systems, Vol. 2, No. 3, 1977.

(4) D. W. Davies, D. L. A. Barber, W. L. Price and C. M. Solomonides, Computer Networks and Their Protocols, John Wiley & Sons, Inc., New York, 1977.

(5) E. J. Feinler, "The Identification Database in a Networking Environment," Conference Record, 1977 Telecommunications Conference, 1977, pp. 21:3-1/5.

(6) J. J. Garcia-Luna-Aceves, "A Study of Computer Mail Services," M. S. Thesis, Department of Electrical Engineering, University of Hawaii, Honolulu, Hawaii 96822, August 1980.

(7) J. J. Garcia-Luna-Aceves and F. F. Kuo, "Addressing and Directory Systems for Large Computer Mail Systems," Proceedings of the First International Symposium on Computer Message Systems, Ottawa, Canada, April 1981.

(8) ISO/TC 97/SC 16-N227, "Reference Model for Open System Architecture," Version 4, International Standards Organization, June 1979.

(9) Y. E. Lien and J. H. Ying, "Design of a Distributed Entity-Relationship Database System," Proceedings COMPSAC 78, 1978.

(10) Y. E. Lien and P. J. Weinberger, "Consistency, Concurrency and Crash Recovery," Proceedings of the ACM-SIGMOD International Conference on Management of Data," 1978, pp. 9-14.

(11) R. R. Panko, "Standards for Computer-Based Message Systems," Report NBS GCR 80-210, National Bureau of Standards, 1980.

(12) R. R. Panko, "A Survey of Electronic Message Systems," Proceedings Pacific Telecommunications Conference, 1981, pp. A3-1/10.

(13) J. B. Postel, "An Internetwork Message Structure," Proceedings of the Sixth Data Communications Symposium, November 1979, pp. 1-7.

(14) J. B. Rothnie, Jr., P. A. Bernstein, S. Fox, N. Goodman, M. Hammer, T. A. Landers, C. Reeve, D. W. Shipman and E. Wong, "Introduction to a System for Distributed Databases (SDD-1)," ACM Transactions on Database Systems, Vol. 5, No. 1, 1980.

(15) D. G. Severance and G. M. Lohman, "Differential Files: Their Application to the Maintenance of Large Databases," ACM Transactions on Database Systems, Vol. 1, No. 3, September 1976, pp. 256-267.

MODELING AND VERIFICATION OF END-TO-END PROTOCOLS

André DANTHINE

Université de Liège, Belgique

Abstract

Using a simple interface protocol as example, finite state automaton and Petri nets are introduced and compared. The concept of an interface machine is discussed and rejected. The problems related to the transmission medium are introduced.

Hierarchical decomposition of computer networks into layers introduced the concept of end-to-end protocols. Such a protocol requests a global model which includes two local models and a transmission medium model. But the local model must be revisited in order to be usable in practical situation. The state variables of an automaton are supplemented by context variables. For Petri nets, predicates are introduced.

After a detailled comparison of these expanded models, the verification problem is introduced and illustrated with one example.

1. INTRODUCTION

A protocol is the set of rules which governs the cooperation between two communicating entities. When the two communicating entities are connected to the same bus, the protocol is based on electrical signals. When the two communicating entities are located in different environments, the protocol is based on message exchanges. Any line protocol is an example of such a situation.

K. G. Beauchamp (ed.), New Advances in Distributed Computer Systems, 125–158.

The designer of a protocol first builds the basic scenario. This involves the simultaneous definition of the basic messages and of the basic sequence of these messages. He then looks into alternative situations and during this process may introduce new messages and new sequences. In general he also has to introduce scenarios for error recoveries. This gives rise to additional messages and increases the list of acceptable sequences.

Except for very simple cases, the protocol reached a level of complexity which requires a systematic approach to be able to validate the design. A description based on a natural language is not adequate. There is therefore a strong need for a formal model at the design level. This formal model must also be usable for formal verification.

We would like to survey the problem of protocol modeling starting with a simple protocol and moving with increasing complexity to the network protocols.

2. INTERFACE PROTOCOL

By interface protocol, we mean a situation where there exists a direct exchange capability between the two communicating entities. A line protocol is a classical example of an interface protocol. They occur also between adjacent layers of a network hierarchy

When dealing with an interface protocol the interest is in general limited to the exchanges taking place in the dashed rectangle of Figure 1.

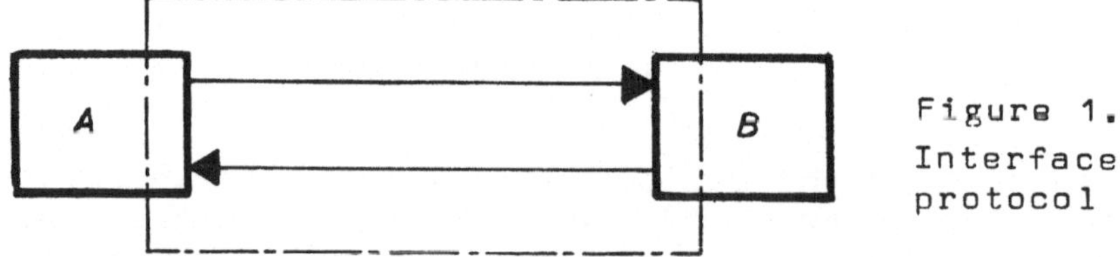

Figure 1. Interface protocol

2.1. A simple example

As an example of such an interface protocol let us consider the flowcharts of Figure 2. This protocol is based on a master-slave relationship between source and destination and the exchange of messages takes place in a half-duplex mode.

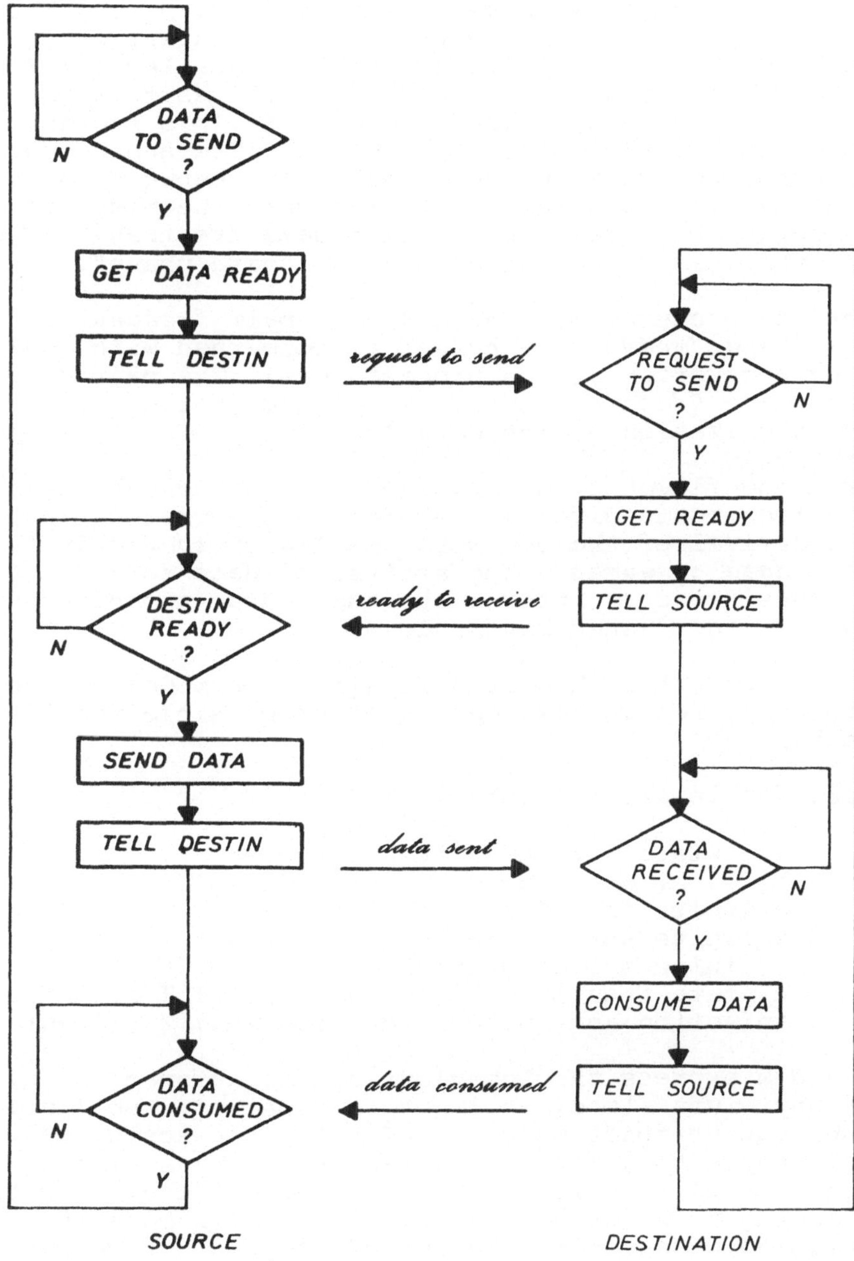

Figure 2. A simple example of protocol

In any protocol, we have to consider two parts. The first one is concerned with the synchronization between the two entities at the "process level". This covers the exchange of messages which has to take place before being able to send data. This exchange is often called the "control phase". The second part is concerned with the synchronization between the two entities at "data level". This covers the message exchange which takes place when data are transferred and which is sometimes called the "data phase".

For the protocol of Figure 2, the pair *"request to send"* and *"ready to receive"* is concerned with the synchronization at the process level. The pair *"data sent"* and *"data consumed"* is related to the synchronization at the data level.

From the flowchart of the source, it is clear that the two events directly related to the protocol are the arrival of the two messages *"ready to receive"* and *"data consumed"*. The arrival of data to send will be considered as internal to the source process but outside the dashed rectangle of Figure 1.

The flowchart of Figure 2 is already a model of the protocol. Let us now look at other possible modeling techniques.

2.2. Finite state automaton

A finite state automaton is a 5-tuple
$< X, I, O, N, M >$ where
X is a finite set of states
I is a finite set of inputs
O is a finite set of outputs
N is a state transition function $(N : I \times X \to X)$
M is an(action and) output function $(M : X \times I \to O)$

N and M express the behaviour of the automaton. If, in any given state, an input is received, the (action and) output function will indicate (the action and) the output to generate and the state transition function will indicate the new state of the automaton. Incoming messages belong to the set of inputs and outgoing messages to the set of outputs but we may have to introduce, in the set of inputs, "internal events" or "null events" which are necessary to model events occuring outside of the specifications of the protocol but in direct connection with its behaviour.

There exist several representation methods for a finite state automaton but the most widely used is the state transition diagram [1-5]. From Figure 2, two finite state automata may be defined, one for the source and one for the destination. They are represented as state transition diagrams in Figure 3.

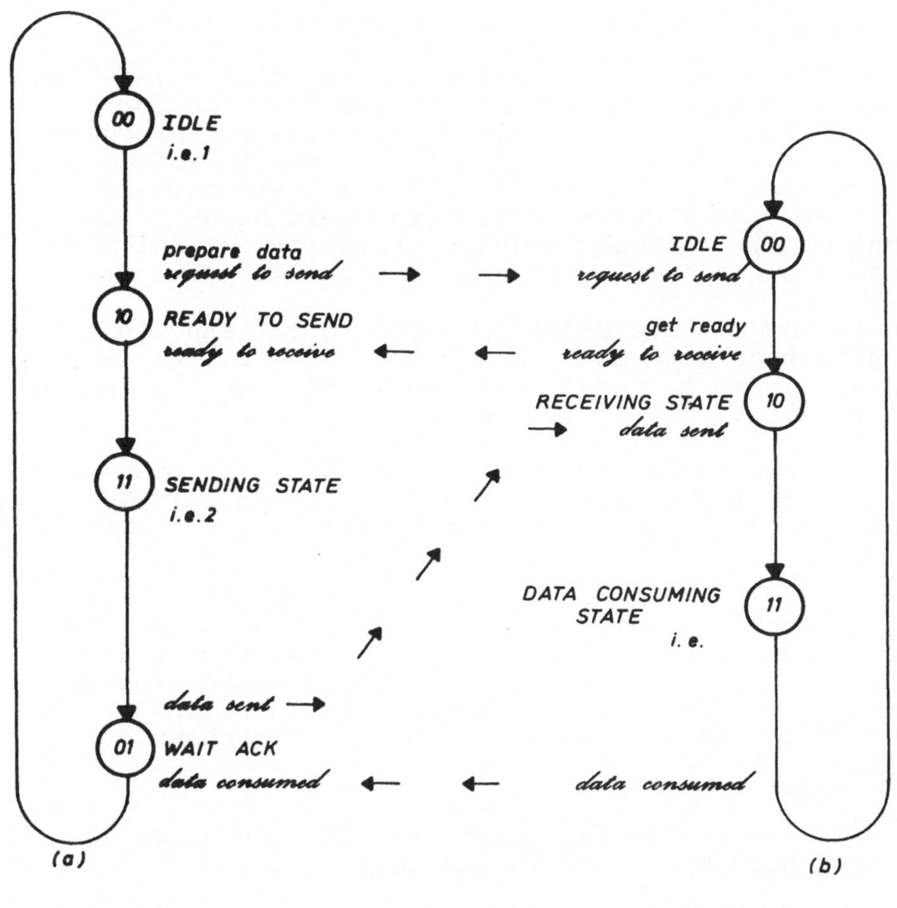

Figure 3. State transition diagrams for the source and the destination (names of states are in capital letters, italic is used for message, lower case is used for local action or internal event)

For the source FSA (finite state automaton) we have as inputs the two messages and two "internal events". The first one (i.e.1.)is related to the arrival of data to send. The second one (i.e.2.)is a true internal event and is related to the end of data transmission. These two internal events model the relationship between the protocol process and its environment. The same thing is true for the internal event introduced in the destination FSA.

From Figure 3, it is clear that our model gives rise to 4 states for the source FSA and 3 states for the destination FSA. It is always possible to introduce additional states. For example, we may introduce between IDLE and RECEIVING STATE a state such as PREPARING TO RECEIVE DATA. This would however imply that we ate not only modeling the protocol but also some details of the processes which implement it.

Up to now we have considered the source and the destination separately but we may take a more global view and try to model the global FSA i.e. the dashed rectangle of Figure 1. The state space of global FSA belongs to the Cartesian product of the two state spaces but is in fact a subset of it. With the 4 and 3 states of the Figure 4, we end up with only 7 states instead of 12 {00/00; 10/00; 10/10; 11/10; 01/10; 01/11; 01/00}.

The state transition diagram is not the only representation which may be used in connection with finite state automata. If we represent inputs and outputs by unit vectors, it is possible to express the state transition and the output functions by matrices whose elements are logical functions of the state [3,6-8]

$\underline{x} = N\ (\underline{x}).\underline{i}$ and $\underline{o} = M\ (\underline{x}).\underline{i}$

For instance, for the source FSA if inputs are represented by the following unit vectors :

i.e.1 = $[1,0,0,0]^T$

ready to receive = $[0,1,0,0]^T$

i.e.2 = $[0,0,1,0]^T$

data consumed = $[0,0,0,1]^T$

and if outputs are represented by the following unit vectors :

request to send = $[0,1]^T$

data sent = $[0,1]^T$

with the following states

$x_1 = [0,0]^T$ = (idle)

$x_2 = [1,0]^T$ = (ready to send)

$x_3 = [1,1]^T$ = (sending state)

$x_4 = [0,1]^T$ = (data consumed)

we have

$$N(\underline{x}) = \begin{bmatrix} p(x_1+x_2+x_3) & p(x_2+x_3) & p(x_2) & p(x_2+x_3) \\ p(x_3+x_4) & p(x_2+x_3+x_4) & p(x_3+x_4) & p(x_3) \end{bmatrix}$$

where

$$\begin{aligned} p(x_i+x_j) &= 1 \quad \text{if state is } x_i \text{ or } x_j \\ &= 0 \quad \text{otherwise} \end{aligned}$$

and

$$M(\underline{x}) = \begin{bmatrix} p(x_1) & 0 & 0 & 0 \\ 0 & 0 & p(x_3) & 0 \end{bmatrix}$$

Another possible representation is based on decision tables. The decision table of Figure 4 corresponds to the source FSA of Figure 2. In a given state and for any possible input we find in the table the next state and the output function if any. From Figure 4 it is clear that in any given state, most of the inputs will give rise neither to a state transition nor to an output generation. This indicates that we do not expect to receive such inputs in this given state. The "unexpected" inputs are not represented in Figure 3.

Input / State	i.e.1.	Ready to receive	i.e.2.	Data consumed
x_1 (00)	x_2 / OI	x_1 / –	x_1 / –	x_1 / –
x_2 (10)	x_2 / –	x_3 / –	x_2 / –	x_2 / –
x_3 (11)	x_3 / –	x_3 / –	x_4 / OII	x_3 / –
x_4 (01)	x_4 / –	x_4 / –	x_4 / –	x_1 / –

Figure 4. Decision table of the source
The outputs are OI :*request to send*
OII:*data sent*
- : no output

2.3. Petri nets

Petri nets [9,10] were used initially to study the interconnection properties of concurrent and parallel activities. It is not surprising that they have been of interest in the modeling of protocols [11-13].

A Petri net (Figure 5) consists of places (nodes, conditions) and transitions (events) which are connected by directed arcs. A directed arc connects either a place to a transition or a transition to a place. The places from which there are arcs incident to a transition are called the input places of that transition. The output places of a transition are similarly defined to be those places which are

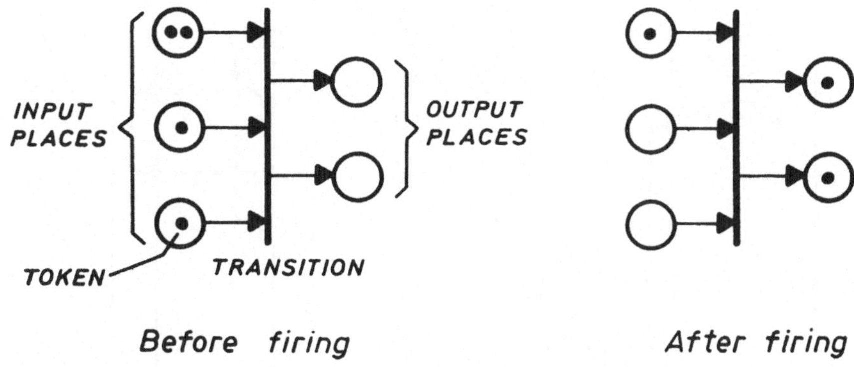

Figure 5. Petri Net Principle

connected to the transition by arcs which originate at the transition and terminate at the place.

A place may have one or several tokens or it may be empty. The transition obeys the following rules of operation

- a transition is said to be *enabled* or *firable* if each of its places contains at least one token;
- the firing of an enabled transition consists of removing one token from each of its input places and adding one token to each of its output places;
- the firing of an enabled transition takes zero time but may not occur instantaneously. The firing of an enabled transition may be considered as depending on an outside authority.

Figure 5 is a representation of the formal definition of a Petri net which is a 4-tuple

$$C = (P,T,I,O)$$

where

P is a set of places (conditions) $P=\{p_1,\dots,p_k\}$

T is a set of transitions (events) $T=\{t_1,\dots,t_n\}$

I is the input function $I(t_k)=\{p_i,p_j,\dots\}$

O is the output function $O(t_l)=\{p_m,p_n,\dots\}$

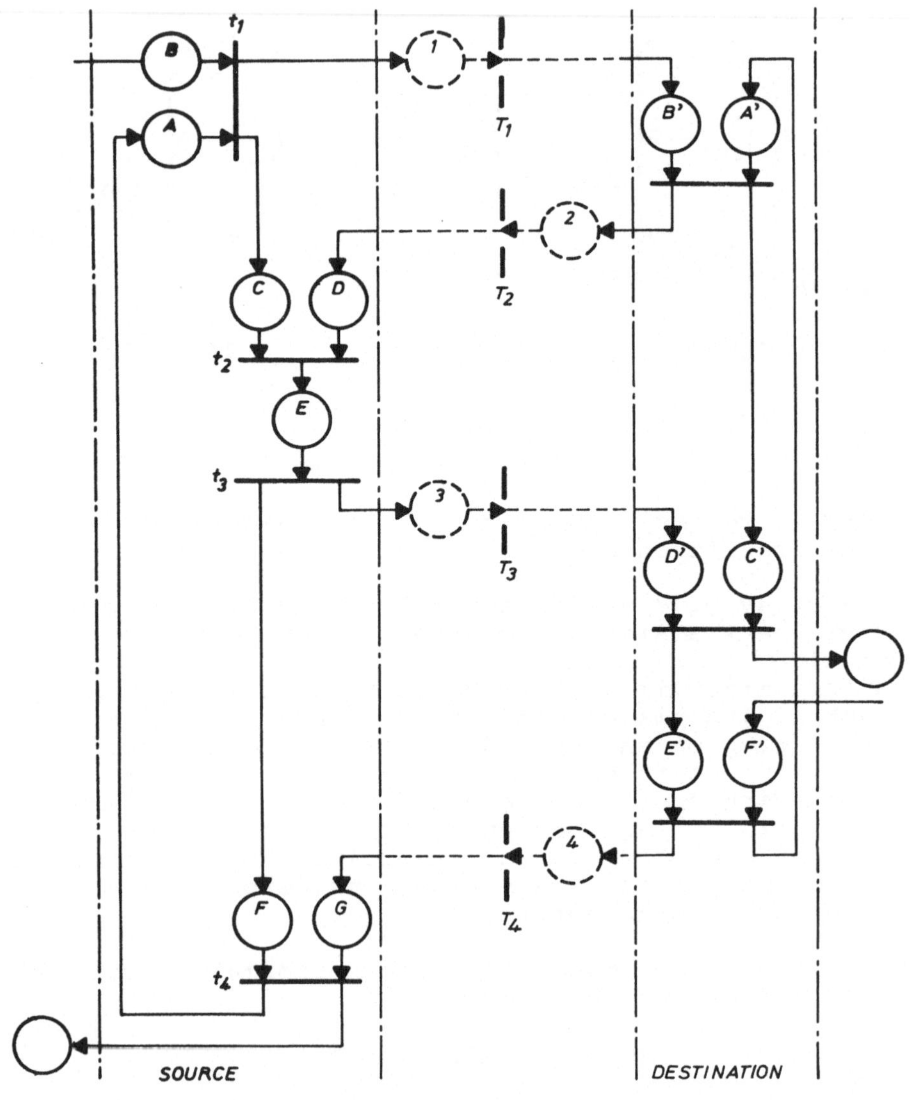

Figure 6. Petri net for the source and the destination
Initial marking: token in A and A'
B = data to send
D = *ready to receive*
G = *data consumed*
B' = *request to send*
D' = *data sent*

For the simple protocol of Figure 2 it is possible to construct the Petri nets of Figure 6 with the Petri net of the source on the left and that of the destination on the right. Our comments about the source FSA and its internal events are illustrated here. The internal event 1 is here the arrival of a token in B and internal event 2 is the firing of t_3. Notice also that the CONSUME DATA of the Figure 2 appears here as an interaction with the environment.

A token distribution amongst the available places in a Petri net is called a marking. From an initial marking of a Petri net, it is possible to construct the set of markings reachable from it. Each marking represents a state of a process and defines a state machine called a token machine [11,12]. The token machine of the source of Figure 6 is represented in Figure 7.

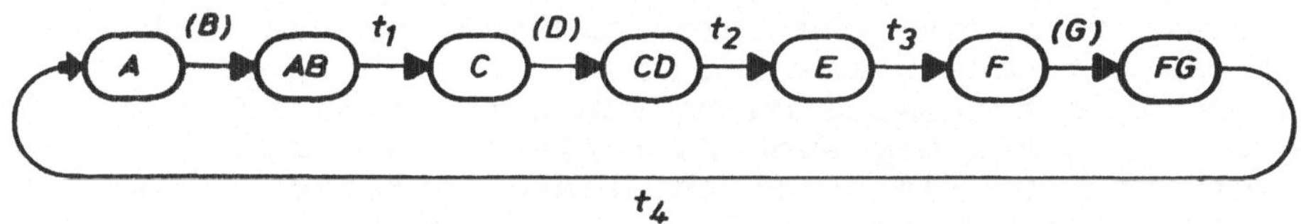

Figure 7. Token machine of the source process

Each arc of the token machine is labeled with the name of the transition that effects it. However we have very often peripheral places i.e. places receiving their token from outside the limits of the process we intend to model. In the source process of Figure 6, B,D and G are peripheral places. The introduction of a token in such a peripheral place results from a transition located outside the process and, in the token machine, the arc is labeled by the name of the place where the token is introduced. The same situation exists with places located inside the model but where the removal of a token depends upon a transition located outside.

Let us compare the Figures 3(a) and 7 which are both derived from the same source process of Figure 2. The markings A,C,E and F are respectively equivalent to the states 00,01,11 and 01. Inputs of the FSA such as *"ready to receive"* and *"data consumed"* appear as additional markings in the Petri net (CD and FG in Figure 7). Finite state automaton and Petri net are not strictly equivalent constructs [9,14], however in most problems they will give the same results.

A Petri net provides a detailled model of the conditions related to the information flow in a process and corresponds, in a more abstract form, to a flowchart or a natural language description. As a token machine and a FSA are equivalent, it means that a FSA is, as a token machine, the result of a transformation. States are derived from conditions. Therefore a FSA is not the best tool to use at the very beginning of the design of a protocol. At this stage, a Petri net model may be very useful while the final design may be presented as a classical and more compact FSA.

2.4. Interface machine

From the figure 6, it is possible to obtain the token machine of the destination process. It is also possible to derive the token machine for the global process located inside the dashed rectangle of Figure 1. A transmission medium with no delay between A and B may even be replaced by a model involving four places (1 to 4) and four transitions (T_1 to T_4) as in Figure 6.

However, if we take into account the master-slave relationship and the half-duplex characteristic of our protocol, it is possible to introduce the idea of an *interface machine*. Such a machine does not exist but it may be a useful conceptual tool. Such an interface machine is located between the two processes A and B of the Figure 1 and receives as inputs the messages generated by A and B (Figure 8a). Figure 8b is the Petri net graph of the interface machine. Figure 8c is the state transition diagram associated with the interface machine and Figure 8d is the token machine.

This virtual interface machine places in evidence the fact that, in the example we analyzed, *the knowledge of the state of the source is enough to know the state of the destination and vice-versa.* Even if the two processes have only local information, our protocol is such that this local information is equivalent to global information.

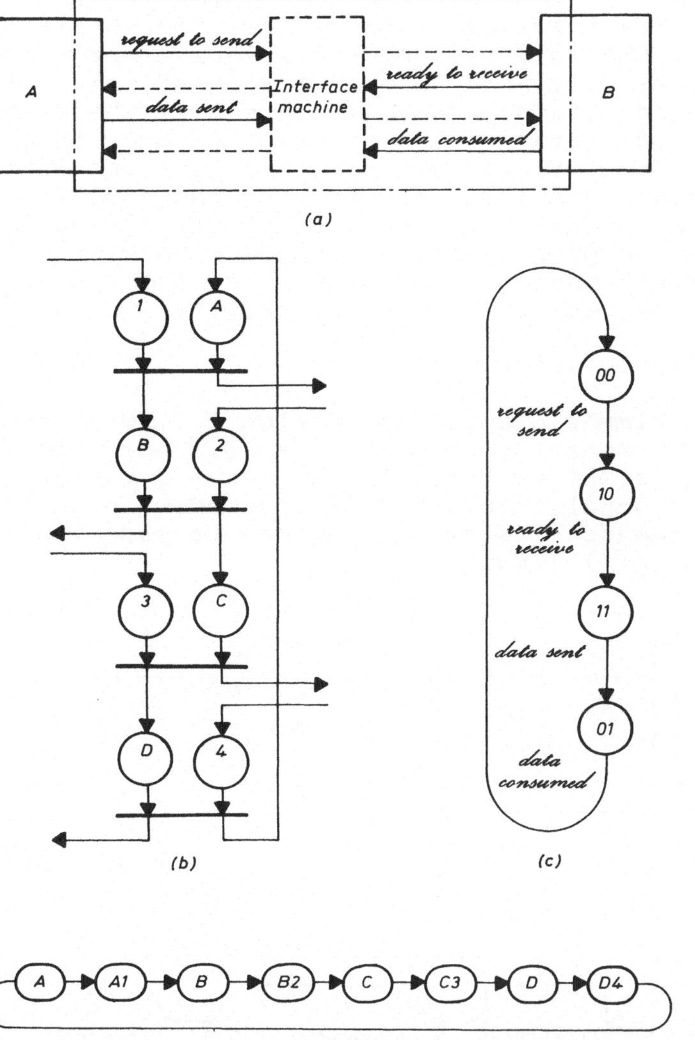

Figure 8. a) Interface machine
b) Petri net of the interface machine
initial marking : token in A
1= *request to send*
2= *ready to receive*
3= *data sent*
4= *data consumed*
c) State diagram of the interface machine
d) Token machine

The FSA of the interface machine involves 4 states (Figure 8c). We mentioned in section 2.2 that the global FSA involves 7 states. Besides the initial states, the only pair of truly equivalent states is the state 11/10 of the global FSA and the state 11 of the interface machine (Figure 8c). The greater number of states of the global FSA comes from the asynchronous character of the transitions of the source and destination processes and also from the internal events introduced in Figure 3. This again raises the problem of the boundary between the specifications of the protocol and the specifications of the processes which implement it.

2.5. Transmission medium

In the simple example of section 2.1. we assumed a perfect transmission medium. We did not consider that a message may be lost. In the general case we will need to model this transmission medium more realistically because its properties may be essential at the design level (Figure 9).

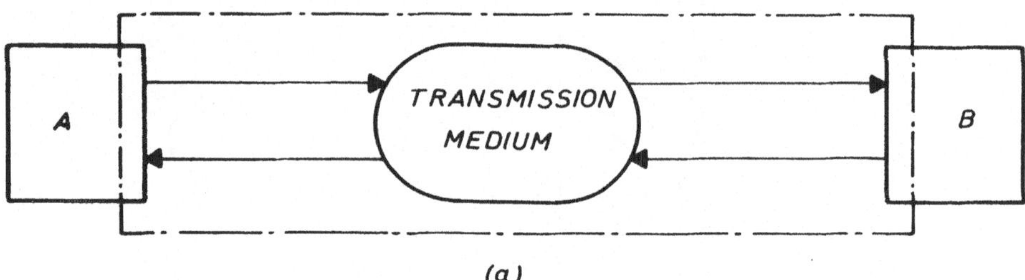

(a)

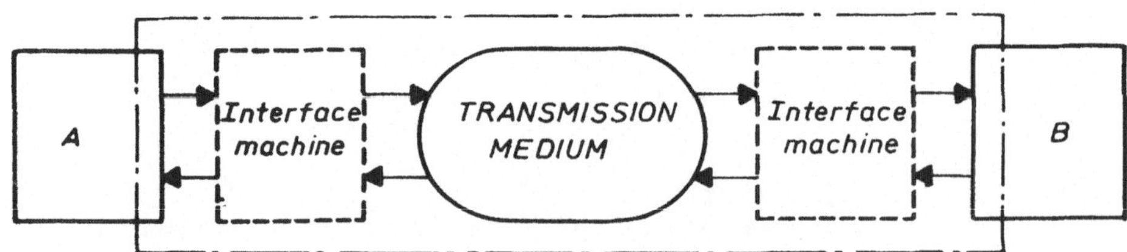

Figure 9. Interface machine with nonideal transmission medium

The transmission medium must be defined in terms of actions on the messages which may take various forms such as variable delay of transmission, loss of a message, duplication of a message, etc. In such an environment, it is no longer possible to introduce the idea of a unique interface machine because the information about one process no longer allows one to deduce the global state. Even with a perfect transmission medium but with a full duplex protocol, the interface machine concept has to be discarded.

It has been suggested that two interface machines be introduced (Figure 9b) each reflecting the view that each side has about the global state. However there may be differences between the state of any interface machine and the real global state due to the behaviour of the transmission medium. Therefore the interest of interface machines is questionable and we prefer to rely on global model which involves two local models and a model of the transmission medium.

2.6. Programming languages and other formal models

Our starting point for introducing our simple protocol was flowcharts and it is not surprising that high level programming languages have been used for modeling the source and the destination processes [15-18].

Let us mention also the use of grammars to model HDLC link protocol [19]. For the simple example of Figure 2 we may introduce the following alphabet:

Σ={*request to send* = A, *data sent* = B, *ready to receive* = C, *data consumed* = D}

The protocol is defined by the grammar : A,C,B,D.

A bibliography containing more references is periodically updated by Day [20].

3. END-TO-END PROTOCOL

A computer network is generally represented as a set of distributed processes organized in a hierarchical structure.

In Figure 10, we used a three layer model with the application layer at the top, the transmission layer at the bottom, and the transport layer in between. The transmission layer may consist, e.g., of a set of private lines, a circuit-switching network or a packet-switching network. The transport layer provides a transport service to the processes located in the top layer. This requires an interface protocol between the upper layer and the transport layer. Another interface protocol is needed between the transport layer and the transmission layer.

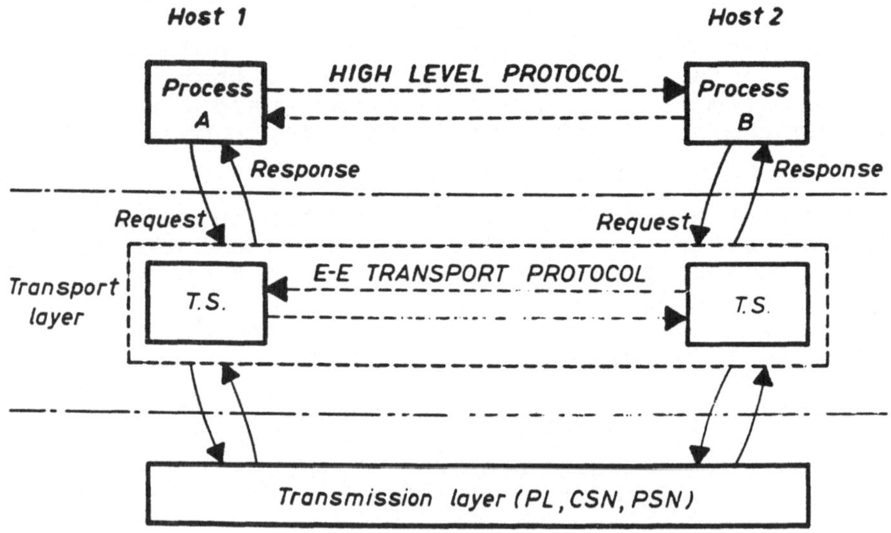

Figure 10. A computer network as a three layer structure

From Figure 10, it is clear that the communication between process A and process B involves a set of chained interface protocols. However, most modern networks have introduced the concept of end-to-end protocols which govern the interaction of processes located at the same level of the hierarchy. In Figure 10, an end-to-end protocol, symbolized by dashed lines, exists between processes located in the upper layer and also in the transport layer. We assumed that the transport service is provided by two distributed processes called transport stations (TS's). The cooperation between these two processes is governed by an end-to-end transport protocol.

As the purpose of an end-to-end transport protocol is to provide a transport service to the processes located in the upper level, it is therefore essential not to limit the model to the "interface" between the two TS but to include the interface protocol between the user process and a TS (Figure 11).

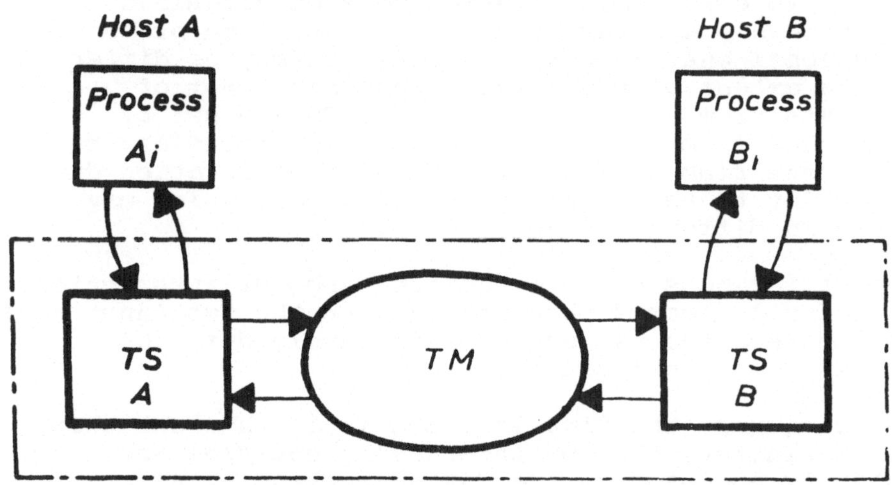

Figure 11. Global model of an end-to-end protocol

The link between the two TS is a virtual one. The characteristics of such a virtual link may depend upon a serie of chained interface protocols and the properties of the transmission medium over the virtual link have to be carefully evaluated.

These considerations eventually lead us to the global model of an end-to-end protocol represented by the dashed rectangle in Figure 11. It involves the model of two TS and the transmission medium model. A TS is a local entity and the model of it will be called a local model.

Any local entity (i.e. the TS in Figure 11) exchanges messages with its peer entity but interacts also with the user process. The user process requests services from the TS which has to respond to these requests. This leads us to introduce for the local model two sets of inputs and two sets of outputs.

The rationale for separation is based on the following considerations. Between peer entities there is no reason to introduce a master-slave relationship and furthermore the set of output messages of one local model must be the same that the set of input messages of the other. Between a user process and the TS which are in two different levels of the hierarchy, it may be essential to introduce a master-slave relationship and furthermore the interface protocol is a local one. In two locations, it may be different in the extent of the services provided. Set of requests from a user to its local TS and set of responses from the TS to the user may therefore be different from one location to another location due basically to the disparity of functions which may exist in different places.

The differences between the two types of interactions call for a clear separation between the two input sets and between the two output sets involved in the modeling of the local process.

If we compare the Figures 1 and 11 it is clear that, in the latter, the two interacting entities are entirely enclosed in the dashed rectangle. All interactions with the environment of a TS are part of the input and output definitions. Here, internal events will now really be internal to the model.

As our global model involves two local models and a transmission medium model, we are back to our initial point i.e. to model a local process. However the additional inputs and outputs raise the problem of the usefulness of the methods presented when the complexity increases. As we will see it is necessary to extend the methods in order not to be limited to academic examples.

4. GENERAL LOCAL MODEL

4.1. State-variables and context variables

When the number of states becomes very large, any type of state machine is difficult to apply and all transition diagrams become unusable. This dimensionality problem may be overcome by decomposition techniques.

In an instance of a program, the status word is only part of the complete state of the process. The information associated to the status word is supplemented by a set of context variables. Following the same approach to overcome the dimensionality problem in protocol modeling, it has been proposed to associate with a set of states, a set of context variables [7,21]. Transitions involving context variables are described by a set of procedures. Using such an extended model, it has been possible to completely model the transport protocol for Cyclades [4] and to verify certain aspects of it [8,22].

During this study two basic questions were raised :
- How do we separate state and context variables ?
- What are the relationships between the procedure execution and the state transition ?

The second question will be considered later on. Let us try to address the first one. The protocols we are interested in are based on message exchanges and the interface with the upper level is based on requests and responses. The general structure of any input to the local process may be assumed to be the following one :

<op. code>< parameter vector>

- the operation code indicates the kind of request or of message e.g. <send letter>,<ack>,etc;
- the parameter vector is used to transmit additional information in connection with this <op.code>. This parameter vector has in general no meaning without the <op.code> and we have therefore a dependency relationship which may be used to separate state and context.

Let us define

$I = \{$<op.code><parameter vector>$\}$

$I_1 = \{$<op.code>$\}$

$I_2 = \{$<parameter vector>$\}$

$I \subset I_1 \times I_2$

The cardinal number of the finite set I_1 of operation codes is in general much smaller than the cardinal number of the finite domain of the discrete parameter vector space. A small cardinal number for the input set is a necessary condition for limiting the number of states of a FSA. Therefore adopting a FSA model with the input set I_1 is a possible way to overcome the dimensionality problem.

To any input from the set I_1 (basic input) will correspond a state or a set of states in the FSA. As the parameter vector has a dependency relationship with its <op.code> i.e. with an input from the set I_1, we may associate to a state or a group of states a set of context variables selected in connection with the elements of the parameter vector. The dependency of the < parameter vector> on the <op.code> will be mapped into a dependency of the context variables on the state variables. Roughly speaking <op.code> will change state variables and <parameter vector> will modify context variables. We will return to this point in section 4.5.

4.2. Petri nets

If the problem of dimensionality of the FSA has to be considered, the same is true for the Petri net. In order to get a tool usable to describe complex and real situations, Nutt [23] introduces the concept of an evaluation net which, like Petri net, is also made up of transitions interconnected by directed arcs to locations (places) but here the transitions obey the following rule : a transition fires if the set of input and output locations *satisfies the definition of that particular transition* causing one token to be removed from each location of a *prespecified subset* of input locations and one token to be placed on each location of a *prespecified subset* of the output locations. Furthermore, the time required for each execution of a transition is part of the specification of the net. This extension allows one to introduce time as a measure of the net performance.

Nutt introduces several evaluation net primitives :

- The T-transition is a transition involving one input location and one output location. The T-transition is enabled if the input location is full (contains one token) *and* the output location is empty (contains no token). Here, the state of the output locations has to be verified before enabling the transition. The F-transition with one input and two output locations and the I-transition with two input locations and one output location also require empty output locations to enable the transition.

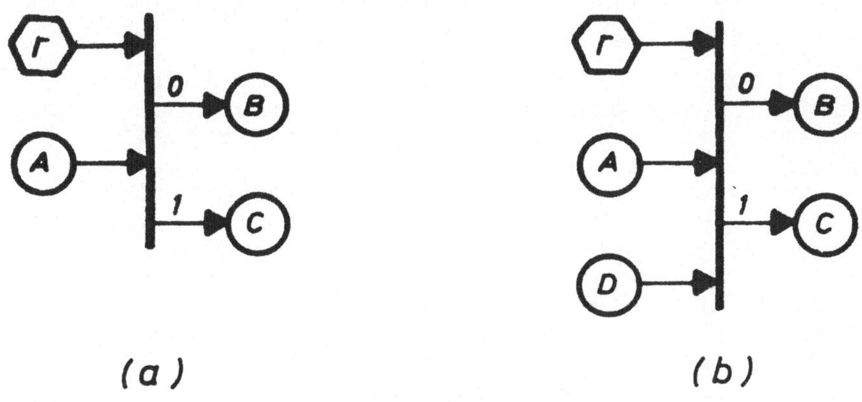

Figure 12. X-transition a) as defined by Nutt[23]
b) as used in [13]

- The X-transition is reproduced in Figure 12. In the X-transition, a hexagon has been introduced. It represents a *resolution location* which is a special type of input location whose status may be 0 (i.e. empty), 1(i.e. full) or ϕ(i.e. undefined). The two leaving arcs have also been marked with a 0 or a 1 and the X-transition definition is the following : the X-transition is enabled if the input location is full (containes one token) *and* the resolution location status is defined (0 or 1) *and* the output location corresponding to the value of the resolution location status is empty. The firing of an enabled transition consists of removing the token from the input location, putting one token in the output location corresponding to the value of the resolution location status and changing the resolution location status to "undefined". As already mentioned the transition time is specified for each transition.

As the resolution location status returns to be undefined state, the complete specification of the X-transition implies a *resolution procedure* which is activated when a token is placed in the input location. One purpose of this resolution procedure is to define the status of the resolution location. The resolution procedure includes an expression of the form

$$M : [p_1 \rightarrow S(r) : = i; p_2 \rightarrow S(r) : = 1-i]$$

where S(r) is the status of r,

i is either 0 or 1, and

p_1 and p_2 are two predicates.

If p_1 is true, the status is set to i and further evaluation of the procedure is discontinued. Otherwise p_2 is evaluated and, if it is true, the status of r is set to 1-i. When both predicates are false, the status of r remains undefined and the procedure need not be evaluated again until one of the arguments of the predicates changes its value.

The purpose of the *resolution procedure* is to prepare the firing of the transition by setting the value of r to 0 or 1. One may also introduce a *transition procedure* which is executed at the firing time.

Another important extension of the classical Petri net introduced by Nutt is the concept of *attribute token*. The token may be a simple token only denoting occupancy or it may be an *attribute token* i.e. a vector representing a set of attributes, some or all of which may change as the token flows through the net. A *transition procedure* may reference *and* alter value attributes of tokens as they flow through the associated transition. They may also reference and alter *environment variables* i.e. global variables that may be accessed by any procedure in the net. A *resolution procedure* may reference but *not* alter token attributes and environment variables.

As in general the execution of a *resolution procedure* is immediately followed by the execution of an associated *transition procedure* we will not here distinguish any further between the two and will only to a *resolution procedure* which in the first step may reference (but not alter) token attributes and environment variables in order to set the value of r and in second step depending upon the value of r, may proceed further and alter token attributes and environment variables.

4.3. Time Petri net

The time Petri net is another extension of the classical Petri net. Introduced by Merlin [11,12], it consists in adding to each transition of a Petri net, two time values. The first time value associated with the transition i will be noted $t*i$ and denotes the minimal time that must elapse from the time that all the input conditions of a transition are enabled until this transition can fire. The second time value associated with the transition i will be denoted $t**i$ and denotes the maximum time that the input conditions can be enabled and the the transition does not fire. We always have $t*i < t**i$ and a Petri net is a special case of a time Petri net with $t*i = 0$ and $t**i = \infty$.

The time concept introduced here is completely different from the transition time of Nutt. Time Petri nets are useful in protocol modeling, for instance to model the discarding of a token received under some conditions (token absorber) or to model a retransmission mechanism based on a time-out.

4.4. Combined Petri net

A combination of the time Petri net with the X-transition of Nutt has been used to model the Cyclades transport protocol in the following way [13]. We already know the general structure of an input (request or message) :

<operation code> <parameter vector>

Therefore the occurrence of an input with a given op-code is not represented by a simple token but by an attribute token. The occurrence of a request or of a message must first be checked with an input condition and therefore the transition will have two input locations (Figure 12b). When an awaited request or an awaited message occurs, the protocol has to check the parameter vector associated with it and to make a basic evaluation. Such an evaluation may be expressed with the help of predicates and the X-transition is just what is needed to model it. Eventually we end up with the basic module of the Figure 13.

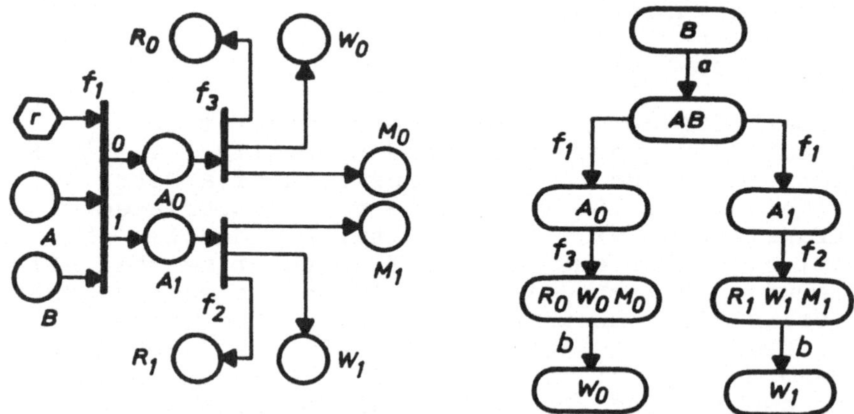

Figure 13. Basic module(left) with its token machine (right)

The occurrence of an input with a given <op.code> will be indicated by a token in location A. Location B will be used for marking the condition under which the request or message will not be discarded.

- If the token is missing in B when a token appears in A, we must provide a mechanism which removes the token in A without firing f_1. Such a mechanism, not represented in Figure 13, involves an arc from location A to a transition f_0 and f_1 we will set (t*0> t** 1).

- If the condition B is not missing when A occurs, the resolution procedure r may be activated and will decide to put a token in A_0 or A_1 or not to fire f_1 immediately. From a conceptual point of view, we may say that the case 1 where a token is placed in A_1 means that the request is accepted and that the case 0 where a token is placed in A_0 means that the request is not accepted. In both situations, the token resulting from the firing of transition f_1 will eventually fire the transition f_2 or f_3. The most that such a firing may do, is
  1) to pass a message to the next unit (token in M_i);
  2) to send a reply to the origin of the request or message received (token in R_i);

3) to set a waiting condition for another incoming message or request (token in W_i).

On the right part of the Figure 13, the token machine associated with the basic module has been reproduced. Transition between states (markings) are due to transition firing (f_1,f_2 and f_3), to token arrival(a) or to token removal (b). For simplicity we assume that if the output location involves a message M_1 and a response R_1, both proceed further at the same time.

In our model we will need the concept of environment. variables that may be accessed by any procedure in the net and we will allow a resolution procedure to alter such variables. Coming back to the Figure 13 and assuming a token in the B location when a token in A arrives, the general structure of a resolution procedure involves
- checking up on available resources;
- setting or updating of environment variables; and
- selection of an output location by evaluation of predicates.

The predicate evaluation may involve the availability of resources, the matching of the parameter vector associated with the input A with the global variables, and so on.

Without environment variables and resolution procedure, the number of places and consequently of markings would be too numerous and the model would not be useable. By keeping the parameter vector information at the environment variable level, it is possible to gain a partial control of the dimensionality problem. To every <op.code> we associate a place like A in Figure 13 and introduce environment variables and resolution procedures to deal with the parameter vectors. As an example, let us mention that for the basic control phase of Cyclades protocol it has been possible to develop a model with 28 places [13].

In section 4.1 and in this section, we introduced very equivalent ideas. With the FSA, we had state variables and context variables. With the Petri net we have the markings and the environment variables. With the FSA, we had a procedure execution and a state transition. With the Petri net we have a resolution procedure and a firing.

To complete this comparison we would like to point out that one of the main interests of Petri net oriented models lies in the insight they give into the mechanisms of the protocol. With the basic module of the Figure 13 a lot of questions are raised when a protocol is analyzed. But such a tool will even be more useful at the design state of a protocol. For instance, if we want to define the way to interpret a request to open a link issued by the user process to its TS (Figure 11) we may use the basic model of Figure 13.

The occurrence of the request <open link> will put a token in place A. A syntactically correct request may be rejected if the source process does not have the access rights to the service or if the TS does not have the resources and does not want to queue the request. If the request is rejected the firing of f_1 will put a token in A_0. The firing of f_3 will put a token in R_0, i.e. will send a negative response to source of the request. Places like M_0 (message sent to the other TS) and W_0 (waiting condition) are not necessary in this exemple. If the request is accepted, the firing of f_1 will put a token in A_1. A message will be sent to the remote TS (token in M_1) and a waiting condition will be set (token in W_1). The introduction of a place like R_1 would mean that the source will receive a reply with the following semantic *"your request has been accepted by your local TS"*. However the designer may prefer to avoid all partial replies and wait until the end of the processing of the request by all distributed entities before reporting back to the source. If so, place like R_1 have very often to be dropped.

In summary, for every received request or message, the designer will tailor the basic module (Figure 13), define environment variables, set the resolution predicates and by the resolution procedure, define the processing of the information included in the parameter vector.

In Figure 14 we reproduce the Petri net related to the request to open a link in Cyclades [13].

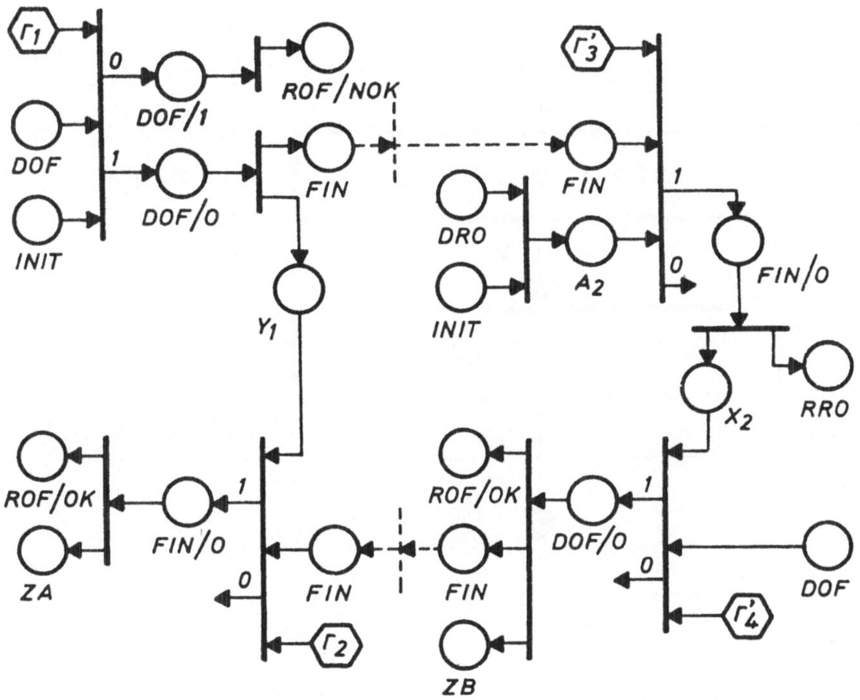

Figure 14. Net of the opening phase

4.5. Finite state automaton decomposition

In our Petri net, we have attribute tokens and states represented by markings *and* environment variables. Furthermore transitions are completed by resolution procedures.

In section 4.1., complex inputs ($i \in I_1 \times I_2$) were introduced, basic state variables were supplemented by context variables and state transitions by procedure executions. We would like to point out that this approach may be formalize. A unique FSA model may be replaced by a two-step process involving two successive FSAs (Figure 15).

The mathematical development may be found in [22,24]. As a FSA may always be represented as an abstract program the context automaton may be described by a set of procedures. This combination of a transition model and of an abstract program is called a *hybrid model*.

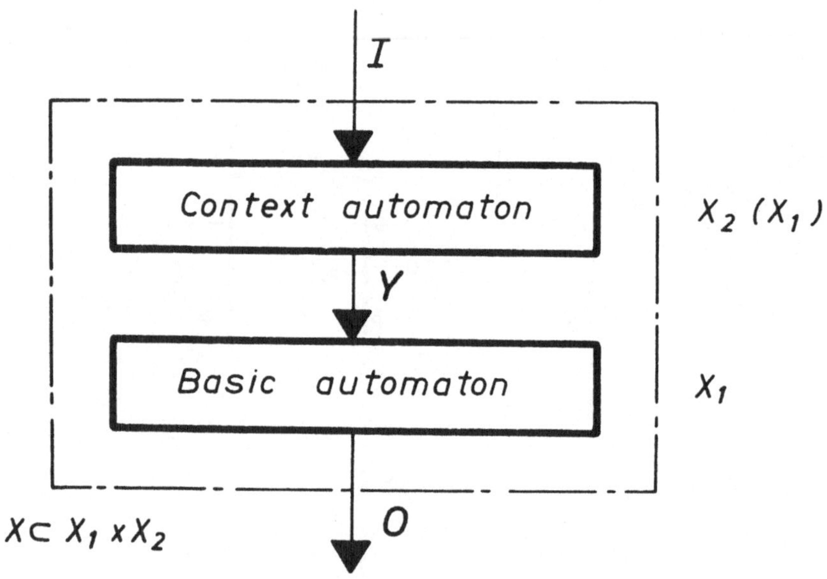

Figure 15. Decomposition of a finite state automaton in a context and a basic automaton

In [25] Bochmann also proposed a hybrid model involving a FSA part with variables associated with the states and concluded that *"since reachability analysis of state machines seems to be more amenable to algorithmic methods than verifying (and finding) program assertions, the above tradeoff may have important implications for future automated methods of protocol verification"*. The hybrid model of [25] is based on Keller's model [26] itself very close in concept to the evaluation net of Nutt [23].

As a FSA and an abstract program are equivalent constructs, it is possible to represent protocols as abstract programs. At the representation level, the dimensionality problem is not anymore an issue but the verification remains a very difficult problem.

In [27] Sunshine raises an interesting objection against the use of the programming language as specification tool : *"a program, even in a 'high level language', is usually not a satisfactory specification because it is impossible to separate the essential features of what the program is supposed to do from the particular way chosen to accomplish those functions"* As almost hybrid models are using at least partially programming languages, Sunshine's objection concerns the whole protocol community. Our personal opinion is that, even with the formal model like FSA, it is extremely difficult to model only the protocol and to avoid completely the process which will implement it. This has already been pointed out in section 2.5.

5. PROTOCOL VERIFICATION

5.1. Global model

As already indicated, the global model of Figure 11 involves the local model of the two peer entities *and* a model of the transmission medium. The nondeterministic aspect of the transmission medium model does not allow the use of the same modeling techniques. The most perfect transmission medium will be modeled as a variable delay device which preser ves the order and the content of the messages going through itself (No loss, no duplicate, no disorder). The most imperfect transmission medium will be characterized by losses, duplicates and disorder. It is clear that this two limiting cases may give very different result at the verification level.

5.2. Verification techniques

Without going into details which may be found in[22,28] let us say that reachability analysis is the basic method for transition oriented models. Between the two TS it may exist deadlock situations i.e. that if one TS is in state x, if the other TS is in state y, if both ara waiting for a message from the other and if there is no message pending in the transmission medium the pair (x,y) represents a deadlock for the global model. The verification techniques consist of first determining all possible deadlock situations and then, by a reachability analysis of testing wether these situations may be reached.

5.3. An example

Let us consider the opening phase for Cyclades characterized by a two-way handshake. If the left user sends an *OPEN* to its local TS (Figure 11), the local TS A will allocate resources to the new association and send a *FIN* (Flow initialization) to the remote TS. The connection establishment will proceed further at the left TS when it will receive a *FIN* from the right TS if all parameters are in agreement. This will allow left TS to complete the establishment of the connection and to send an *OPEN* to the left user as a positive response to the original request.

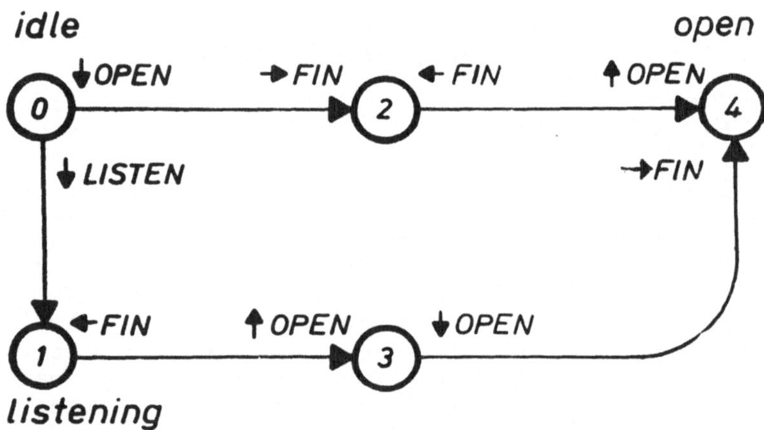

Figure 16. State diagram of the opening phase for two-way handshake

The Figure 16 reproduces the state diagram of the opening phase. Circles represent states and we use the following conventions

↓ *XXX* : request *XXX* received from the process
↑ *YYY* : response *YYY* sent to the process
→ *ZZZ* : message *ZZZ* sent to the remote TS
← *WWW* : message *WWW* received from the remote TS

The exchange of request, response and messages during a two-way handshake opening is reproduced at Figure 17. On this figure it is possible to follow the sucession of events and exchanges. The first request comes from the process in Host B which issues a *LISTEN* and the TS B goes from state 0 to state 1. When the request *OPEN* from the process in Host A is received by the TS A, a *FIN* is sent to the TS B and the TS A goes from state 0 to state 2. And so on...

In the connection opening of Figure 17 the process in Host A initiates the connection while the process in Host B is ready to be called. The two-way handshake may also be successful if both processes try at the same time to open a connection. Such a situation is reproduced Figure 18.

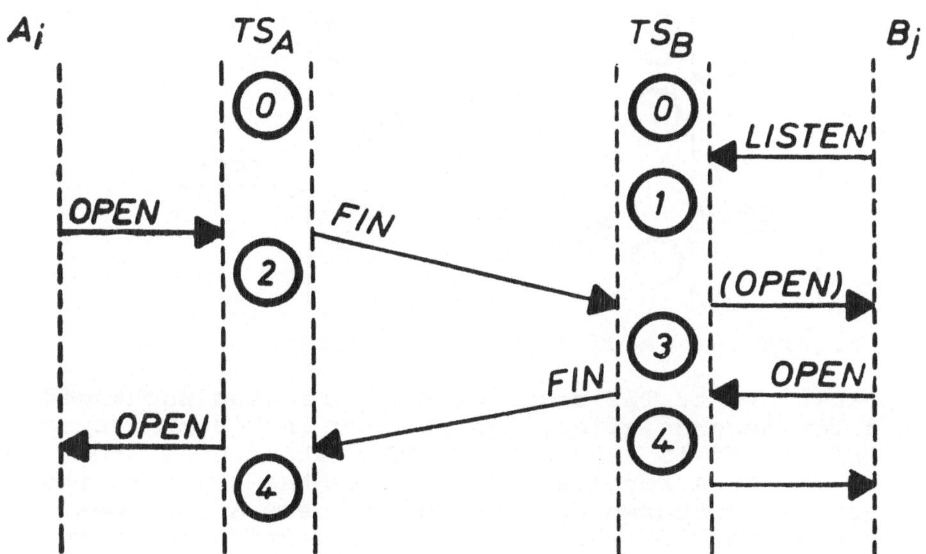

Figure 17. Two-way handshake opening

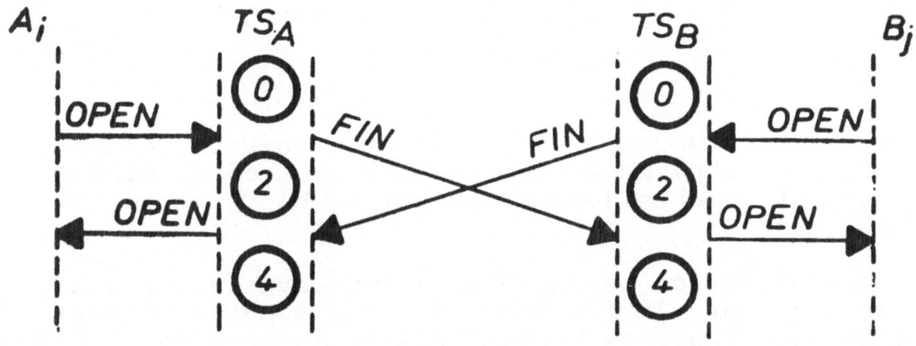

Figure 18. Symmetrical opening

However there exist situations where the protocol is not able to avoid deadlock.

The Figure 19 is concerned by an attempt of a symmetrical opening which fails due to a race condition. The process in Host A is the first to try to *OPEN*. When the *FIN* sent by the TS A reaches the TS B, it is discarded because in state 0 (i.e. idle) a *FIN* is not an

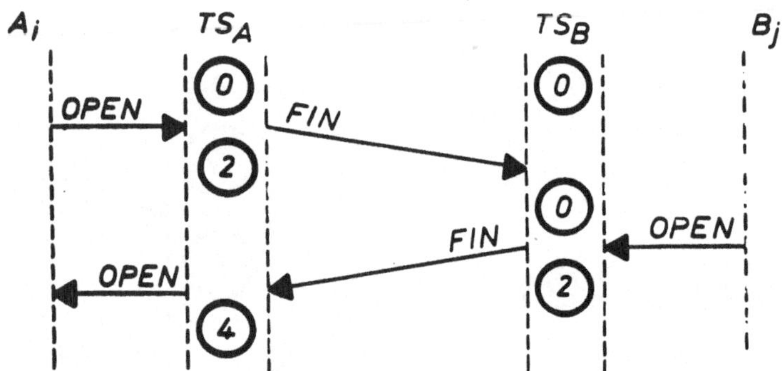

Figure 19. Deadlock due to a race condition

acceptable event. The state of TS B remains unchanged. When a few seconds after, the process in Host B issues its *OPEN*, a *FIN* is sent and the TS B state goes from 0 to 2. The TS A receives this *FIN* as the reply to its *FIN*, sends a response *OPEN* to the process A and goes to state 4. Now for the TS A, the connection is in the open state (Figure 8) but for the TS B the connection is still in an opening state. Such a deadlock will be eventually recovered either by the failure to transmit data or by a time-out on state 2.

6. CONCLUSION

The problem of modeling protocols in order to validate or verify or implement them has now reached a level of development which allows practical problems to be solved.

The generalization of the hybrid approach is certainly not an accident. In the long run, the superiority of reachability analysis versus program assertions or the converse will probably depend upon their relative capability in terms of automated methods.

At the modeling level, there are quasi-equivalent tools starting from Petri net and evolving to programming languages through automata theory, grammars and formal languages. The past experience of every people will probably decide of his choice. It is therefore more important to stress upon the similarities than to focus on the differences and we hoped to have contributed to this goal.

REFERENCES

1 D. BJORNER, "Finite State Automaton - Definition of Data Communication Line Control Procedures", *AFIPS Proc.*, Vol. 37, FJCC, Houston, November 1970, pp. 477-491.

2 H. KAWASHIMA, K. FUTAMI, and S. KANO, "Functional Specification of Call Processing by State Transition Diagrams", *IEEE Trans. Comm. Tech.*, Vol.COM-19, October 1971, pp.581-587.

3 R.E. RUSBRIDGE, and A. LANGSFORD, "Formal Representation of Protocols for Computer Networks", *Report AFRE-R-7826*, UKAEA, Harwell, England, December 1974, 20 p.

4 A.A.S. DANTHINE, and J.J. BREMER, "An Axiomatic Description of the Transport Protocol of Cyclades", *Rechnernetze und Datenfernverarbeitung*, Aachen 1976, Springer-Verlag, pp.259-273.

5 G.V. BOCHMANN, "Finite State Description of Communication Protocols", *Proc.Computer Network Protocols Symposium*, Univ. of Liège, February 1978, pp. F3-1 to F3-11 and *Computer Networks*, 2,4/5, October 1978, pp.361-372.

6 R.W. STUTZMAN, "Data Communication Control Procedures", *Comput. Surv.*, Vol.4, N° 4, December 1972, pp. 197-220.

7 A.A.S. DANTHINE, and J.J. BREMER, "Communication Protocols in a Network Context", *Proc. ACM Interprocess Comm. Workshop*, Santa Monica, March 1975, pp. 87-92.

8 A.A.S. DANTHINE, and J.J. BREMER, "Modelling and Verification of End-to-End Protocols, *SART 77/11/13*, Third European Network User's Workshop, IIASA, Laxenburg, Austria, April 19-20 1977, 17 p.

9 J.L. PETERSON, "Petri nets", *ACM Computing Surveys*, Vol. 9, N° 3, September 1977, pp. 223-251.

10 R.C. CHEN, "Representation of Process Synchronization", *Proc. of the ACM SIGCOMM/SIGOPS Interprocess Communications Workshop 1975.*

11 P.M. MERLIN, "A Methodology for the Design and Implementation of Communication Protocols", *IEEE Trans. Comm.*, Vol.COM-24, N° 5, June 1976, pp. 614-621.

12 P.M. MERLIN, and D.J. FARBER, "Recoverability of Communication Protocols. Implications of a Theorical Study, *IEEE Trans. Comm.*, Vol. COM-24, N° 9, September 1976, pp. 1036-1043.

13 A. DANTHINE, "Petri nets for Protocol Modelling and Verification", *Proc. of the Computer Networks and Teleprocessing Symposium*, Budapest, Hungary, October 1977, Vol. II, pp. 663-685.

14 P.M. MERLIN, "Specification and Validation of Protocols", *IEEE Trans. Com.*, Vol. COM-27, pp. 1671-1680, Nov.1979.

15 G.V. BOCHMANN, "Logical Verification and Implementation of Protocols", *Proc. 4th Data Comm. Symp.*, Québec, October 1975, pp. 7-15 to 7-20.

16 N.V. STENNING, "A Data Transfer Protocol", *Computer Networks*, Vol. 1, N° 2, September 1976, pp. 99-110.

17 D. BRAND, and W.H. JOYNER, "Verification of Protocols using Symbolic Execution", *Proc. Computer Network Symposium*, Univ. de Liège, February 1978, pp. F2-1 to F2-7.

18 J. HAJEK, "Automatically verified Data Transfer Protocols", *Proc. Int. Comp. Comm. Conf.*, Kyoto, September 1978, pp. 749-756.

19 J. HARANGOZO, "Protocol Definition with Formal Grammars", *Proc. Computer Network Protocols Symposium*, Univ. of Liège, February 1978, pp. F6-1 to F6-10.

20 J.C. DAY, "A Bibliography on the Formal Specification and Verification of Computer Network Protocols", *Proc. Computer Network Protocols Symposium*, Univ. of Liège, February 1978

21 A.A.S. DANTHINE, and J.J. BREMER, "Définition, représentation et simulation de protocoles dans un contexte réseau", *Journées AIM Mini-Ordinateurs et Transmission de Données*, Liège, janvier 1975, pp. 115-126.

22 A. DANTHINE, and J. BREMER, "Modelling and Verification of End-to-End Transport Protocols", *Proc. Computer Network Protocols Symposium*, Univ. of Liège, February 1978, pp. F5-1 to F5-12 and *Computer Networks*, 2, 4/5, October 1978, pp. 381-395.

23 G.J. NUTT, "Evaluation Nets for Computer System Performance Analysis", *AFIPS Conf. Proc.*, Vol. 41 Part 1, 1972, pp. 279-286.

24 A. DANTHINE, "Protocol Representation with Finite-State Models" *IEEE Trans. Com.*, Vol. COM-28, pp. 632-643, April 1980.

25 G.V. BOCHMANN, and J. GECSEI, "A Unified Method for the Specification and Verification of Protocols", *Proc. IFIP Congress*, Toronto, 1977, pp. 229-234.

26 R.M. KELLER, "Formal Verification of Parallel Programs", *CACM*, 7, 1976, pp. 371-384.

27 C. SUNSHINE, "Formal Techniques for Protocol Specification and Verification", *IEEE Computer Magazine*, August 1979, 21 p.

28 C.A. SUNSHINE, "Survey of Protocol Definition and Verification Techniques", *Proc. Computer Network Protocols Symposium*, Univ. of Liège, February 1978.

WHY USUAL TRANSMISSION PROTOCOLS ARE NOT APPROPRIATE FOR HIGH SPEED SATELLITE TRANSMISSION

HUITEMA Christian

Centre National d'Etudes des Télécommunications (CNET)

Satellite links offer interesting possibilities : broadcasting and very high bit rate. However they also have a long propagation delay (300 ms) and not very well known error rate characteristics.

Even protocols such as HDLC are not convenient for high speed transmission on satellite. Efficiency decreases with the mean number of errors during the propagation delay. Very accurate tunings have to be done on flow-control parameters in order to keep the transmission going smoothly.

First studies conducted with NADIR have shown that protocols less sensitive to errors or flow-control tunings can be designed to operate high speed file or message transfers on satellite links.

1 - INTRODUCTION

A transmission protocol aims at ensuring a "safe" exchange of Data Units (e.g. messages, frames, ...), in an efficient way.

A safe exchange means that D.U. are delivered without errors, in the right order, at the right speed. An error detection/correction procedure allows to recover any transmission error. A flow-control procedure allows the receiver to slow, if necessary, the production of D.U. by the sender.

An efficient protocol will not make any undue consumption of ressources (e.g. transmission bandwidth, CPU, memory size). It will keep the delivery delay as short as possible.

K. G. Beauchamp (ed.), New Advances in Distributed Computer Systems, 159–165.

When using a local area network, very simple protocols, based on positive acknowledgments, can be designed. More sophisticated protocols, such as ISO HDLC, have designed to be used on long haul ground lines.

We will discuss HDLC error control and flow control mechanisms, compute their efficiency in terms of bandwidth consumption, memory size needed and mean delivery time, and, as a conclusion, try to explain the best way to design efficient transmission protocols to be used on satellite lines.

2 - HDLC

HDLC is the standard line-level protocol defined by ISO. It is bit oriented and flags are used for frames delimitation. A 16 bits CRC transmitted along with each frame is used for error detection. Frames in error are ignored by the receiver.

Informations frames are numbered, either modulo 8, or modulo 128 (extending numbered). The sender can transmit up to X unacknowledged frames, X being a number called the "frame-window",(X is less than 8 or than 128, depending on the numbering space).

The receiver can detect lost frames if the difference between the number of two frames received consecutively is not equal to 1.

By sending a "Reject" (REJ) command, the receiver can force the sender to retransmit all unacknowledged frames. By sending a "Selective Reject" (SREJ) command, the receiver will force the sender to retransmit the frame number N; this command also acknowledges all frames received before the frame number N.

All unacknowledged frames are retransmitted after timeout.

The "Receiver not Ready" (RNR) command allows to prohibit the sender from transmitting any information frame until a following "Receiver Ready" command is received.

This mechanism, with the "frame-window", can be used for flow-control purpose

3- ERROR CORRECTION PROCEDURES AND EFFICIENCY

3.1. REJECT

If the "Reject" error recovery mode is selected, the receiver will oscillate between two states :

a - accept frames until the next error

b- send "REJ" and wait until the error is corrected.

Let's quote D the links nominal throughput , BER the link bit error rate (between 10 E-8 and 10 E-5 for a satellite link), and T the roundtrip delay (roughly 600 ms for a satellite link). One may observe that the mean number of bits received during an "accept frames" period is 1/BER, while T*D bits could have been received during the "wait for correction" period.

Thus, the efficiency ratio (number of received bits / number of sent bits) of the REJ procedure is :

R = (1 / BER) / (1 / BER + T*D)

... = 1 / (1 + BER*T*D)

The product of the bit error rate by the nominal throughput by the roundtrip delay is the mean number of error during a roundtrip delay.

3.2. Selective Reject

A "Selective Reject" command for frame number "N" acknowledges all frames received before the frame number "N". If the receiver detect an error after having sent a "SREJ", he must wait until correction of the previous error before sending any other SREJ.

Thus, no more than one error can be corrected during a roundtrip delay. The efficiency ratio will be lower than :

R = 1 / BER*T*D

3.3. Optimisation

A protocol allowing to correct as many frames as necessary during a roundtrip delay can be designed. Efficiency will then be limited by the error detection scheme : if an error modifies a single bit in a frame, the whole frame must be discarded. Thus, if "L" is the length of a frame (number of bits), the efficiency of an optimal protocol will be :

R = (1 - BER)**L

i.e. the probability that no error occurs on a given frame.

4-- FLOW CONTROL AND MEMORY SIZE

The "RNR" command allows the receiver to stop the emission of frames by the sender.

If the receiver reaches a "saturation" state, he will send a RNR command. This command will be received by the sender after a propagation delay (300 ms) ; the latest frame emitted by the sender before reception of the RNR command will be received by the receiver after another roundtrip delay.

The total amount of memory needed to store incoming frames after a RNR command is M = T*D (product of the nominal throughput by the roundtrip delay). If a 2 Mbit/s satellite link is used, M = 2 E+6 * 0,6 = 1.2 Mbit (150 Kbytes).

The sender must store a copy of all unacknowledged frames. If "REJ" commands are used, an amount M of memory is needed. If "SREJ" , or an optimised protocol , is used, the amount must be 2*M, in order not to slow the sender during the correction of an error. Another amount M of memory will be needed at the receiver side to perform sequencing.

If full duplex transmission over a 2 Mbit/s satellite link is operated using a "REJ" protocol, 300 Kbytes of memory will be needed at both sides. Using "SREJ", 600 Kbytes will be needed!

5 - SEQUENCING AND DELIVERY DELAY

The delivery delay is the time spent between the first transmission of a packet and its delivery to destination. When using high speed satellit links, this delay is not usually very different from the link's propagation delay, i.e. 300 ms.

It can be greater in two cases. If an error occurs during the transmission of a packet, it will have to be retransmitted ; the delivery delay will then be(roughly) three times the propagation delay, i.e. 900 ms. If the packet is correctly received, but if a previous error has not yet been corrected, the packet will be kept until correction of all pending errors, so that packets are delivered in the right order ; the delivery time will then be between 300 and 900 ms.

If packets are short (1000 bits) and if the BER is low (less than 10 E-6), only 1 packet out to 1000 will be in the first case. But if the throughput increases, the probability of the second case grows. Assuming that arrival of errors is poissonnian, that the data flow is steady, and that the first case is neglectible, the mean delivery delay can be easily computed :

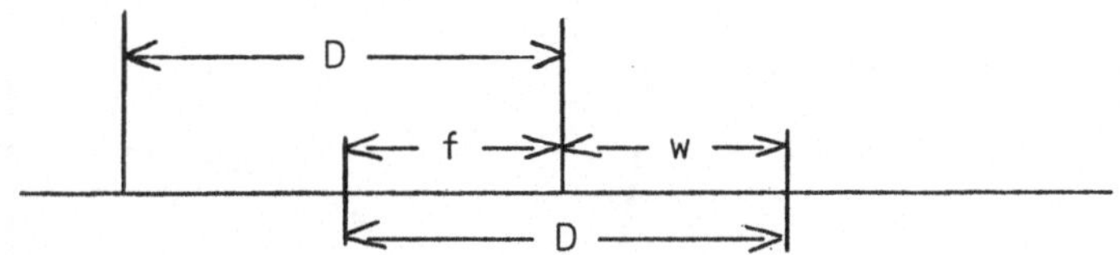

Bit error rate : BER
Throughput : V
Roundtrip delay : D

$$P(f \geq t) = \exp(-BER.V.t)$$

Mean value of w :

$$E(w) = \int_0^D (D - t) \, . \, BER \, . \, V \, . \, \exp(-BER.V.t) \, dt$$

$$= \left[-(D - t) \, . \, \exp(-BER.V.t) \right]_0^D$$

$$+ \left[\exp(-BER.V.t) \, / \, BER.V \right]_0^D$$

Mean number of error during a roundtrip delay = X = D.V.BER

$$E(w) = D \, . \, \left(1 \, . \, (1 - \exp(-X)) \, / \, X\right)$$

Mean delivery delay = D/2 + E(w)

Computation leads to the formula :

$$W = D/2 + D \, . \, \left(1 - (1 - \exp(-X))/X\right)$$

Where D is the roundtrip delay and X is the mean number of error during a roundtrip delay.

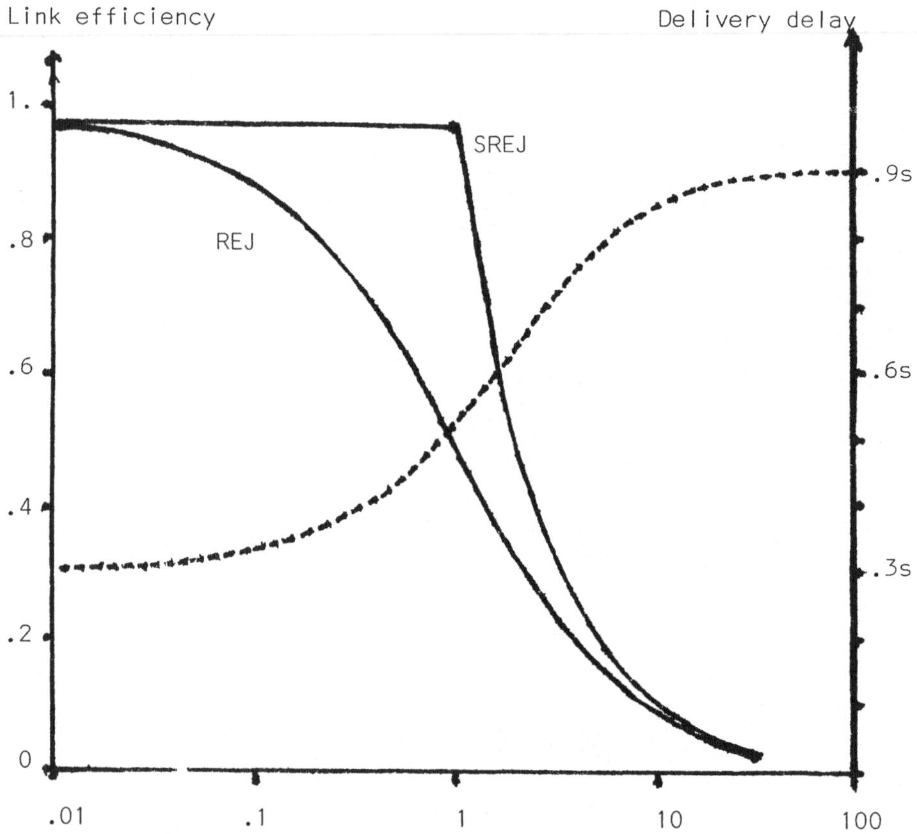

Figure 1 :

Plain curves show the efficiency of HDLC REJ or SREJ as a function of X, the mean number of error during a roundtrip delay.

Dotted curve shows the mean delivery delay of a single frame, also as a function of X.

6 - CONCLUSION

Figure 1 shows that both efficiency and delivery delay are a function of the mean number of error during a roundtrip delay, "X". When X is big (>1), efficiency is low and delays are long. Two kinds of traffics have then to be considered : file transfer and message transfer.

When transfering huge files, the delivery delay "per frame" is not a problem. A very efficient protocol should be designed; i.e. a "super SREJ".

When transfering messages, e.g. for local area interconnexion, the mean delivery delay should be optimized. Figure 1 shows that the only way to do it is to keep X low. X is the product of the throughput per the roundtrip delay per the bit error rate. Using forward error correction would lower the bit error rate. Using end to end error correction would lower the throughput.

RESER, A LOCAL AREA NETWORK USING A FIBER-OPTIC RING

S Miège, R Baduel, G Even, A Khalil

E.S.E. (Supelec), Antenne de Rennes

Abstract

During the last two years we have developed a local area network. It is based upon a fiber-optic ring topology, empty slot access method and operates at a serial transmission rate of 10 Mbps. Following an introduction to the background to this project, we present a brief description of its basic principles, and describe the ring-and-cluster philosophy of the network.

We have implemented a parallel packet oriented high throughput interface. Microsystems may concentrate up to 12 data connections or 8 voice communications on each one of the seven addressable parallel interfaces of each node.

This paper presents the first results of contacts being developed with a manufacturer in order to evaluate the industrial viability of these concepts, as well as the current state of our experimental network. We will soon be using the system in an experimental office environment for evaluation and training purposes.

K. G. Beauchamp (ed.), New Advances in Distributed Computer Systems, 167–179.

1 BACKGROUND OF THIS PROJECT

1-1.- What is Supelec?

Supelec, familiar name for Ecole Supérieure d'Electricité (E.S.E.), is what in France we call a "grande ècole", specialized in graduating engineers in the field of electrical engineering. While the main part of Supelec is in Orsay near Paris, a small part has been settled in Rennes, Britanny. Our laboratory is a part of this Rennes branch. We have developed a good competence in the field of computer science, especially for computer interconnection, microcomputers and word-processing. Two other labs exist here : one is specialized in industrial automation problems, the other one in the design of integrated circuits and the processes to manufacture them.

1-2.- Why the need of a local area network (lan)?

A short analysis of our present and future activity leads to a decomposition in four points:

- in order to teach programming methods with standard languages such as a microprocessor assembly language or Pascal, we need small and available interactive systems, managing small source and code files: individual microcomputers are now a better answer than a time sharing computer.

- to produce specific pieces of software for various equipments, which are requested by the applied research part of our activity, we need access to large computers with specific development tools and languages; these computers may exist or not locally.

- simulation and CAD applications need access to programs and systems whose power ranges from medium to very large scale. In most cases we have to use external computing systems, accessed through local terminals, for example graphic intelligent terminals.

- we have a lot of texts to prepare and print for students, lectures, etc... word-processing activity also includes mailing to teachers and students, timetables edition, etc...

One can see that we need to interconnect CRT terminals, personal computers, printers, intelligent terminals, mainframes and teleprocessing gateways to distant powerful computers.

1-3.- The ancestor of our local area network

Four years ago, most of our connections were established between two minicomputers and about fifteen terminals. As our lab is spread over two floors, we developed a network of cables with manual switching.

At that time, we chose a standard for connecting plugs, derived from the V24 recommendation and intended to supply a wired flow control not only between a computer and a terminal, but also between two computers or two terminals. This solution, which is still operational as a back-up of RESER, has some qualities, but is not quite satisfactory about the following points:
- manual switching does not allow us to share an equipment for short transactions (data base, printer...), unless we accept to afford a time-sharing processor attached to each of these equipments, and several cables linked to them.
- in order to connect a new equipment, we have to modify its input-output interface. In the cas of a mainframe, a software set of protocoles sould be much more flexible.
- The transmission rate is limited to 9600 bps.

1-4.- A few words about other French lans

In Rennes, there are other important laboratories interested in the field of local networks; one of them, the C.C.E.T.T., a research organization shared by the French Telecom Administration (CNET) and the TV Broadcasting Authority (TDF), is developing CARTHAGE, a lan first intended to be an extension of a PABX; it mixes telephone circuit PCM frames, circuit switching for data and packet switching for data, on a fiber-optic ring. The 8Mbps are dynamically shared, likely as follows: 6Mbps for circuit switching and 2Mbps for packet switching.
Near Paris, a pilot project, KAYAK, helped to develop 2 lans, TARO, a medium speed ring with twisted pair links, and DANUBE, a 1 Mbps coxaial bus network based on about the same principles as those of Ethernet.
We just mention experiments made by the car manufacturer RENAULT; others by the university of Toulouse, and this brief list does not claim to be complete.

2 NETWORK DESCRIPTION

RESER is a local area network for integrated data and voice transmission. Though primarily intended for use in such areas as distributed data processing, terminal access, office automation, and other applications requiring economical connection to a local communication medium carrying bursty traffic at high data rates as well as steady circuit-like traffics for voice transmission, its fiber-optic links allow industrial applications requiring high protection against induced electromagnetic noise.

2-1.- general structure of RESER

Ring Stations are connected by fiber-optic links. Computers can be connected to each ring station through Ring Plugs. Terminals, personal computers and vocal sets can be clustered by dedicated microcomputers, the Connection Processors.

Thus RESER provides simple interfaces with a wide variety of equipments: standard RS-232-C serial interfaces or TTL-level parallel interface are provided for data terminals and small computers;
Fast computers are directly connected to a ring station with a small packet parallel interface.
CODEC ICs sample and restore voice channels.

This clustering in two levels seems to offer a rather cheap acces. This point will be discussed later.

2-2.- Access control

We chose the empty slot control technique, which seems to be very convenient in the case of ring lan
Control mechanisms for lans being well known, we will give only a brief description of this particular one. Our slot is longer than that of the Cambridge Ring: 18 bytes. We call it a Station Frame. It is composed of a command byte, two address bytes and 15 data bytes.

Each station shifts four bytes of the incoming frames and retransmits them on the outgoing link. Therefore, a frame may be shared by 5 stations at most, or by less than that, depending on the number of bits being travelling on fiber-optic links (about one bit every 30 meters). One or more S-frames, depending on the number of stations, continuously move around the ring

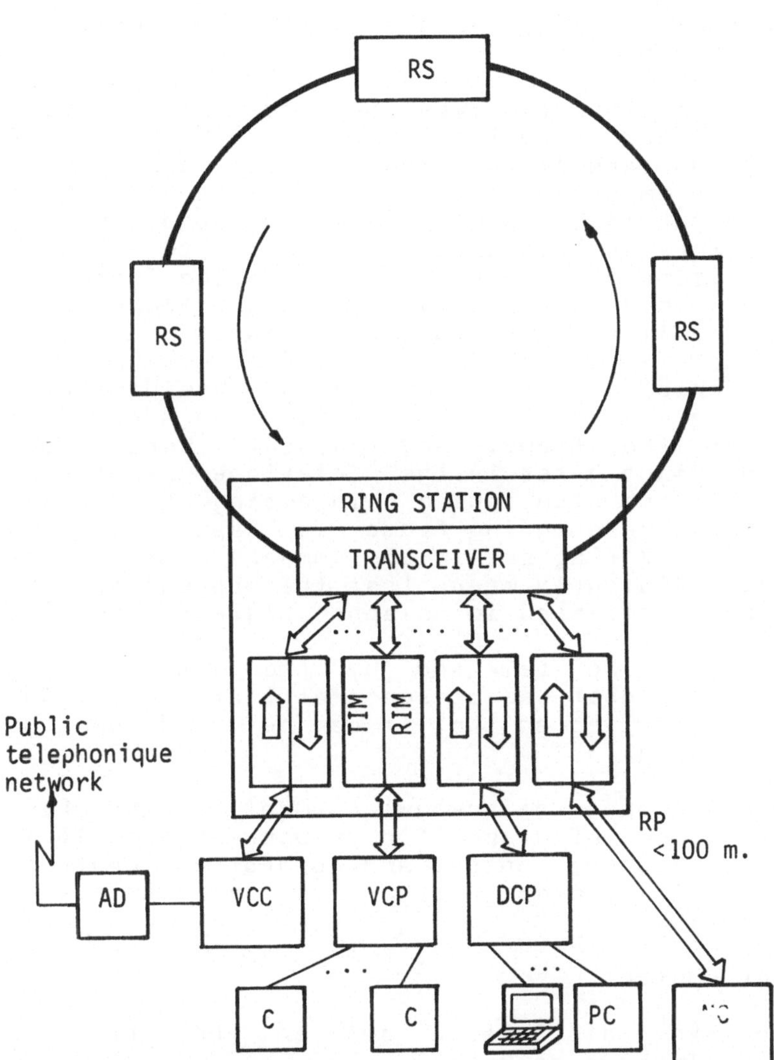

RS · Ring Station
TIM : Transmite Interface Memory
RIM : Receive Interface Memory
RP : Ring Plug
DCP : Data Connection Processor
VCP : Voice Connection Processor
VCC : Voice Communications Controller
PC : Personal Computer
C : Voice Channel (CODEC)
AD : Automatic Dialer
MC : Mini Computer

and may be full or empty, according to an indicator inside of the header.

When any ring station receives the header of an empty frame, it may mark it full and replace the other 17 bytes by the address of the ring plug (if any) which asks for transmission and which has the highest piriity, and the 16 bytes of a packet found in the Interface Memory of this plug (the first byte of which being the address of the correspondent). The frame is transferred unidirectionally through links and stations until it reaches its destination.

Upon reception of a S-frame marked as beeing full, every ring station analyzes the destination address in the header of the frame, and possibly copies the enclosed remaining bytes in the Interface Memory of the addressed ring plug, while reinjecting the frame into the ring. The outgoing frame is kept untouched except for an ACK flag set in the command byte, in the case of copy. That only means that the content of the frame has been delivered in an appropriate memory.

The frame will be released by the sending ring station, which cannot use it again at that moment. The ACK flag is memorized so that it can be read on the sending ring plug.

This simple mechanism of acknowledgement is very fast and avoids the use of a specific response frame. It is useful for both error control and congestion control. Upper levels will decide whether to retransmit the packet or not.

2-3.- Error Control

We have to make a distinction between the error on data and the loss of packets.

The bit error rate on links and stations is estimated to be better than 10-9, so we chose not to apply error control on the 15 data bytes of station frames.
But this very small error rate must be taken into account when applied to a frame moving around without stop. For example, at 10 Mbits per second, a 10-9 bit error rate produces an error about every 100 seconds. An error in the packet header would cause awful consequences such as bad addressing of the packet or modification of packet state (for example: change from full to empty or from empty to full).

Therefore an error control on the header of station frames is provided. When a ring station detects an error in the frame header , it changes it into an empty one.

To sum up, a packet transmitted in the ring may not arrive at its destination because :
- an error is detected in the packet header, or,
- the interface memory of the receiver is full.
One can see that a loss can occur either before or after the passing of the frame trough the receiving station; in the latter case, the frame may have been placed in interface memories but the sender will not see the corresponding ACK flag.
Upper levels must be conscious of this fact, and protocols will detect and discard packets that may have been repeated wrongly.

2-4.- Congestion Control

As a frame is released after one revolution, even if there was no place in the addressed memory to copy its content, congestion cannot occur on the ring itself.
When station A has filled a frame, next station B will see an empty slot, so no station, or couple of stations can monopolyze the ring bandwith.

2-5.- addressing scheme

A plug address is composed of:
-3 bits to select one of the 7 possible ring plugs in one ring station (the value 7 asks for a broadcasting to all of the plugs),
-5 bits to address one of the 31 possible stations on one ring (the all-ones value asks for a broadcasting to all the stations).

In this way, three different kinds of broadcasting are allowed:
* to all plugs of a station, and by this way all the equipments clustered in some place,
* the ith plug of all stations (for example all voice concentrators, as plug assignment is not a random process, since there is a precedence among plugs),
* all plugs in all stations.

The set of these two addresses is contained in one byte and allows addressing 7 ring plugs distributed in 31 ring stations, thus 217 plugs and associated computers.

2-6.- Fiber optic links

As mentioned above, the transmission medium of RESER is a fiber-optic of broad bandwidth which provides high data exchange speed, immunity to electromagnetic interferences and very small bit error rate (10-9 to 10-11 at 10 Mbps NRZ). We chose Hewlett Packard kits (HFBR): they were available and easy to use: TTL-S levels, 5V DC power supply, automatic gain control, they accept to transmit a 10 Mbps NRZ coded signal.

2-7.- Ring Stations

The task of a Ring Station is to transfer data between S-frames and ring plugs, while ensuring the progression of S-frames along the ring.
It is composed of:

- the Transceiver which receives data from the fiber optics and drives them into it. It also performs the operations of data serialization/deserialization and decodes the Ring Station address field and the Ring Plug address field.
The transceiver is reponsible for managing the command byte, its error control as well as its full/empty flag and acknowledgement flag.

The part of the transceiver specialized in ensuring ring continuity can be isolated from the rest of the transceiver. Then the ring station is isolated from the ring. This feature improves the reliability of the network.

- the Interface Memories organized as first-in-first-out queues. They are used as a buffer between the transceiver and the ring plugs. There are two interface memories for each Ring Plug, one of them (TIM) can store one packet for transmission, and the other one (RIM) can store severals received packets (5 in our first experiment).

Packets in IM, called plug packets, are 16 bytes long. The first byte is the address of the correspondent plug (i.e. the receiver address in a packet for transmission (in a TIM) and the sender address in a received packet (in a RIM) : the 15 last bytes are the data bytes of S-frames, transmitted unchanged to the addressed plug.

2-8.- Ring Plugs

Ring Plugs are the interfaces between a Ring Station and digital equipments using the network. It is a high speed 8 bit parallel bus operating at 1 mega bytes per second. RS422 drivers and receivers allow up to one hundred meters between station and connected computers. The transfer via this bus is bidirectional and controlled by four control signals:

- WRITE DATA in the interface memories,
- READ DATA from the interface memories,
- SEND ORDER to or READ STATUS from the interface memories.

Due to the use of FIFO memories, 16 bytes of one packet are transmitted asynchronously, at the rate that each connected processor can afford.

STATUS informations:

-TIM is full; when this flag switches to 0, the transmit memory starts emptying and will be empty in a few micro-seconds.

-ACK received; this flag is reset when you write data in the TIM. It will be set when a full frame is received by the ring station, with this plug address as plug sender, and the ack flag set.

-RIM is empty: when this flag switches to 0, a first packet starts being received and reception will be completed in a few micro-seconds (16*8/10=12micro-seconds).

2-9.- Connection Processors

They operate a "second layer" in clustering. Many terminals are not able to manage the packet interface offered by a ring plug, or to manage high level communication protocols, or cannot take advantage of the high transmission rate offered at ring plug level. Connection processors are dedicated intelligent devices, such as microcomputers, intended to allow these terminals to access the network.

In fact only fast computers or dedicated computers can directly use ring plugs.

2-9-1.- Data Connection Processors

A data connection processor is a microcomputer specialized in managing communications in order to provide clustered low or medium speed end users with simple protocols of connection. It contains:

- the presentation software necessary to provide the

users (terminals, personal computers, etc...) with an access to the session software,
- the session software that mix several low speed channels into transport communications.
- the transport software necessary to deal with data communications.

2-9-2.- Voice Connection Processors

Though there is no difference between data packets and voice packets when at the firts level (loop and interface memories), voice packets are processed in special connection processors, with an appropriate hardware and software architecture, in order to manage the caracteristic features of voice transmission.
Voice is first sampled by CODEC ICs: after a filter stage, the analog signal is sampled according to a 8Khz clock, and the 12 bits resulting of this A/N conversion are compressed with a logarithmic law to one byte. A local clock drives Codecs so that a sample is made every 125 micro-seconds. Samples are put in a packet and sent onto the ring. Receiving voice CP will put a sample in the addressed Codec every 125 micro-seconds.

A real time transmission mechanism is necessary in order to guarantee the presence of the next sample for the receiving codec.

3 PRESENT HARDWARE CONFIGURATION

- -6 ring stations, four of which are only minimum transceivers,
- -2 MI cards, each comprising up to four sets of TIM and RIM FIFOs,
- -3 data connection processors, composed of a Z80 CPU on one card, and a 8 USARTs on another card,
- -2 voice connection processors, composed of a 8X300 micro-processor on one card and 8 Codecs and filters (INTEL 2911) on another one.

Speaking about packaging, we have designed Ring Boxes, 4 U, 19" cabinets in which are stuffed: one ring station, one MI card, one voice CP and two data CP, and of course connectors and power supply.
This access device offers 8 voice channels, 16 data channels (8 of which are asynchronous ones, the other eigth can be either serial asynchronous or parallel

interfaces), and one direct acces to a ring plug. Ring boxes are powered via a standard 220V-AC distribution cable that runs along the ring.
We estimate that one connection of this configuration should cost about 3000FF in case of industrialization.

An experimental automatic dialer designed by Supelec for the KAYAK pilot project allows one voice channel to be extended by our PABX.

3-1.- A few words about protocols

In an experimental environment, and because of the various kinds of data that can coexist on a local area network, it is very important that several types of protocols may coexist.
With RESER this is very easy, due to the fact that the lowest level of ring plug is very basic, and several connection processors can be connected to a ring station, each of which supporting one kind of protocols.

Until now we implemented only very simple softwares:

- for data communications, virtual circuits without flow or error control, only allowing for transparent data links.
A more elaborate software is being developed: The OSI reference model is taken into account, although some layers are almost empty:
* a presentation layer manages standard terminals,
* a session layer multiplexes data packets from presentation layer and from computers connected in packet mode. Flow control is made at this layer by allowing the transfer of only one session packet at a time.
* a transport layer cuts session packets and put them into ring plug packets. It performs error control. Error recovery is made at session level.
*low level layers are almost empty, thanks to the packet oriented ring plug interface.

- for voice transmission, samples are put into packets and sent without any error control or flow control. A first version allowed for testing of codecs and processors, but could not withstand an even moderate load of the network. A new version enforces a small anticipation of about 5 milliseconds, and then guarantees that a precise number of voice communications can be treated as circuits (with 6 stations, one voice CP in each station, the 48 codec entries may be simultaneously active, without any influence from data transmission required by data CP). This implies that voice CPs are connected to the first ring plugs of stations, i.e. those with highest precedence.

A special processor on the network deals with setting and clearing of voices communications. Broadcasting from one codec is possible, and a special voice CP allows for conversation between several users. This processor responds to requests made through a CRT or by an application program according to standard data protocols, and controls voice CPs according to voice protocols.

4 FIRST CONTACTS ABOUT RESER

A French manufacturer is now reproducing our network in order to evaluate what part of our solutions may be of interest to him. It is too early by now to give a final conclusion about this operation, but the first feeling is cheerful.

In another respect, RESER will be the kernel of an experimental office designed by Supelec and an Organization whose interest is to improve the productivty of administrative tasks performed in French Insurance Companies.
A few work stations are interconnected by RESER, in order to measure the impact of such applications as electronic mail, common electronic diary, etc... Our local area network will allow us to put new applications into play, thanks to its flexibility and high transmission rate.

5 CONCLUSION

Our experiment is still going on, but RESER will be used by students after next october. A good deal of work remains in front of us, especially studies and implementation of various protocols, testing and evaluation tools.

We have found that the use of fiber-optic links is very easy, that a 10Mbps transmission rate is something you can afford with standard components, that voice and generally speaking real-time circuit-like transmissions are compatible with a packet switching network, provided there is no contention and that you can bear an ounce of centralization for circuit setting and a trifle of priority.

We wanted to use existing hardware for our design, simple and not expensive solutions. Clustering appears to be the good solution, for it gives a cheap individual connection (the communication software that must exist may be shared by several terminals) and it suits the organization of buildings, divided into floors, departments..., each of these entities being served by a cluster, clusters being linked together by a fiber-optic ring.

BIBLIOGRAPHY

N.Abramson and F F Kuo
"Computer Communication Networks" Printic-Hall, 1973

Reference Model of Open Systems Interconnetion"
ISO/TC 97/SC 16.

"Architecture et Services pour l'Interconnexion de Réseaux Locaux"
Afnor, CF/TC 97/SC 16/CE 1 F 426, mars 1981

"Project 802, local Area Network Standards, Functional Requirements Document"
IEEE, Computer Society Decembre 1980

N.Naffah
"TARO, un système de transmission en boucle pour réseaux locaux"
"Description fonctionnelle du réseau expérimental DANUBE"
INRIA, projet pilote KAYAK, REL 2.501 et REL 2.514.1,

"Réseau local d'entreprise multiservice CARTHAGE"
Afnor, réseaux locaux, nro 2

C.Macchi, J.F.Guilbert,
"Téléinformatique" DUNOD, 1979

SESNET - LOCAL AREA NETWORKS FOR SOFTWARE ENGINEERING DEVELOPMENT

M. J. Norton and P. P. Sanders

British Telecom
Systems Evolution and Standards Department

1) INTRODUCTION

This paper describes the rationale behind the development of Local Area Data Networks (LANs) for the British Telecom (BT) System Software Engineering Centres (SSECs). The paper describes the facilities which the authors believe such networks should offer. It indicates some of the design criteria used and describes some of the problems of quickly implementing, a system which is at the limit of current practice.

It must be noted that the LAN's described in this paper are data networks, confined within single buildings or sites. They are not local networks as normally recognised by British Telecom, which would serve a telephone exchange area of several miles radius.

2) THE SYSTEM SOFTWARE ENGINEERING CENTRES

In planning the transfer from a national telecommunications system based on electromechanical switching to a system based on stored program control digital exchanges, BT made estimates of the number of "Software Engineers" required to design, commission and maintain these new services. The estimates lead inexorably to the view that BT would be unable to acquire the quality and quantity of system software staff required by normal internal career development, or by external recruitment. Thus was born the idea of specialised centres (or software houses) with the aim of recruiting both internally and externally, staff with good technical backgrounds and training them in real-time software and

K. G. Beauchamp (ed.), New Advances in Distributed Computer Systems, 181–200.

systems techniques. In order to tap previously unused talent and to reduce staff turnover or wastage, it was decided that these SSECs would be sited regionally. London and existing major BT or computer industry centres would be avoided. The first two sites selected were Belfast in Northern Ireland and Newcastle upon Tyne. Staff are currently being recruited and trained for these centres which should commence full operation at the end of 1981.

The provision of good internal and external communications was seen as crucial to the success of the SSECs. The effect of their geographic remoteness can be almost totally overcome by the use of suitable audio, video and data links. BT is probably uniquely equipped in the U.K. to provide these facilities. This paper is concerned with the facilities for data communication planned for the SSECs and at present being evaluated by BT at the London base of the parent division. These network developments are purely for BT internal use; it is not intended to offer them as BT products or services in the marketplace.

3) SSEC NETWORKS

The role of an SSEC as a "Real Time Software House" places two distinct requirements on its support network :

A) Domestic / office automation functions.

B) Operational / software development functions.

Examining these in more detail, category A divides into:

A1) Text processing / report generation.

```
A2) Electronic mail / message.  )
A3) Electronic filestore.       )-- Electronic In-tray
A4) Forms replacement.          )

A5) Diary management.           )
A6) Meetings scheduler.         )--Executive Service
A7) Whereabouts list.           )
```

Category B divides into :-

B1) Transparent access to local and remote software development computers.

B2) Provision of backing store and printer services to microprocessor development projects, etc.

B3) Provision of special emulations (eg HASP) to access

remote mainframes.

B4) Access to remote target machines for software loading, monitoring and retrieval.

It would be nonsensical and uneconomic to provide two separate networks to support these functions. The concept evolved of providing a transparent data communications system to support the category B items; on this would be mounted the value-added-services to fulfill the category A requirements.

It was felt that a suitable transparent system would be an integrated network, comprised of Local Area Networks (LANs) at each SSEC linked by BT services, (eg NPSS and Teletex), into a complete Wide Area Network (WAN).

4) SSEC LOCAL AREA NETWORKS

Consider first the local area network part. The contending technologies of "Rings" (Cambridge and Token Passing), "Buses" (Ethernet and various byte parallel or frequency division multiplex systems) and centralised circuit switches were evaluated. See references, (1), (2), (3). The following items were considered important factors:

4.1) High speed interfaces and transmission media.

4.2) A large percentage of the total system bandwidth should potentially be available to an individual user thus providing the optimum response time.

4.3) Flexible configuration, allowing a gradual increase of users and services.

4.4) Distributed peripherals and services.

4.5) Reliability, (i.e. a very low undetected error rate).

4.6) Maintainability, (i.e. automatic detection and location of network faults).

4.7) Reasonable cabling costs.

4.8) Availability / Cost effectiveness.

Space in this paper does not permit a full discussion of the merits of the contending technologies, but the conclusion reached was that the two best approaches were the Ethernet and the Cambridge Ring Mk. 2. A more detailed assessment of these, against twenty criteria, is given in a later section of this paper. In the end, product availability determined that an Ethernet product (Ungermann Bass NET/1) be purchased for the prototype system. It is also our intention to purchase and evaluate a Cambridge Ring Mark 2 system, so an in service comparison of the two systems can be undertaken.

5) SSEC LOCAL AREA NETWORK PROTOCOLS

This section deals with the enhancements required to the Ungermann Bass NET/1 to make it suitable for SSEC use. The proprietary transmission medium is seen to provide layers one and two of the ISO Open Systems Interconnection (OSI) architecture, (4). The NET/1 does not support (at present) a user multiplexing protocol. Thus only one virtual circuit (permanent or switched) can be supported on each physical (V24) interface. This is adequate for driving terminals but makes connection to multiple user service. machines very expensive. To overcome this problem the authors propose to mount a subset of the X25 protocol on the user interfaces of the NET/1 network interface units. Application processors are available for which the software may be written and these can then support multiple logical channel interfaces to service machines over single physical interfaces, (parallel Direct Memory Access). Thus the OSI layer three (the network layer) will be represented in part by X25, (5).

It is envisaged that the upper part of OSI layer three and layer four will be provided by the PSS Study Group Three Network Independent Transport Service, (Yellow Book), (6). At layers five and six there will be a Virtual Terminal Protocol and a Virtual File Protocol. Whether the file protocol should be based on the UK "Blue Book" proposals, (7) or on the CCITT Teletex standard (8), is still under review.

6) SSEC WIDE AREA NETWORK PROTOCOLS

As has been stated earlier, the objective of the SESNET project is not merely to provide internal services to an isolated SSEC, but to provide a fully integrated system for all SSECs. Each SSEC requires communication with :

i) Other SSECs.

ii) The controlling headquarters in London.

iii) Other computer resources of British Telecom.

iv) Centres of excellence outside British Telecom.

The basic modes of communication are available, typically NPSS, Telex, Prestel, PSTN dial-up and the proposed Teletex service. The difficulty is that merely providing a means of transfering bits, bytes, or packets does not ensure useful communication. The missing element is a widely agreed package of "High Level Protocols" and procedures. The best solution will be the ISO high level protocols, these are under development but will need at least two more years to complete and agree. The interregnum has been filled in the U.K. by the "Yellow" and "Blue" book protocols (see section five). These will be used initially in SESNET, to allow access to other networks, eg those of the Science Research Council and the National Physical Laboratory.

The following sections of this paper are concerned with a more detailed description of the SESNET LAN transmission system and discussion of the control and server functions required to support the network.

7) SESNET TRANSMISSION MEDIA

This is the ISO OSI layer 1 equivalent and may include all or part of layer 2. The following comments compare the "Cambridge Ring" and "Ethernet" approaches. The weight given to each comparison point will vary according to the particular design criteria of the specific network. The comparison was produced to aid the choice for SESNET.

7.1) Wide addressing range

The addressing range depends on the type of physical addressing scheme used:-

a) Universal...Each node on every network anywhere has a unique address. There are administration problems in the allocation of addresses and the large address has also to be carried as a transmission overhead.

b) Local...The address has local significance only and the total address is built up on a hierarchical structure.

The Ring's limited address capability of 255 sources and destinations seems adequate. The Ethernet's 96 bit address overhead on every transmission seems of questionable benefit.

7.2) Adequate data rate

What is considered to be an "adequate" data rate is very

much open to debate. The value needed depends upon current and future requirements. Assuming a required response time of less than 0.2 secs. and a typical small application program, eg.mail, editing etc. using approx 20k bytes, then a user will need a burst data rate of 1Mbit/sec, to down-line-load this into his local terminal. Both systems can exceed this rate on a single user basis. The Ethernet has a better data rate when predominately large packets are used.

7.3) Simple fault finding

A network fault and it's position should identify itself quickly, (eg failed node or broken transmission link). An automatic mechanism is most desirable. Here the Ring has a major advantage over the Ethernet. The Ring's monitor station will log and locate transient faults, locate cable faults and detect hogging.

7.4) Well-defined standard

This makes life simpler and gives the transmission scheme a much better chance of success. The standard should be in order of preference: international, national, defacto. Both systems have become de facto standards. There is a better prospect for international standardisation of Ethernet than of Cambridge Ring.

7.5) Ease of adding extra users

This function should be capable of being performed while the system is live. The physical connection and disconnection of nodes should have no visible effect on an individual user. Ethernet nodes can be added easily. Although the Ring has to be broken, this is not necessarily visible to the users so long as tapping points have already been installed, (as in the latest Logica implementation); otherwise the Ring has to be stopped.

7.6) Reliable transmission

A low error rate is highly desirable, ie. less than 1 in 10,000,000 bits. Good common mode rejection is implied. The rigorous approach required goes right down to the type of connector chosen, which should not deteriorate with age. An inbuilt error detection and correction mechanism would be an advantage. Reliable transmission is helped by the environment of the installation, (i.e. kept away from strong electrical interference and extreme atmospheric conditions). The error detection mechanism of the Ethernet is necessarily better, as it can send much larger packets. There is no inherent automatic correction mechanism in either system.

7.7) The ability to carry speech

The transmission system should not preclude the sending of real-time digitised speech. The support of many simultaneous speech channels would be an advantage. The Ring's worst case access time can be calculated, but the effect on speech channels of a heavily loaded Ethernet is still the subject of much academic debate.

7.8) Suitability for fibre optic transmission

The increased usage and advantages of fibre optic technology makes its inclusion highly desirable. The Ring is well suited to fibre optics. The current Ethernet is for all practical purposes, unable to use this medium.

7.9) Commercial potential

This is a significant advantage, especially in an open systems environment. Units will be manufactured with the interface already integrated into the design, with consequent cost benefits. Factors which will influence the level of commercial interest in the two systems include the arrival of integrated interfaces, government backing and adoption by influential companies. At the moment, Ethernet appears to be ahead internationally.

7.10) Availability of purpose-designed interface devices

This is related to the previous point. The greater the commercial potential the more likely it is that integrated interface devices will become available, and be used in products. The integrated Ring interface is now being tested. It may be the middle of 1982 before significant quantities of the Ethernet equivalent are available.

7.11) Transmission confirmation

An indication that one's message has been received is an added benefit, especially if it is not an overhead on the transmission protocol. The Ring gives automatic confirmation, Ethernet does not.

7.12) Good overload characteristics

When the traffic on a system reaches saturation, the system should not experience a decline in overall throughput or exhibit instability. The Ring will reach a maximum throughput and be stable. Xerox claim that Ethernet does the same, (9).

7.13) Broadcast facilities

The ability to broadcast a single message to all local addresses is a very valuable feature, and its absence could impose severe limitations on a system. Both systems have broadcast facilities.

7.14) Not dedicated to any specific manufacturer

The system should be manufacturable by anyone. No single organisation should have the sole distribution and/or production rights. This is true of both systems.

7.15) Ease of installation

The new system should be simple to install. If existing wiring can be used eg. telephone, all the better. Co-axial cable should be of a type that does not have severe restrictions on its bending radius. There is little to choose between the two systems, although the Ring needs a monitor station and power distribution that the Ethernet does not.

7.16) Large local-user base

This can help overcome the problem that the chosen system may not be the leading standard. The Ring is widely used in UK universities, with some private company interest. Ethernet has found moderate support from both private and public UK companies.

7.17) Long transmission distances between nodes

The transmission distance between nodes affects the type of installation possible. A minimum of 100 metres is probably adequate for intra building operation with a 400 metre minimum for inter building working. The Ring's inter-node distance is dependent on both transmission speed and the type of cable used. The Ring in theory can cover a wider area than the Ethernet. The Ethernet can transmit further without repeaters.

7.18) Ease of interfacing to stations

When a system is installed, the physical entry points to the network (eg tapping points), will probably be fixed. The interface or workstation should be able to plug easily into the network and not have arbitary restrictions on position. A station on a local spur of the network should have all the attributes of the rest of the network.

7.19) Network Reliability

It is desirable that no single fault shall stop the system functioning. Any transfer of network control, due to failure, should be transparent to the user. With both systems, certain single faults can cause total failure. In principle the ring is more vulnerable, due to the greater number of active devices in the main transmission path. Neither system has an automatic fault isolation and recovery mechanism.

7.20) Increasable Network Speed

The mode of operation should be capable of enhancement for higher data rates when new technology makes this economic. The Ring has a distinct advantage in that increases to >100 Mbits/s are feasible with current ECL technology. To date there has been no mention of a higher speed Ethernet and this could probably only be achieved by the use of multiple, parallel Ethernets on a frequency division multiplex system.

8) NETWORK MANAGEMENT

Why is it needed?

Although mechanisms are available by which nodes can communicate with each other, the problem of which nodes need to be connected is still present. This is where NETWORK MANAGEMENT provides the mechanism for connecting users and applications. A whole range of functions and protocols are needed to accomplish this, the types required depend upon the management architecture chosen.

Network management consists of a number of functions, e.g. resource allocation, security, error handling and monitoring. These functions are independent of the network management system chosen. The architecture used influences the manner in which the functions are implemented. Two network management architectures are discussed below, viz DISTRIBUTED and CENTRALISED.

8.1) Distributed

Two advantages of the distributed management system are:

a) that it does not need independent processors dedicated to controlling the network.

b) by virtue of (a) the system is more robust. System operation does not cease if a node goes down, (assuming it does not take transmission system with it !).

These factors make the distributed approach attractive when a small, low-cost system is needed. The disadvantage comes in locating resources on a large system, since a broadcast facility is almost obligatory, polling each node to find the required resource is time consuming. This could be overcome by a table of resources and users, or access to a multiple copy file which contains the information.

8.2) Centralised

With the centralised system, control is vested in a dedicated network controller. Its functions can be legion; resource control, translation, resource allocation, first line security, monitoring, etc. The main advantage is that all system information is held together, saving the unnecessary duplication at each node.

The main drawbacks are :

a) A processor will have to be allocated to run the network controller and backup processors must be available.

b) Changeover of central controllers will need to be transparent to the network users.

The main advantage is that as all network control goes through one point, tight control can be maintained. Extra features can be added without burdening each node.

8.3) SESNET control

It was decided that SESNET would use centralised network control; the reasons being:

a) Although initially SESNET could be controlled by either method, its ultimate size and diversity will dictate that only a centralised control system is sensible.

b) Costs savings over a distributed control method for a large network are unlikely.

The above are based on several premises

i) That there should be a large number of workstations, e.g. 200, (hopefully approaching one/user).

ii) A workstation is a terminal with input (keyboard), output (display), local processing and a network interface. The interface is capable of holding all the protocol information and data buffering.

iii) The total size of the protocol programs is likely to be large. There would not be sufficient space to include an ever growing database of features and users.

The following is a description of a typical call connection.

RM = RESOURCE MANAGER

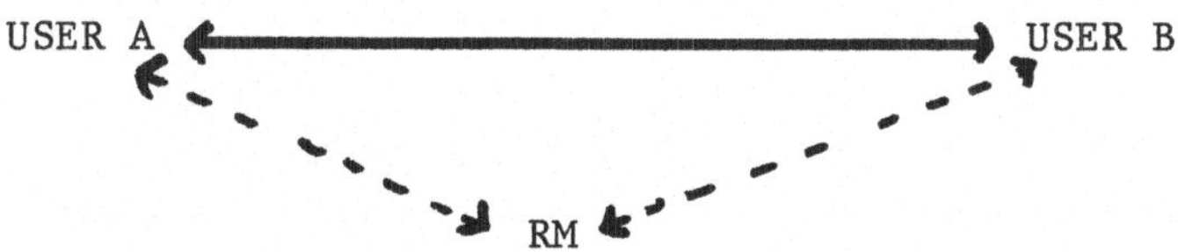

Fig. 1. Elements of a call

USER A -> RM a request is made for connection to USER B

RM changes logical names to physical addresses. Checks security to see if connection is permissible. Checks for busy, if so inform user A; not busy, allocates spare channels X and Y.

RM -> USER B connect to USER A via your channel X.

RM -> USER A connect to USER B via your channel Y.

USER A -> USER B confirmation of connection by calling party.

When a call is finished.

USER A or USER B - RM request to disconnect

RM -> USER A disconnect logical channel y

RM -> USER B disconnect logical channel X.

The RM has many other functions. If a user or resource is requested which is not available on the local network, the RM should recognise this and pass the request to the GATEWAY server.

A request for a local service is validated by the RM. The RM determines whether the function should be performed in the user's intelligent terminal, or a general purpose processor. If, for example, the function MAIL is typed and it can be run locally, the RM should set up a call between the FILE server and the user and provide the name of the correct applications file. This would be loaded into the user's workstation and the function MAIL should then be available to the user. If the user's terminal could not support this locally, then a spare general purpose machine will be allocated, it will be loaded with the application and given the address of the user. The only visible effect for users that are running an application remotely will be a slower response time.

9) SERVICES

Services on the network divide into four categories.

9.1) Services needed to run the network.

e.g. RESOURCE, BOOT, ADMINISTRATION, ERROR.

9.2) Services which add to the networks capabilities but are invisible to the user.

e.g. GATEWAY, EMULATOR, CONCENTRATOR, PRINT, TRANSMISSION.

9.3) Services designed to support specific user functions.

e.g. MAIL, REPORT, FORMS, DEVELOPMENT, VOICE, MESSAGE, DIARY, MEETINGS.

9.4) Services that help the network operate and are visible to the user.

e.g. FILESTORE, TIME, ARCHIVE, ENCRYPTION.

The services are identified as separate items, but may be grouped together on particular physical computers or indeed dynamically assigned to processors on demand.

The following is a description of the services listed except those defined under item 9.1.

9.5) Gateway

This has access to one or more communications services (eg NPSS, Teletex, Telex, PSTN, Prestel, etc). It is used as a window into other networks or systems. If a request is made for a user or resource which is recognised but does not exist on this local network, then the request is placed with the gateway. The gateway will endeavour to set up the connection. The information needed is:

a) Who or what has to be connected ?

b) Can the connection be sensibly made ?

c) What routing is needed ?

d) Where is the routing and addressing information ?

e) What service is to be used ?

f) When is it to be sent ? (This may be carried out by another server concerned with transmission cost optimisation).

The call can be set up on a cost/facility/urgency basis and alternative routing could be applied if difficulties arise. The gateway is also the place where incoming calls are examined. It would check that the person or facility being called is on the network, and that any necessary protocol conversion (eg Transport Service, Virtual File, or Virtual Terminal) can be carried out; before completing the local network connection.

9.6) Emulator

This permits connection of the Local Area Network to existing mainframe computers. BT currently use ICL, IBM and Burroughs mainframes. It is envisaged that the LAN would masquerade as a remote job entry terminal for each mainframe, as required. Appropriate emulations, (eg HASP for IBM), will need to be developed.

9.7) Concentrator

This permits the economic connection of peripherals since currently, node connection costs are high. A concentrator should cope with a minimum of four devices. An attached terminal should be able to work at it's desired speed without degradation of service, independent of the number of active users on the concentrator.

9.8) Print

This performs the spooling action for files to be printed. It routes the file to the appropriate type of printer, eg high speed line-printer, general purpose dot-matrix printer, or letter-quality printer. Text is transformed to match the printer and character set specified. The server will allocate a printer in the cluster nearest to the user's work station.

9.9) Transmission

Bulk files that need to be moved from one location to another can be spooled. The files will be transmitted, depending upon their urgency, in the most cost effective manner, (eg. during the off peak hours).

9.10) Text

The following group of services deal with text processing in its various forms. For clarity they are split up, but some services may be combined at a later date.

9.10.1) Mail

This will be the most important service of any local network. There are many facilities to be provided and the following description indicates some of the basic ones. Incoming mail, if addressed correctly, would go to the recipient; otherwise it would go to a central point, where the headings are read and an operator addresses them to the correct destination. Mail, internal to the network, should always be sent to the recipient directly. It should be possible to broadcast mail to many users. Any mail not capable of being delivered electronically by the network is converted for transmission via the appropriate medium. Mail has a set of time stamps; a time-stamp is applied when mail is delivered, (analagous to postal franking), and then when it is read. Alarms can be set for when the mail should have been read, and/or when a reply should have been received. It is also essential that mail cannot be fraudulently sourced, or changed retrospectively and that it is archived.

9.10.2) Report

This will have many "word processor" type features. Special features that are needed are :

a) Diagram insertions --- these could be linked to files and printed as a dot matrices.

b) Reformatting of text into two columns on one A4 page.

c) Sample printout of selected pages, (i.e. there is no need to have the whole report printed just to check a page).

d) "Boiler plate" construction with other reports showing both parts on the screen at the same time.

e) Automatic indexing, Paragraph numbering, Contents list generation etc.

9.10.3) Forms

There are two parts to this application, form definition and form filling. At the moment, there seems no alternative but for each form to be printed when filled out. The old practice of having specially printed forms is not generally practical, as human intervention is required to load the paper stock onto the nominated printer. The form definition part should be presented on a typical display thus allowing most users to fill in the form when needed. Form filling should be made easier as the entry points can be protected fields and "help" facilites can be appended to the form for beginners. Standard information (e.g name, dept, number, etc.) can be entered automatically from system parameters. The printers selected for form printout should be capable of printing dots, this allows lines to be defined giving a similar appearance to the forms that are being replaced.

9.10.4) Development

Program and system development can take on a whole new perspective since the range of accessible machines widens, transfer of files is made much simpler and workload can be shared. Another advantage is that it is possible to transfer files throughout the wide area network. Thus if a problem needs special attention, files can be transferred to a different location where personnel and facilities may be better equipped to find a solution.

9.10.5) Message

This is similar to Mail. The main differences are that messages are not automatically archived and that they have a finite life time.

9.10.6) Diary

A person's office diary (for appointments, movements, engagements etc), is an essential item to be held on the network. This gives a greater flexibility than the paper type, (eg. Alarms, reminders, appointments etc. can be arranged).

9.10.7) Meetings

Meeting scheduling is a function that can be performed in the diary service. Common dates of availability can be arranged and if agreed put into the diary. Prompts for papers, notes and files (paper and electronic), can be associated with this. It should be noted that this facility does not mean that private diaries must be accessible to other users, merely to the meetings scheduler software.

9.10.8) Voice

The potential for voice applications is large. One use is for an external user to leave a verbal message, similar to a tape-based system. Editing of documents could be done by leaving a voice message attached to a text file, after which a secretary would perform the required functions. Future applications include real time voice communication between users and voice recognition.

9.11) Filestore

A filestore is an arrangement of backing store (generally disc) controlled by a processor. The aim is to give a large number of users fast access to many files, both shared and private. To achieve this, overheads in information access must be minimised and certain features, usually options on an operating system, now become essential. Typically these would be:

a) No overlays in the filehandling software.

b) All or the most significant part of the directory is held adaptively in main memory.

c) The block or fragmentation size of each file must be set according to its special requirements.

d) There can be many levels in the directory structure, but the minimum number of levels must be small.

e) It can be specified that a file can be stored in an unfragmented manner.

f) The ability to transfer part of a file.

g) The ability to retain an audit trail of file changes.

Hardware can also help disc access by providing a cache, reducing the number of head movements. As future storage requirements are difficult to estimate, the ability to build up to a large storage capacity is advisable. This assumes that current trends in disc size increase and cost reductions will continue.

9.11.1) File access

The following description shows how traditional operating system file access rights, might have to be modified when using a machine as File Server. Three things must be considered:

Security File duplication Access rights

Security, This is similar to that used in conventional time sharing computer systems with additions to allow encryption and a less coarse allocation of security classes.

Duplication, the number of files carried by the file system can easily escalate. For example , a message that is to be broadcast can be either copied to all recipients, or a pointer put in each mail box. Having one master message does save file space, but it poses access right problems. A file should be able to be read by more than one user simultaneously, although a lock has to be imposed if any write operation is to be done. Who can gain access to a file is not an easy item to specify if the network contains a large number of users. Strict friend/foe control is not enough. One possible method is that the file directory is structured as :

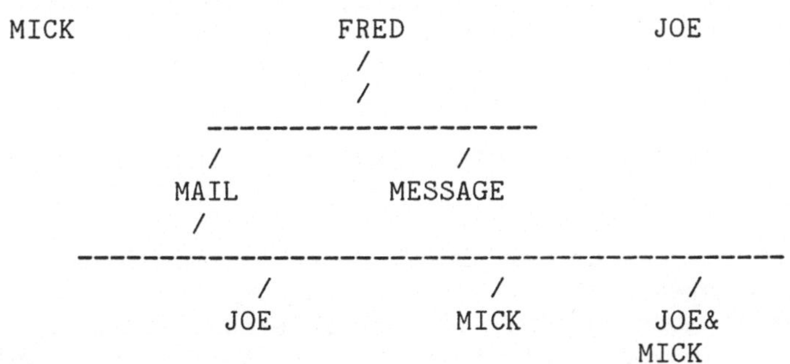

Fig.2. Directory structure

Figure 2 shows the directory of user FRED with MAIL and its sub-directory. Access rights are granted by virtue of the sub-directory name. Thus the mail created for MICK by FRED is deposited under the heading MICK, and is not visible to JOE. Mail deposited under MICK&JOE is only viewable by those two users of FRED's MAIL files.

9.12) Time

This is more a piece of specialised hardware than an application. It provides a time and date facility locked to an encoded time and date radio transmission, backed up by an internal frequency standard. Its integrity is vital to the validation of Mail, Messages etc.

9.13) Archive

Files are backed up automatically. The service does not require an operator; information archived but not in main store may take much longer to find. The structure of some storage systems, (e.g. write only once, laser discs), may cause updates to be done on a two-stage basis; a list of changes to an old backup and then a complete new backup.

9.14) Encryption

This is a service for increasing file security. The parameters needed are a key and the file name. The same key will encode and decode a file. It is envisaged that the United States Data Encryption Standard (FIPS 46) algorithm would be used. This server is used mainly to protect data on the backing store. It can protect data in transit on the network only if the relevant terminals are capable of encryption and suitable key distribution systems are in use. When VLSI implementations of the RSA public key crypto-system become available, it is envisaged that all network terminals will use this for MAIL and other private items.

10) THE IMPLICATIONS OF OPEN SYSTEMS INTERCONNECTION (OSI) STANDARDS

The implementation of SESNET poses all the imcompatability problems associated with any system put together from various manufacturers' modules. OSI, if available now, would make SESNET much easier to build. The multitude of terminals , operating systems, gateways, etc. increase the complications of interconnection.

As the OSI standards are still some years from completion, only the principles can be applied. The upper layers of the SESNET software will be defined in accordance with the OSI requirements of layer independence and precise definition of interfaces. The following protocols will be required :

A Virtual Terminal Protocol(VTP) for transforming I/O into a network and terminal-independent data stream.

A Virtual Filestore (VFS) presenting a network-independent filing system for opening and closing files, transfering files and error messages.

11) SESNET - THE CURRENT POSITION

The first stage of implementation for the prototype LAN is now in progress. The basic transmission system, (an Ungermann Bass NET/1 Ethernet), has now been installed, serving about thirty offices spread over two floors. The NET/1 currently supports about sixty V24 interfaces and an upgrade to provide another 20 interfaces is about to occur. A contract has been placed for a Norsk Data NORD 100 minicomputer with 210 Mbytes of disc storage, to act as the FILE SERVER. This machine is scheduled for installation in September 1981. Software development is in progress to provide the GATEWAY SERVER on an existing Ferranti ARGUS 700 minicomputer. Various other "server" developments, (eg. TIME and ENCRYPTION), are under way. It is planned to offer an initial service from October 1981. This will allow unintelligent VDUs on each desk cluster to access applications packages on the Nord 100, (at first only packages providing the REPORT and MAIL services). The VDUs will also be able to make and receive calls from NPSS via the GATEWAY. Several other existing service machines will be connected at this stage.

The initial service will use the normal single channel V24 mode of the NET/1 and thus there will be contention for physical ports on all the attached service computers. The Nord 100 will initially be equipped with thirteen V24 ports on the NET/1. Resource management for this first service will use the NET/1 distributed name server mechanism, allowing multi-port resources to be accessed by a single symbolic name.

12) SESNET - FUTURE AIMS

The first major development will be to provide X25 (level 3) as an access protocol to the NET/1. This will be carried out as an "in-house" software development to run on the NET/1 interface application processor boards. These are 64 kbyte Z80A micro computers. The objective is to provide multi-channel interfaces between the NET/1 interface units and large servers eg. the FILE SERVER, the GATEWAY SERVER and the RESOURCE MANAGER. A hardware development is already in progress to permit data transfer between these servers and the NET/1 units at a maximum of 1 Mbit/second. It is anticipated that this phase will be completed by Summer 1982.

The second stage will be to shed many of the application

packages from the FILE SERVER into smaller, micro-processor based servers and intelligent terminals. This will allow the FILE SERVER to concentrate on it's original purpose of acting as a general file resource. The intelligent terminals are likely to be 64 kbyte Z80A machines and a significant development effort will be required to generate appropriate implementations of high level protocols, (X25, Transport and File Transfer Protocols). This stage should be complete by Spring 1983. Subsequent projects will cover the other servers discussed earlier in this paper.

13) ACKNOWLEDGEMENT

The views expressed here are the authors' personal views and do not represent British Telecom policy statements.

Acknowledgement is made to the Senior Director of Technology Executive for permission to make use of the information contained in this paper.

14) REFERENCES

1. The Ethernet (Data Link Layer and Physical Layer Specifications). Version 1.0, September 1980. IFIP WG 6.4 Local Computer Networks, December 1980.

2. M. V. Wilkes, Communication Using a Digital Ring. Proceedings of the PACNET Conference, Sendai Japan 1975.

3. N. B. Meisner, D. G. Willard, G. T. Hopkins, Time Division Digital Bus Techniques Implemented on Coaxial Cable. Computer Networking Symposium. December 1977, pp 112-117.

4. Open Systems Interconnection Reference Model ISO/TC 97 SC 16/N227.

5. CCITT Recommendation X25, Geneva 1980.

6. A Network Independent Transport Service, British Telecom PSS Users Forum, February 1980.

7. A Network Independent File Transfer Protocol, UK Data Communications Protocols Unit February 1981.

8. CCITT Recommendations S60, S61, S62, S70, F200. Geneva 1980.

9. J. F. Shoch, J. A. Hupp, Measured Performance of an Ethernet Local Network, Xerox February 1980.

THE INTERCONNECTION OF LOCAL AREA NETWORKS VIA A SATELLITE NETWORK

C J Adams, J W Burren, C S Cooper and P M Girard

Science and Engineering Research Council
Rutherford Appleton Laboratory
Chilton, Didcot
Oxfordshire OX11 OQX

1. INTRODUCTION

During the past few years a number of potentially cheap and reliable network technologies have been developed which provide high speed, low error rate packet-switched communication on a distributed basis. In these technologies the communications channel is a single cable running round an area with host computers connected to the cable by 'T-junctions' (see fig.1). Addressed packets are sent on the cable to allow all-to-all communication between the hosts and use of the channel is divided among the hosts that wish to transmit by some time-division scheme. Two well-known examples of this type of network are the Xerox 'Ethernet' (ref 1) and the 'Cambridge Ring' (ref 2). These particular network technologies are only applicable over a limited geographical area, say within a radius of a few kilometers and because of this are now usually referred to as 'Local Area Networks'.

The emerging use of these local area network technologies is to interconnect computer based office equipment, personal computers, special purpose minicomputers and mainframes within the confines of single buildings, company sites or university campuses. A comparison of the number of hosts capable of being supported by a single local area network with the number of offices or pieces of equipment requiring interconnection within a site leads to the natural requirement to connect individual local area networks to each other in such a way as to preserve their essential properties.

K. G. Beauchamp (ed.), New Advances in Distributed Computer Systems, 201–210.

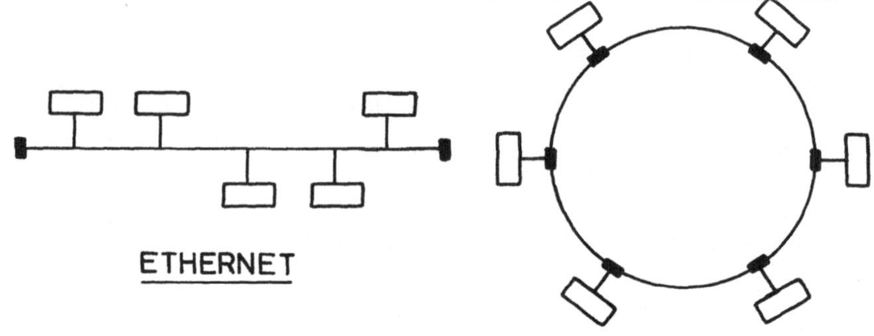

Figure 1. Local Area Network Technologies

Of equal importance to the need to interconnect LANs for the purpose of covering a single site is the requirement to inteconnect sites separated by distances ranging from tens to thousands of kilometres. Recent experience with broadcast communication at 2 Mbit/sec. using a satellite suggests that such a system has a number of properties in common with local area networks. The purpose of this paper is to describe an architecture for interconnecting a set of local area networks to cover a wide area, in which the wide area links are provided by a broadcast satellite network: we term such an interconnected set of networks an 'internet' (see fig.2).

We begin by introducing the idea of a 'one-hop' network as being a computer network in which there is no storage of information within the communications subnetwork that connects the host computers together, apart from storage induced by propagation delays. The most familiar example of a network of this type is the public dialled telephone network. Local area networks and broadcast satellite networks also have this property, although in the latter case the amount of information in transmission at any given time is large because of the long (1/4 sec) propagation delay.

This paper is concerned with the interconnection of networks of this 'one-hop' type. The general architecture envisaged is that a given site would have one or more local area networks connected together, with the sites themselves connected via a satellite communications network. It will be assumed that the constituent networks give high-speed transmission (of some number of Megabits/sec) with very low transmission error rates (better that 1 in 1000 million). The interconnection of networks is via computers that are hosts on two or more networks of the internet: we shall call such interconnection points 'bridges' on the internet. While they have some of the features commonly associated with network gateways, such as knowledge of addressing on several networks, in the internet architecture their role is perhaps nearer to that normally taken by an exchange in a packet switched network.

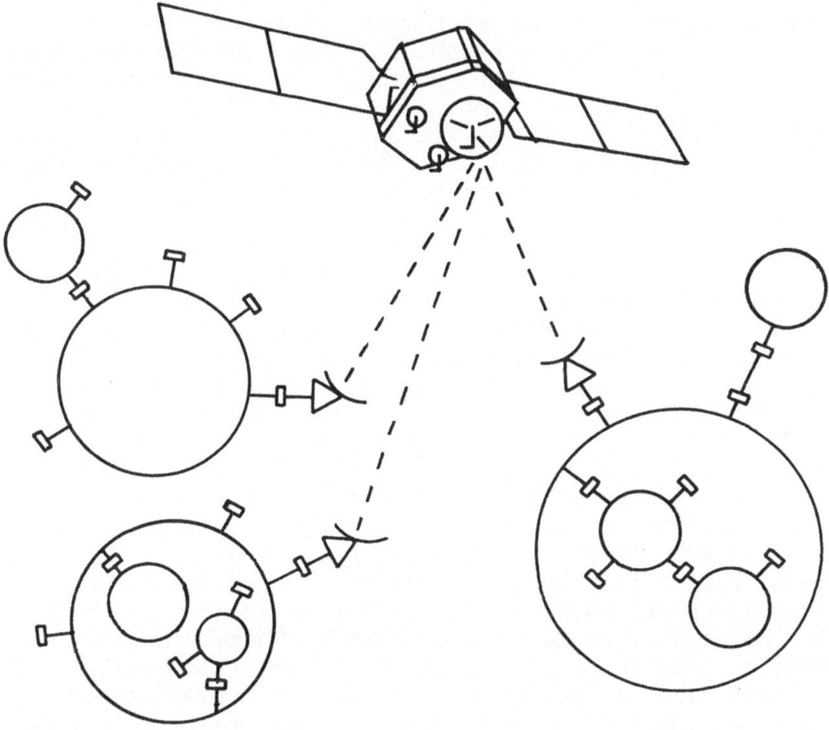

Figure 2. Local Ring Networks connected via a Satellite

The approach to network interconnection outlined in this paper was developed by members of the Network Development Group at the Rutherford and Appleton Laboratories in the UK. An approach on very similar lines has been developed independently by I M Leslie and R M Needham of the University of Cambridge (ref 3).

2. GENERAL APPROACH

The basic objective of the proposal outlined in this paper is that the bridges between networks should extend the features of a single 'one-hop' network into the internet configuration. A second objective is to minimise the need for protocols that require universal agreement, that is for protocols to which all the networks forming the internet must adhere. In the proposal the networks are required to be able to carry out a small number of basic functions, but the protocols associated with these functions, with one exception, do not require universal agreement. This is achieved without significantly increasing the work required from the bridges over and above that required to interconnect identical networks.

The most important feature of a single 'one-hop' network is that, at the communications level, it is highly transparent. To take a block of information between any pair of hosts on the network it is only necessary for fields containing the source and destination host numbers to be at known positions within the block. Above this lowest level protocol the hosts may decide to build all types of service, datagram or call based, error-free or noisy, layered or monolithic. We shall follow Cambridge and take as the basic transfer mechanism the exchange of blocks with fields defining the source host number, the destination host number and the destination port number. These port numbers provide a first level of multiplexing and can act as a key to the type of protocol or service being carried by the block. In fact it is essential to have these port numbers at this level to allow the network to be used transparently. These numbers can, of course, be used in many ways, ranging from fixed 'well-known' numbers providing a standard service to numbers being dynamically assigned for the particular requirements of a single user.

A second feature of the constituent networks that we wish to preserve in the internet is that of sequentiality, that is that blocks of information can only arrive at their destination in the same order in which they are sent by the source. This feature is guaranteed in a single network since there is no storage within the network and as many existing higher level protocols assume that this is a feature of the communication network we propose to keep the property of sequentiality in the internetwork configuration.

For communication on the internet we are proposing that a virtual circuit mechanism be used. The intention however, is to provide minimal function within the virtual circuits, in fact to provide little more function than would be provided by a datagram service.

The mechanism proposed is as follows. Suppose that HOST A on network P wishes to communicate with HOST B on network Q. HOST A sends on 'opening' block containing the full address of HOST B. The purpose of this block is to create a fixed route to HOST B through an appropriate set of intervening bridges. Thereafter the route may be used to transmit data in either direction between HOSTS A and B. The form of addressing to be adopted depends essentially upon the naming domain structure anticipated for the internet. The general attitude towards addressing is that described in the UK Transport Service (ref 4) and it is assumed that individual physical networks such as satellite, ring or ethernet are wholly contained within one naming domain; the latter may however contain several networks. If the internet forms a single domain then particularly simple forms of (Transport Service style) addressing may be agreed, such as unstructured unique addressing, hierarchical addressing or source route addressing. In the proposed configuration of a single satellite network, joining a number of sites with clusters of local networks, the addressing could take the particularly simple form of a universal hierarchical address such as Site/Network/Host/Port number. On receiving an 'opening' block a bridge joining two networks will create a 'channel' through itself to open the route. Thus blocks arriving from a particular source on network R addressed to a port number associated with the channel will be routed through the channel and sent to a fixed destination on network S and similarly in the opposite direction. The channels must be designed to preserve sequentiality. A mechanism for closing the route is also required, if only to protect resources in the bridge, and this will be discussed in the next section.

Once a route has been set up, the port on the local bridge serves as a 'front' for the port on the remote computer at the end of the route. Apart from time-delay considerations the end-user may converse with the port on the local bridge as if it were in fact the remote port.

3. DETAILED MECHANISM

3.1 Basic transmission of data

Within each 'one-hop' network a mechanism must be provided for transmitting blocks of data from a source host to a destination host. Each block will carry the destination host number, a destination port number, and a source host number. The port number provides internal routing information for the receiving host. A port number to which replies to the source host are to be addressed may be present as part of a higher level protocol. Port numbers may be fixed and 'well-known' or may be dynamically assigned. In many circumstances the 'well-known' port numbers will only be used indirectly via a 'name-server' machine, that is a machine which, when given a name, responds with a host number and port number.

The basic block may carry a CRC or checksum to guard against transmission errors and normally a maximum length will be defined for blocks. The length of blocks will usually be defined in terms of an element containing more than one bit. Blocks must be an integral multiple of elements. Typically the length of blocks will be given in terms of bytes.

The exact format for these basic transmission blocks does not need to be universially defined, but is a matter for each 'one-hop' network. As the parameters, host numbers, port numbers, and CRC are not propagated transparently across bridges but are always mapped, even the field sizes of these parameters do not require universal definition. It is, however, desirable for the basic block to carry one field that is universally defined for the internet. A single bit is required to indicate that a block contains a message that has been generated by the internet. Messages of this type do require universal definition, but no mandatory actions would be defined for hosts receiving these messages. As port numbers are mapped as a block passes through a bridge, it may be useful to carry a small sub-port number field transparently across the internet. Such a field would require universal definition.

3.2 Opening a route

To open an internetwork route, a host will send an 'opening' block to a known port on the appropriate bridge on its own network. This opening block will be similar to an opening block that might be sent to any host on the local net, but, in addition to the normal data that would be contained in such a block, the block will contain the internet address of the destination, the internet address of the source and the port number that the source host wishes to use for its own end of the route. The latter port number would normally be present in an opening block destined for a local host. Since it is not the purpose of this paper to discuss the intricacies of naming and addressing, we shall assume that we are dealing with a single naming domain in the sense of reference (4).

Assuming that the source host (#Ø) is on network A and that the first bridge (#1) joins networks A and B (see fig.3), the bridge will remember the source host number and reply port PA0 on the source and will allocate a port PB1 on network B for the route. It will then forward the block on network B, giving PB1 as the reply port, to the known port on the next bridge (#2) towards the destination, in exactly the same way as the original source sent the block to the first bridge. The procedure is repeated until the destination is reached. At this stage the opening block has reached its destination and has caused a return path to be set up from the destination back to the source.

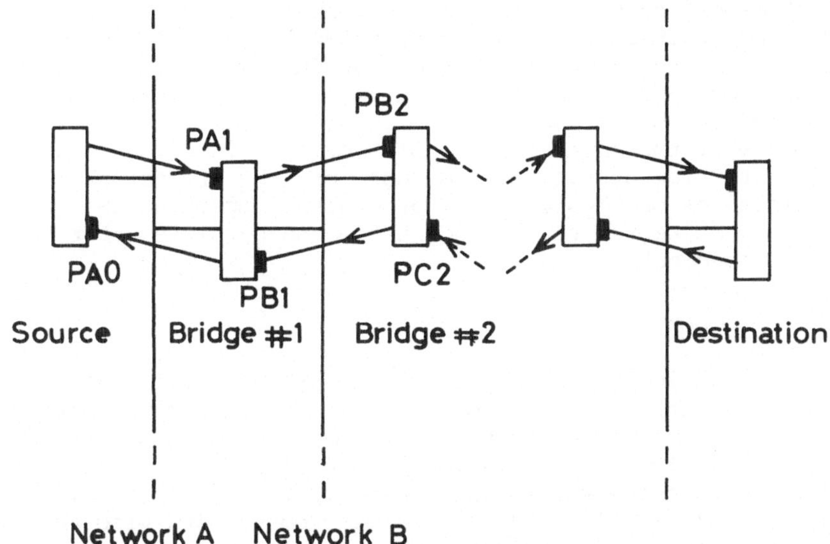

Figure 3. Routes and Ports

The destination host may now send a 'reply' block back along this path. This block may contain a reply port number which the bridges can use to set up the forward path from the original source to the destination host in the same way as the opening block was used to set up the backward path. Thus when the block reaches the first bridge (nearest the source), this bridge will allocate a port PA1 on network A as the reply port and forward the reply block to the source host. A bi-directional route has now been set up between the two hosts and thereafter this route can be used transparently by the two hosts, with the original source host treating port PA1 on its local bridge as if this port was in fact the real destination. Time delays on any conversation on the route would of course be those appropriate to the whole route, not the single network.

Two variations of this scheme may be used. If no reply port (or a zero reply port) number is given with the opening block then one has a simple datagram service. If no reply port is given with the reply block, then no forward path is created and one has a simple transaction process, that is a request followed by a reply after which the conversation is terminated.

3.3 Closing a route

A basic time-out mechanism must be available within the bridges for closing routes. This would be triggered by the absence of data flow along the route. The time-out interval could either be a fixed constant or be given as a parameter when the route is opened. The end hosts would not be informed by the bridges of the closure of the route. They would normally inform each other of their intention of terminating use of the route at some higher level of protocol. It is, of course, important that port numbers are not reused for as long a time as possible after they have been timed-out.

3.4 Maximum block size

In general each individual network and bridge will define a maximum size for transmission blocks and clearly the maximum size allowed for a given route will be the minimum of the maxima. The value of this maximum block size for a route could be indicated to the hosts in the open process. There is no packing of small blocks into larger ones, or vice versa, although the means of transferring the blocks on particular networks could include fragmentation and reassembly.

3.5 Transmission errors

Transmission errors may be detected within the networks that make up a route. In general it will be assumed that such errors are sufficiently infrequent that they may be dealt with by the end-users. If a transmission error is detected the block involved will normally be discarded. If a particular network has a high error rate it would be expected to operate internally some retransmission scheme for basic blocks.

3.6 Flow control

The bridges along a route will not provide any flow control for the hosts using the route. If hosts require flow control protection, they will have to provide their own arrangements on an end-to-end basis. Bridges will have to provide protection for themselves against congestion. Normally they will be able to provide themselves with this protection by limiting the number of routes that they accept. However bridges must still be able to protect themselves from situations such as flooding of a route by a host, congestion or failure of a network to which the bridge is connected, or congestion or failure in other bridges or an end-user. In these circumstances a bridge may have to discard blocks on particular routes until the problem has cleared. Network messages may be sent along the route to inform both ends that information has been discarded. Messages may also be sent to warn of potential congestion.

3.7 Parameters set when a route is opened

The basic task of the opening procedure for a route is to deal with addresses and thus to create the route. Addresses themselves are rather complex parameters and there is thus a temptation to add further parameters into the open request to describe in general the quality of service required on the route. Clearly a great number of options could be introduced into the open protocol, however the general line followed in this paper would be to keep options to a minimum and to allow bridges where possible to adapt to the actual flow of information along the route. It may, however, be necessary to include billing and authentication information in the open request.

4. CONCLUSION

In the coming years very large numbers of computers and computer driven devices will be deployed in offices, factories, shops and laboratories. To harness these computers together to work collectively on tasks, it will be necessary to connect these computers together into large networks for the purpose of information exchange. Local area network technologies of Ethernet and Cambridge ring type (refs 1 and 2) will be very strong contenders for providing the networking structure at the local level. It will be necessary to link many of these networks together over a wide area to form large internets. The purpose of this paper has been to propose a possible networking architecture for this internet. The aim has been to outline a very light-weight 'carrier' protocol scheme for the internet, similar to those used on single one-hop networks, and to avoid the need for detailed agreement on universal protocols. It remains to be seen whether the proposed architecture can make efficient use of a satellite channel without introducing unacceptable congestion and delay in the bridges. This needs to be tested by experiment.

References

1. The Ethernet. A local area network. Digital, Intel, Xerox Version 1 September 1980.
"Ethernet: Distributed Packet Switching for Local Computer Networks by R M Metcalfe and D R Boggs. Communications of the ACM 197 July 1976.

2. The Cambridge Digital Communication Ring by M V Wilkes and D J Wheeler Computer Laboratory, University of Cambridge, England. Proceedings of the LACN symposium May 1979.

3. A High Performance Gateway for the Local Connection of Cambridge Rings by I M Leslie, Computer Laboratory, University of Cambridge. June 1981.

4. A network independant transport service. SG3/CP(80)2 prepared by Study Group 3 of the British Telecom PSS User Forum. February 1980.

REPORT ON THE LOCAL AREA NETWORKS FFM-MCS AND MICON

Ludwig Rössing

Forschungsinstitut für Funk und Mathematik
D-5307 Wachtberg-Werthhoven, Königstr. 2
F.R. Germany

ABSTRACT

This report gives a brief description of the two experimental local area networks FFM-MCS and MICON. The philosophy of communication between autonomous tasks only through messages, which is used in these networks, is reported.

INTRODUCTION

This paper reports the research activities at the FFM (Forschungsinstitut für Funk und Mathematik - Research Institute for Electronics and Mathematics) on the subject of distributed computing with local computer networks. The project startet 1974 at the FFM, having its origin in experiences gained in managing multicomputer installations for controlling a real time system (phased array radar) and in former projects in automatic air traffic control with distributed radar sites. These projects were extraordinarily good applications for distributed computing. Problems of such kinds can be naturally distributed in independent tasks (processes) which can run concurrently on the different computers. These typical applications for distributed processing were the starting point for systematically investigating computer networks and the software for distributed systems and for the different programming techniques in an experimental local connected multicomputer system.

K. G. Beauchamp (ed.), New Advances in Distributed Computer Systems, 211–221.

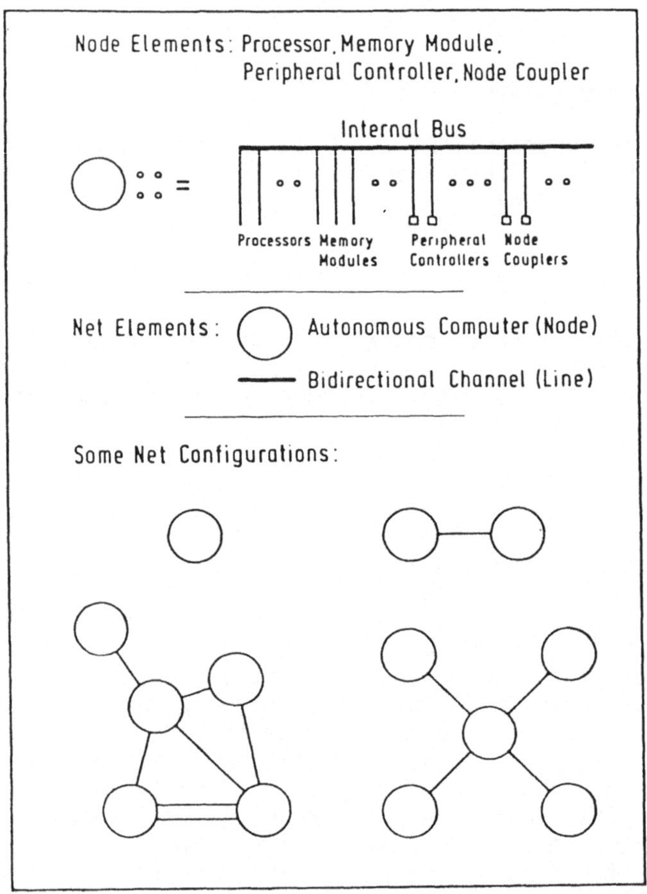

Figure 1: The building block system

THE FFM-MULTICOMPUTERSYSTEM (FFM-MCS)

In 1975 the FFM-MCS was built up as such an experimental system based on minicomputers. (1,2,3,4) It was designed as a building block system, adaptable to a variety of applications and requirements. Figure 1 shows the elements of this building block system. The FFM-MCS consists of autonomous computers called nodes and bi-directional channels (lines) for node connection. With these network elements arbitrary network configurations can be constructed. There are no restrictions concerning the number of nodes or the ways of connecting the nodes. Whereas the nodes are provided with a bus architecture, the network has point-to-point connections (see Figure 1).
A special defined configuration for an application consisting of nodes with peripheral equipment and node connections is called a 'structure'. The local operating systems (= OS-Kernels) including the corresponding device drivers are produced by a generator and

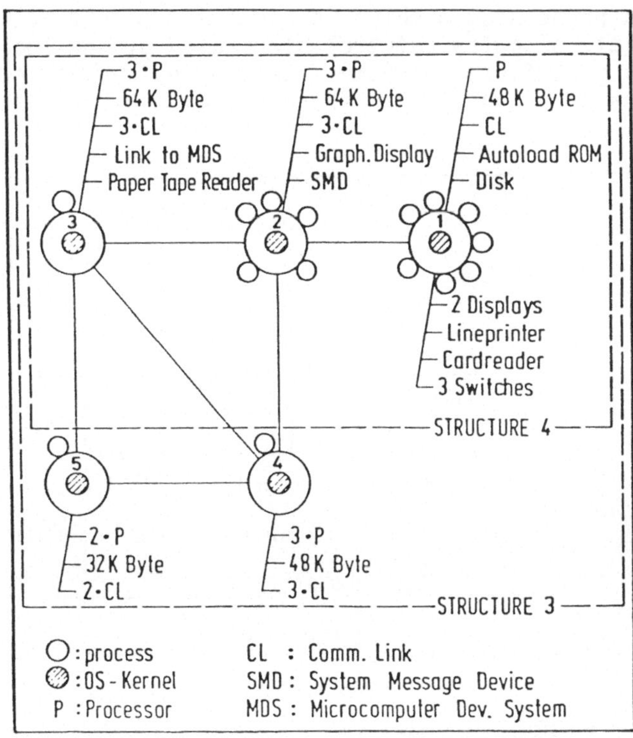

Figure 2: Some configurations of the FFM-MCS

are stored in a file with the corresponding structure name. The system is bootstrapped by loading the local operating systems according to the structure from this file. Figure 2 depicts two structures and shows the nodes with all their equipment.
The system is based on the 16 bit SUE-minicomputer from Lockheed Electronics (5) because its bus architecture was already prepared to be equipped with more than one processor. Moreover, it offers the chance to develop special customer interfaces. Originally it was not designed for distributed data processing, so a fair amount of hardware and software development had to be done. (2,6,7)
The software should not narrow the flexibility of the hardware, so the software was also built in a modular manner. The whole basic operating system of the network is partitioned in local operating systems which run on the nodes of the network. So every autonomous computer has its own local operating system.(8) The local operating system consists of a part identical in each node (processor management, supervisor call (SVC) handling, message handling) and a part that is depending on the equipment of the node (device monitors, net dependent lists).
A given data processing problem is partitioned into autonomous tasks. A task consists of a task control block used by the local operating system, message buffers, private data and instruction code. The address space of a task is not shareable with other task There-

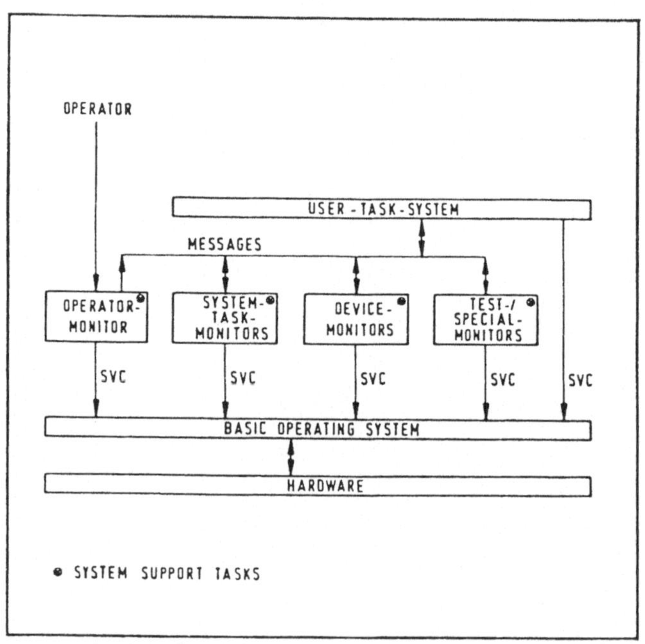

Figure 3: Logical system structure

fore no common memory areas and no global variables exist. Every task is assembled separately. For this reason the tasks could be located anywhere in the network. But the device monitors (which are also normal tasks) have to be located on that nodes where the devices are connected to. A task is identified by its name which therefore has to be unique in the network. This kind of task mentioned here is similar to the well known 'sequential processes' in Hoare's CSP system. (9)
Figure 3 displays the logical structure of the system and the communication paths between the different layers. Tasks communicate with tasks only through messages and with the (local) operating system through supervisor calls (e.g. I/O-instructions within device monitors). As can be seen, the hardware is completely invisible to the task system. The user need not care about the hardware configuration the task system is running on (see Figure 1). Therefore it is possible to optimally accomodate the hardware configuration to a processing problem without redesigning the software.
The intertask communication is performed by the transfer of messages. Messages between tasks in different nodes are transported via the 'node couplers' (6) (see Figure 1) by direct memory-to-memory access, eventually using intermediate nodes for the transfer.

Type	SENDER		RECEIVER
I	W(MB,L,TNAME,I) Sends message to task 'TNAME'	⟶	R(MB,L,TNAME,I) Expects message from task 'TNAME'
II	WU(MB,L,TNAME,I) Sends message to task 'TNAME'	⟶	RU(MB,L,I) Expects message from any task
III	WB(MB,TNAME,I) Sends 1-word-telegram to 'MAILBOX' at task 'TNAME'	⟶	No command; 'MAILBOX' is automatically updated without notifying the receiver

Figure 4: Message types for intertask communication
MB: message buffer; L: message length
TNAME: task name; I: error indicator

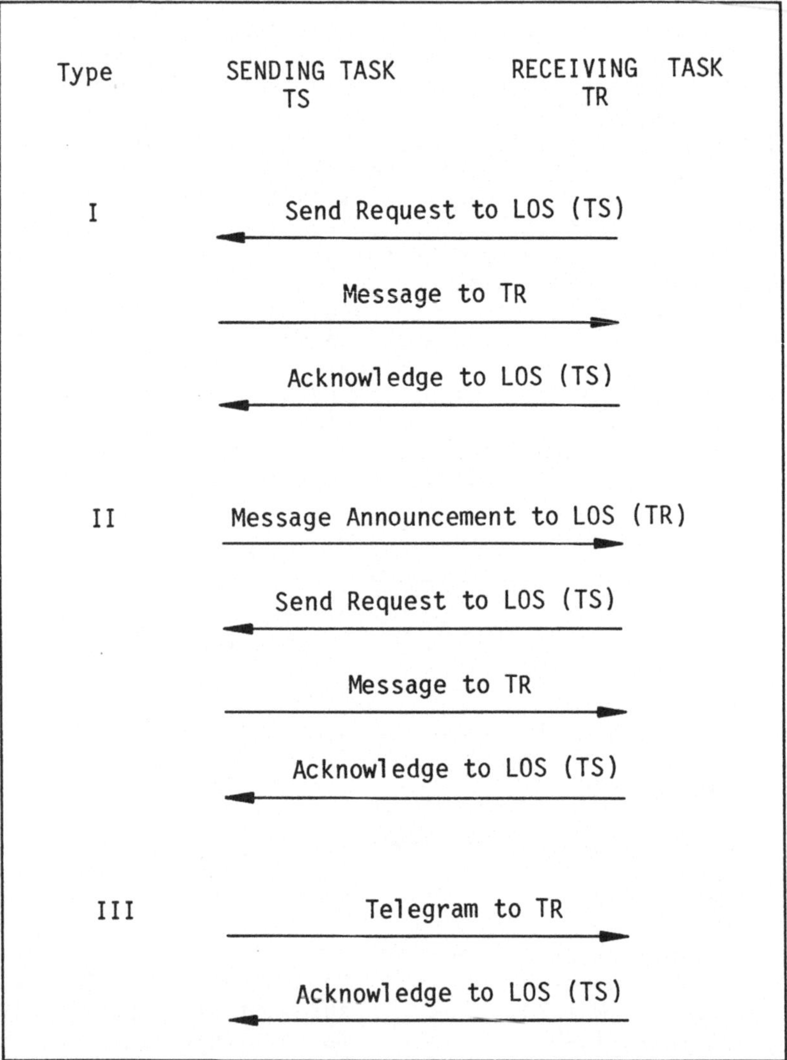

Figure 5: Message transfer protocols
LOS (T): task T's Local Operating System

The FFM-MCS knows three types of messages, which are necessary for efficient and safe operation of a distributed network (see Figure 4)

1) 'private message': Sender and receiver know each other by name and location and expect only this partner to communicate with. This message concept is used in Hoare's CSP
2) 'open input message': The receiving task accepts messages from any task of the system. This concept is to be used in the monitor tasks (e.g. printer monitor). It is realized in ADA (10)
3) 'telegram': a telegram is a message sent to a predefined buffer (mailbox) of the receiver without notifying the receiver. Former telegrams will be overwritten.

Figure 5 illustrates the protocols used for these three types of message transfer. Messages of type I and II result in a synchronization of sender and receiver via the rendezvous, whereas the type III communication does not require any synchronization. It has only a two step protocol (see Figure 5) and therefore it is the fastest communication type. It is used when the user situation does not demand any synchronization, for instance acceptance of sensor data, where only the latest message is to be taken for evaluation.

In distributed data processing it is an important question, whether the deadlock problem caused by the complicated communication structures of major task systems can be controlled in a satisfactory way. A deadlock arises from a circular waiting condition of several tasks, for instance raised by their communications. But as it became apparent in the FFM-MCS, even in complex program packages with about 40 tasks running on the system, it is possible to avoid deadlock situations by mapping the communication structures into a Petri net and proofing the freedom of deadlock by performing the 'token game'.

In order to get experience and a better insight into this network, many measurements were executed to find out which performance can be reached by increasing the number of processors, in which cases the multicomputer system is superior to a single computer, how the distribution of the tasks influences the performance of the system, and how the communication overhead affects the performance. Another aim was to trace the bottlenecks of the FFM-MCS.(11)

THE FFM MICROCOMPUTER NETWORK (MICON)

With all the experiences obtained with the FFM-MCS, an improved network design was devised trying to avoid as many drawbacks of the FFM-MCS as possible. The objective of this new system as well as the way to attain this objective (e.g. the retention of the principle of a building block system with as few different elements as possible) is similar to those in the previous one, though in the realization there are many differences, strongly influencing the capabilities of the system. In the following it is attempted to point out some differences between the FFM-MCS and the MICON.

It is desirable to have the possibility of adapting the microprogram of the processors to the conditions of the problem the system should deal with. This offers the opportunity of transferring complex functions into the microprogram to decrease bus contention within the nodes and augment the safety of that parts of the software which have to be protected against damage. So, in MICON microprogrammable microprocessors were used, built up in bit-slice technique (with the Am 2900 family of AMD (12)). The microprocessor, which was developed in our laboratory (13), is called JANUS because it has two identical I/O ports, but only one ALU and one register file. Besides its function as a normal processor (14), JANUS can be used in the following ways as shown in Figure 6:

- as processor with private memory (e.g. to store data which are heavily accessed, so getting a better protection of the data and a reduction of bus contention)
- as an intelligent peripheral controller
- as communication link.

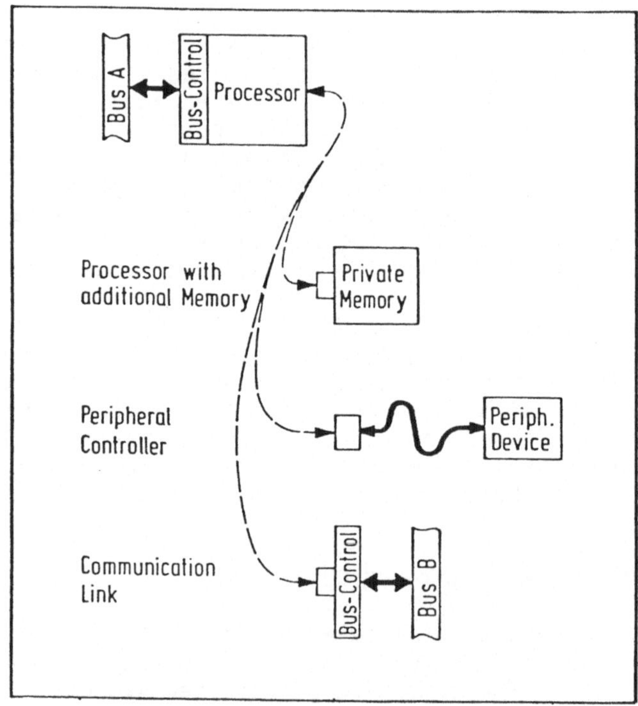

Figure 6: The JANUS processor

As communication link, JANUS is connected to the buses of two nodes. It acts as an intelligent coupler, but can be used by either side as a normal processor as long as there are no communication requests in one of the two nodes. Exploiting the possibilities of JANUS will reduce the number of different node elements to two: JANUS and memory modules.

One major disadvantage of the SUE-system used in the FFM-MCS is the master-slave relationship of the processors: only one processor (master), defined by its position in the crate, can be interrupted by external signals and so all operating system activities, including message handling and task scheduling, has to be done by this processor. In MICON this master-slave relationship (causing a bottleneck as clearly shown by the measurements and representing an undesired central part in a node) is avoided. The structure of the local operating system allows the concurrent processing of system code by all processors to the point where access to common lists forces mutual exclusive access.(15) This mutual exclusion is efficiently supported by microprogram and hardware.(15)

Also a second central part of the SUE system, the bus controller, has been changed in the new system in that way that every JANUS and every memory module has its own bus controller.(16) This leads to some hardware redundancy, but protects the node against complete failure if the bus controller fails.

Besides the message types used in the FFM-MCS (cf.Figures 4 and 5) some new variants are provided in MICON.

In addition to the open input message described earlier, there is an 'open output message' for sending information to any task requesting it.

Within these open message conceptions there is the possibility (e.g. for improved safety) to dynamically confine the total set of tasks to a subset by explicitely naming the wanted communication partners.

If it is not determined a priori which communication is ready for a rendezvous, the user provides a list with wanted communications. From this list the communication will be executed which is ready for transmission first. If there are several pending rendezvous, the position in the list determines which communication will be executed.

The so-called 'multicommunication' is introduced to decrease unnecessary communication waiting times. In this case the user provides a list of more than one communication order. The local operating system collects these orders and tries to accomplish them as fast as possible, irrespective of the order sequence in the list, so avoiding unnecessary waiting times and order sequence dependent deadlocks. The task will be blocked as long as all requests of this multicommunication will be executed.

With these communication variants mentioned so far and implemented in MICON, all essential communication demands of autonomous tasks in distributed systems can be handled.

Whereas in the FFM-MCS fixed routing tables are used, in the MICON adaptable failsafe distributed routing protocols will be installed, as for instance specified in (17), for efficiently supporting recovery management.
In the FFM-MCS the tasks have to be written in assembler language, adopted from the SUE system. In the new system 'DISTRIBUTED PASCAL' can also be used.(18) DISTRIBUTED PASCAL in PASCAL-P4, extended by standard procedures for communication between autonomous PASCAL programs.

In the future MICON will be used for further research on distributed networks and is considered to be a step on the way to reliable, fault tolerant, fail safe, and damage resistent systems.

ACKNOWLEDGMENTS

This report presents the results of the research work which was done at the FFM by many colleagues of mine. I want to appreciate my gratitude to all of them, especially to W. Grünewald and G. Neumann.

REFERENCES

(1) H. von Issendorff, "Modulare Multicomputersysteme mit Mini- und Mikroprozessoren", FFM-Bericht Nr. 246, Werthhoven (1976) (in german)

(2) W. Grünewald, "Modulare Programmierung von Mehrrechnersystemen", FFM-Bericht Nr. 277, Werthhoven (1979) (in german)

(3) H. von Issendorff, W. Grünewald, "An adaptable network for functional distributed systems", Proc. 7th International Symposium on Computer Architecture, May 1980, La Baule (France), pp. 196-201

(4) H. von Issendorff, "Experiences with the experimental FFM-MCS", Proc. AGARD Conference, No.303, Roros, Norway, 22-26 June 1981, pp. 10-1 - 10-8

(5) Lockheed Electronics Company Inc., "SUE Computer Handbook", Los Angeles, California 90040, 1973

(6) W. Langer, "Der Rechnerkoppler des FFM-Multicomputersystems", FFM-Bericht Nr. 278, Werthhoven, 1979 (in german)

(7) W. Jansen, "Funktionsmodule des FFM-Multicomputersystems", FFM-Bericht Nr. 282, Werthhoven, 1979 (in german)

(8) W. Bauer, "Der Betriebssystemkern des FFM-Multicomputersystems", FFM-Bericht Nr. 276, Werthhoven, 1979 (in german)

(9) C.A.R. Hoare, "Communicating Sequential Processes", Comm. ACM 21, No. 8, 1978, pp. 666-677

(10) Reference Manual for the Ada Programming Language, United States Department of Defense, July 1980

(11) G. Neumann, R. Ackermann, W. Grünewald, "Messungen zum Kommunikationsaufwand für Prozesse in einem lokalen Rechnernetz: Erfahrungen aus einer Implementation", Ber. z. German Chapter of the ACM, Bd. 7, Kaiserslautern, Febr. 1981, pp. 186-208 (in german)

(12) Advanced Micro Devices, Inc., "The Am 2900 Family Data Book With Related Support Circuits", Sunnyvale, Calif. 94086, 1979

(13) W. Langer, "Ein mikroprogrammierbarer Prozessor für Mehrrechnersysteme", FFM-Bericht, Werthhoven, (to appear) (in german)

(14) W. Grünewald, N. Haak, W. Langer, U. Skupin, "MUMI-PROZESSOR Befehlshandbuch", FFM-Bericht Nr. 306, Werthhoven, 1981 (in german)

(15) G. Neumann, "An operating system for a microcomputer network", Int. Conf. on performance of data communication systems and their applications, Paris 14-16 Sept, 1981 (to be published)

(16) K. Kittmann, "Ein dezentraler Systembus als Baustein für Mikrorechnernetze", FFM-Bericht, Werthhoven (to appear) (in german)

(17) P.M. Merlin, A Segall, "A Failsafe Distributed Routing Protocol", IEEE Trans. on Comm., Vol. COM-27, No. 9, Sept. 1979, pp. 1280-1287

(18) W. Grünewald, "DISTRIBUTED PASCAL", NTG/GI Fachberichte "Struktur und Betrieb von Rechnersystemen", Ulm 1982 (to be published) (in german)

THE CAMBRIDGE RING

R Banerjee and W D Shepherd

University of Cambridge, England

This paper describes the Cambridge Ring, a communication system with a raw data rate of 10 Megabits/sec, and outlines how it is used to provide a Distributed Computing System in the Computer Laboratory, University of Cambridge.

INTRODUCTION

The Cambridge Ring provides a system which allows the interconnection of computers and similar devices within a building and is capable of operating at a data rate substantially in advance of that provided by conventional techniques. It is used to support a local area network within the Computer Laboratory at the University of Cambridge and is used for a similar purpose in a number of other Universities and Industrial Concerns [1,2,3].

In the first part of this paper we describe the structure of the Cambridge Ring and the low and high level protocols implemented on it. We then give a brief description of how it is used to provide a Distributed Computing System in the Computer Laboratory at the University of Cambridge. Finally we give a brief indication of possible future developments.

RING ORGANISATION

The Cambridge Ring is based on the empty slot principle. The bandwidth is subdivided into slots which circulate continuously round the ring either carrying data or ready to accept data. The main aims of its designers were [4,5]:

K. G. Beauchamp (ed.), New Advances in Distributed Computer Systems, 223–237.

1. To preserve the conceptual simplicity of a ring system in the practical design.

2. To include a minimum of protocol at the hardware level.

3. To prevent the possibility of hogging of the ring by one station.

The designers felt that the bandwidth available over short distances was more than sufficient, and therefore making best use of the bandwidth was not one of the aims.

At first they thought that a register insertion mode of operation would best satisfy the design aims. However, its unreliability led the designers to consider an empty slot scheme. In its simplest form an empty slot system suffers from hogging, but it was realised that by allowing the minipacket to make one complete revolution and not mark it empty until it had passed the source, this defect could be overcome. Therefore, the empty slot principle was chosen as the basis of the design.

The structure of the ring is given in Figure 1.

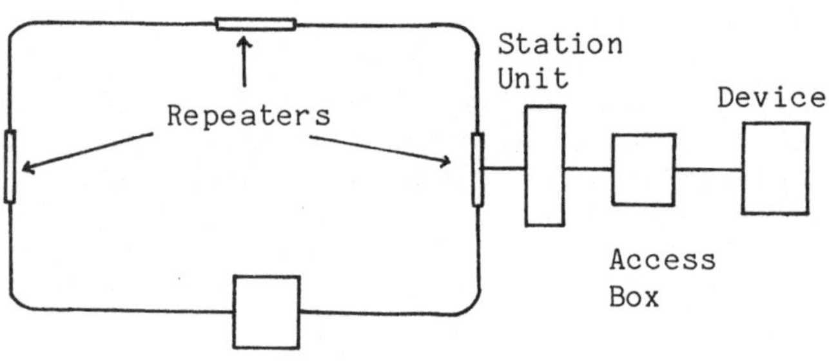

Figure 1 Ring Structure

A station consists of a repeater and a station unit which are identical for all stations, apart from a plug specifying the address of each station. In addition, an access box provides the interface between the device and the station. A unique station called the monitor station, is used to create the slot structure on initialisation, to clear corrupt minipackets and to collect error and general ring statistics.

The minipacket structure shown in Figure 2 is designed to minimize delay at the transmitter and receiver, and to maximize timing tolerance.

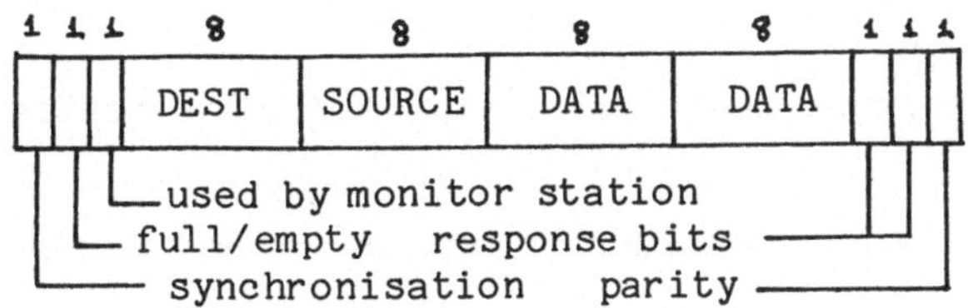

Figure 2 Minipacket Format

The first bit is always one and is used by the framing circuitry to synchronise with the "start of minipacket". The second bit indicates whether the minipacket is full (1) or empty (0). The monitor station uses the third bit to remove erroneous minipackets. This will be dealt with later. Next comes the destination and source bytes, allowing 255 different stations, followed by two data bytes. The response bits are used to convey response information and will be mentioned later. The parity bit is used to help with error detection and will be discussed in the section on error recovery. The minipackets follow each other head to tail up to the capacity of the line, the "train" being terminated by at least two zeros called the gap digits. The current ring has four minipackets circulating.

STATION ORGANISATION

The station is shown in greater detail in Figure 3.

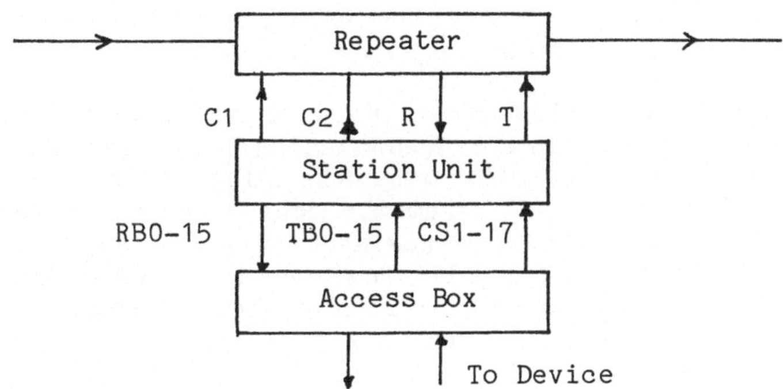

Figure 3 Station Structure

The repeater is designed to operate autonomously from the station unit and is used to regenerate the signal at each node, as well as providing access to the ring for the station.

The interface between the station and the repeater consist of four lines:

R: This passes the data stream continuously from the repeater.

T: This enables the station unit to pass digits to the repeater for transmission.

C1: This line carries control waveforms instructing the repeater to either pass on the digits from the ring or insert digits from the station unit.

The fourth line is a further control line, used only at the monitor station to initialise the ring.

The station unit contains three registers:

1. Transmission Shift Register

2. Receive Shift Register

3. Source Select Register

It also contains framing circuitry, parity checking circuitry, independent transmission and reception circuits, and circuitry for detecting and interrogating returning minipackets.

The Source Select Register is used to limit access to a station. When this register contains all ones, the station will listen to a minipacket from any source; if it contains zero, the station is deaf to all sources; otherwise the station will receive from the source whose address is in the register.

When the transmission circuitry detects an empty minipacket it will form the data in the Transmission Shift Register into a minipacket and transmit this onto the ring. This circuitry is invoked by a transmit signal from the access box and assumes that the access box has filled the register with the destination and data bytes. The Transmitting Shift Register is circular and retains a copy of the minipacket, allowing the minipacket to be retransmitted, if necessary, without reloading. On transmission the response bits are set to "11".

On the receiving side, a circuit continuously monitors the data stream for minipackets addressed to the station. On detecting one, provided the receiving register has been cleared after the previous transactions, the digit stream will flow into the

Receiving Shift Register. If the register has not been cleared the response bits are marked "00" (station busy). In the meantime the source is checked against the Source Select Register. If the station is deaf to the source the response bits are set to "10" otherwise shifting ceases and the response bits are set to "01" (minipacket accepted). On acceptance the minipacket remains in the Receiving Shift Register until it is cleared by a signal from the access box.

The returning minipacket detection circuitry uses its knowledge of the number of minipackets in the ring to detect a returning minipacket. The control bits of the minipacket are interrogated and this information is made known to the access box, which then takes the appropriate action depending on the protocol. The possibilities are:-

Response Bits	Interpretation
11	destination absent
01	minipacket accepted
10	destination deaf to this source
00	destination busy

The interface between the station unit and the access box consists of two local 16 bit data busses, TB0 to TB15 and RB0 to RB15, and 17 control lines, CS1 to CS17.

TB0 to TB15 pass data in parallel to the Transmission Shift Register, whereas RB0 to RB15 accept data in parallel, from the Receiving Shift Register.

The control lines, CS1 to CS17, are the low level instructions provided by the station unit to the access box for writing protocol control programs.

CS1 sets the 8 bit Source Select Register from RB0-7

CS2 gates the first data byte from the Receiving Shift Register to RB0-7

CS3 gates the second data byte from the Receiving Shift Register to RB8-15

CS4 gates the source address from the Receiving Shift Register to RB0-7

CS5 gates the Source Select Register to RB0-7

CS6 gates a one to RB6 if the station has rejected a minipacket from an unselected source at any time since the Source Select

Register was last set

CS7 echoes the logical OR of CS1-6 and CS16 (handshake response to access box)

CS8 sets the first data byte in the Transmitting Shift Register from RB0-7

CS9 sets the second data byte in the Transmitting Shift Register from TB0-7

CS10 sets the destination byte in the Transmitting Shift Register from TB0-7

CS11 gates the response bits for the last minipacket returned to TB0-4; these bits are not valid unless CS14 has gone up

CS12 echoes the logical OR of CS8-11 and CS13 (handshake response to access box)

CS13 causes the minipacket in the Transmitting Shift Register to be transmitted or retransmitted (transmit command)

CS14 falls at the end of CS13 and rises again when the transmitted minipacket has returned (minipacket returned signal)

CS15 falls at the end of CS16 and rises again when a minipacket has arrived in the receiving register (minipacket received signal)

CS16 discards the information in the Receiving Shift Register and prepares the station unit to receive (receive command)

CS17 signals appearing on this line indicate the ring is up and working.

The above constitutes the hardware on which the layered protocols can be built.

To prevent congestion of the system with minipackets being repeatedly marked busy, this happens if devices with varying speed characteristics are interconnected, a further algorithm has been incorporated into the hardware. If a minipacket is returned twice marked busy, a delay in transmission is invoked by delaying the appearance of the CS14 signal from the access box. This delay is dependent on traffic load and is 2 x ring delay x traffic density. Further busies increase this delay to 16 x ring delay x traffic density.

RING MAINTENANCE AND ERROR RECOVERY

Additional hardware has been included to shorten error recovery time by localising the fault. A paper on this subject was produced by Hopper and Wheeler [6].

A simple scheme is provided by the monitor station to remove erroneous minipackets. This is done by unsetting the monitor bit on transmission. This bit is then set at the monitor station. The monitor station will then mark as empty any minipackets which reach it with the monitor bit already set. Therefore, full minipackets can circumnavigate the ring once, and once only.

In addition use is made of the parity bit in each minipacket. Each station has two parity circuits, one to check the incoming minipacket , and one to produce a parity bit for the outgoing minipacket. Each station continuously computes the parity of passing minipackets and simultaneously senses and overwrites the old parity (this avoids increase in delay). If the old parity does not match, an error has occurred since the last active station. The station proceeds to launch a reporting minipacket to a logging station. A stream of reporting minipackets from a given station will indicate a malfunction, otherwise a spurious noise signal can be assumed. Empty minipackets are also monitored in this way, and therefore a continuous check is carried out on the ring.

Breaks in the ring can also be detected. The leading digit in a minipacket is always one, so an unbroken series of zeros cannot occur. If this does happen a station will emit a continuous stream of reporting packets.

It should be noted that in the scheme above, erroneous minipackets are allowed to continue. There is no error recovery at the minipacket level. This is left to the packet level of protocol (see later) and will normally be backward error recovery (retransmission) on receiving an erroneous checksum.

HARDWARE

The ring is built using TTL technology and operates at 10MHz, with a maximum distance of 200 metres between repeaters. The signals are transmitted along two twisted pairs. To increase reliability the repeaters are powered directly from the ring. On the current ring the power is injected at five points. Therefore, repeaters continue to work even if a station loses power.

The use of a two channel/four wire system allows a simple self-clocking modulation technique to be used. In each pulse interval, a change on both pairs indicates a one and a change in one pair indicates a zero, each pair being used alternatively for signal changes.

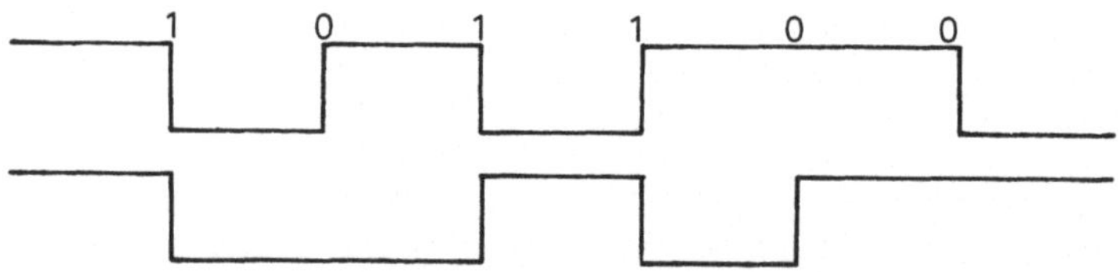

Figure 4 Four Wire System of Modulation

This scheme was considered superior to the more common phase or frequency modulation techniques. The advantages of the four wire system include: no half pulse ambiguity, it is fully balanced, little encoding or decoding delay, and it is easy to detect some errors (e.g. no transition in a given pulse interval).

The delay in the repeater is just greater than a bit, and the electronic delay in the cable is approximately 6 bits per 100 metres. This gives a ring delay of approximately 5 microseconds in a system with 8 stations, 100 metres apart (disregarding probable need for an additional shift register).

In the prototype system one segment of the ring uses fibre optic cables. These are 200 metres long but because of their low signal/noise ratio could easily be extended to several kilometres without the need for additional repeaters.

HIGHER LEVEL PROTOCOLS

Although communication on the ring can be done using the minipacket protocol described above in practice most transmitting is done using protocols above this level. At present there are three higher level protocols used on the ring. These are the Packet Protocol (PP), the Single Shot Protocol (SSP) and the Byte Stream Protocol (BSP). BSP allows the setting up of a virtual circuit for communication between transmitter and receiver, whereas the other two support datagram type communication.

Packet Protocol

The PP normally sends data in larger quantities than can be held in a single minipacket and has the additional advantage that the data can be directed to a notional "port" on the destination station. The block sent is known as a basic block or packet. In this paper we will refer to it as a packet.

A packet has the structure shown in Figure 5(a) and commences with a header of the form shown in Figure 5(b).

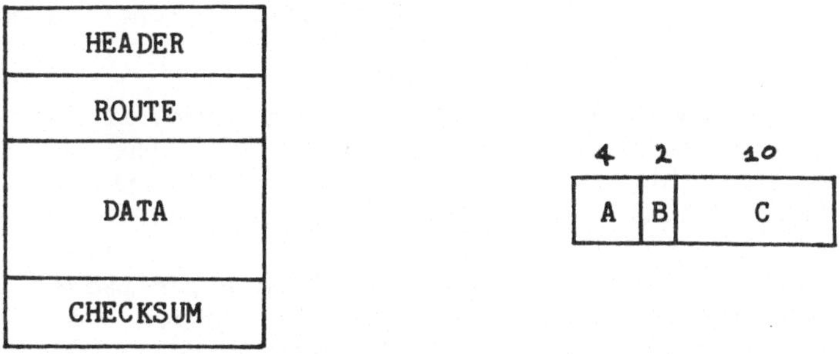

HEADER
ROUTE
DATA
CHECKSUM

a) Packet

A (4)	B (2)	C (10)

b) Header

Figure 5 Packet Structure

Field A is the binary pattern 1001
Field B is 0 long packet with checksums
1 long packet with checksum field set to zero
2 this packet consists of this single minipacket carrying data C
3 reserved for further expansion

A long packet consists of: route packet, C+1 data minipackets and a checksum-minipacket.

A route minipacket consists of a port number in the bottom 12 bits, the packet being notionally directed to that port at the destination station.

The C+1 data minipackets conform to the protocol that is currently agreed to be in use at the port identified in the route part.

The checksum minipacket for type 0 packets consists of a 16 bit end-around- carry checksum over the entire packet commencing with the header minipacket up to and including the last data minipacket. In type 1 packets, the notional checksum packet is sent as zero, and checked to be zero.

The method of reception is as follows [7]:

1. While a station is totally unable to receive anything it keeps its select register to zero.

2. When a station is potentially capable of receiving input it sets its select register to 255.

3. It then listens for a valid header minipacket ignoring anything else.

4. When a valid header minipacket has been found, if the station wishes to receive from the station from which the header came, then the receiving station sets its select register to that source, thus rejecting input from any other source.

5. The receiving station must operate either a per-packet timeout or a per-minipacket timeout in order to recover from a packet being sent shorter than the header minipacket suggested. The timeouts commence with reception of the header minipacket. If the timeouts expire at any time henceforth, the input thus far accumulated is ignored, and the station is reset to state 2 above, ignoring the incoming packet.

6. The next minipacket after the header is the route minipacket. If interpretation of this leads the receiver to believe that it cannot process the remainder of the packet (e.g. specific port not active) then it may reset itself to state 2 above, ignoring the incoming packet.

7. On reaching the end of a packet the checksum minipacket is received and checked. If, for type 1 packets, the checksum is incorrect, or for type 2 packets, it is non-zero, then the entire packet must be ignored as if it had never been received.

8. After reception of a packet, the selection register may be restored to 255 if more input is possible, otherwise zero.

9. As an alternative to resetting immediately to state 2 if a partially received packet is to be rejected, the selection register may be set to zero for a short time in an attempt to cause the transmitter to stop sending. The selection may either be for a fixed time, or until the station hardware indicates that a minipacket has been rejected "unselected". When the latter strategy is used, a timeout is also required.

For transmission:

1. When transmitting the first minipacket (the header) of a packet, due allowance must be made for the possibility of the receiving station being busy or unselected owing to it being in the process of receiving a packet from another source. Attempts to transmit the header should be maintained at least as long as the longest possible packet can take at the reception station. Any other ring failure can be regarded as fatal.

2. Having successfully transmitted the first minipacket (header), allowance may have to be made for certain reception stations to perform certain set-up operations for the packet, during which time the station will reject as busy.

3. After that, the number of busy rejects that may be expected per-minipacket should be very low, as the receiver is supposed to be concentrating on one source only. It will be necessary for transmitting stations to have a timeout or repeat count on a per-minipacket or per-packet basis in order to recover from a reception station crashing in the middle of a packet. A timeout is also necessary to recover from certain ring errors (such as power off) which result in a minipacket never returning to its sender.

Single Shot Protocol

A number of transactions on the ring consist of a single packet "request" to which the response is a single packet "reply". Such transactions do not require the setting up of a virtual channel as provided by the Byte Stream Protocol (BSP), see below. The Single Shot Protocol (SSP) provides a standard way of handling these simple transactions. It provides for data to be sent and results received; in addition, a return code can be sent in the reply to indicate any error that might have occurred.

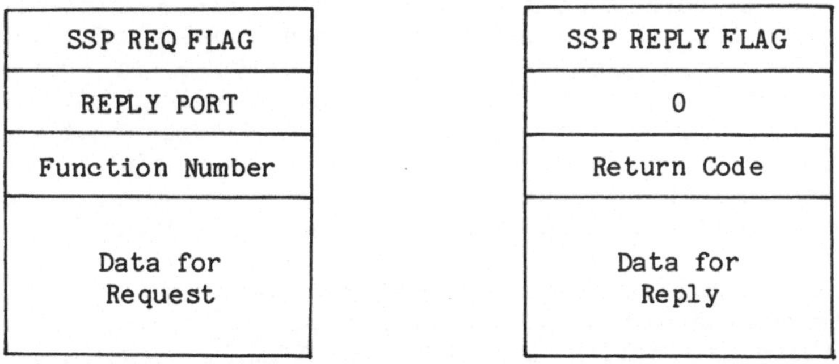

SSP REQ FLAG
REPLY PORT
Function Number
Data for Request

SSP REPLY FLAG
0
Return Code
Data for Reply

Figure 6 Packet Layout

The SSP defines the return code to be zero if and only if successful and it is not used to carry results. It is important to note that SSP's must be repeatable.

Byte Stream Protocol

The byte stream protocol[8] is built on top of the Packet Protocol. It provides a pair of synchronised byte streams, and corrects all errors detected by the Packet Protocol. It assumes that undetected errors will be sufficiently rare to be ignored, although there are facilities for resetting to a standard state if otherwise unrecoverable errors do happen. Acknowledgements are used to ensure data integrity. All erroneous packets are ignored, and the timeout mechanisms of the byte stream protocol repeat unacknowledged packets.

The acknowledgements also provide flow control, in order to ensure that a transmitter does not send more data than the recipient has committed himself to accepting. In order that the timeout mechanisms should not lead to futile communication during periods when there is no data to send or nowhere to put it, it is possible for either the transmitter or the recipient to stop the traffic (and assume responsibility for restarting it).

The protocol is designed in such a manner that simple machines can have two fixed buffers, one for transmission and one for reception. On receiving a packet into the reception buffer, the command it contains may be processed independently, and the appropriate parts of the transmission buffer updated. During this operation, it is decided whether or not the updated packet should be transmitted or not. This technique may lead to some unnecessary repetition of non-essential elements, but this is defined to be harmless.

Each packet that is sent consists of: a command referring to the reception of data, a command referring to transmission of data, and the data itself. Alternatively it can consist of a control command referring to the transaction as a whole. In order that repetitions of packets can be identified as such, commands are given sequence numbers. For example RDYn indicates that packets with sequence number less than n have been successfully received and that the recipient is now ready to receive packet n. An example of a control command is CLOSE which winds up the transaction.

In the implementation of BSP at Cambridge the initial connection consists of sending a single packet in each direction. These packets are called the OPEN and OPENACK packets. An OPEN packet is sent by the originator of the connection and it contains the port number to be used in reply and a number of BSP parameters. These parameters indicate the maximum packet size that the originator is prepared to receive and the largest packet it will send. The OPENACK packet is directed to the station from which the OPEN came and contains a port number for the connection, a return code which will be zero if successful, and a number of BSP parameters. The parameters again indicate packet sizes that will be used.

THE CAMBRIDGE DISTRIBUTED COMPUTING SYSTEM

The Cambridge Ring is used to support a local area network within the Computer Laboratory at the University of Cambridge. The network consists of a number of committed computer systems such as the CAP computer, that are ordinary time sharing computers and have their own internal resource control mechanisms outside the control of the network. There are a number of shared servers on the ring including: a nameserver, a printing server, and a fileserver [9]. These servers provide a general service for other devices using the ring.
There is also a terminal concentrator that allows a number of terminals to be connected to the ring and is capable of making connections between terminals and computers.
In addition there is a processor bank which consists of a number of uncommitted microcomputers, most of which have no local peripherals except for a connection to the ring [10].
Part of the system is shown schematically in Figure7 .

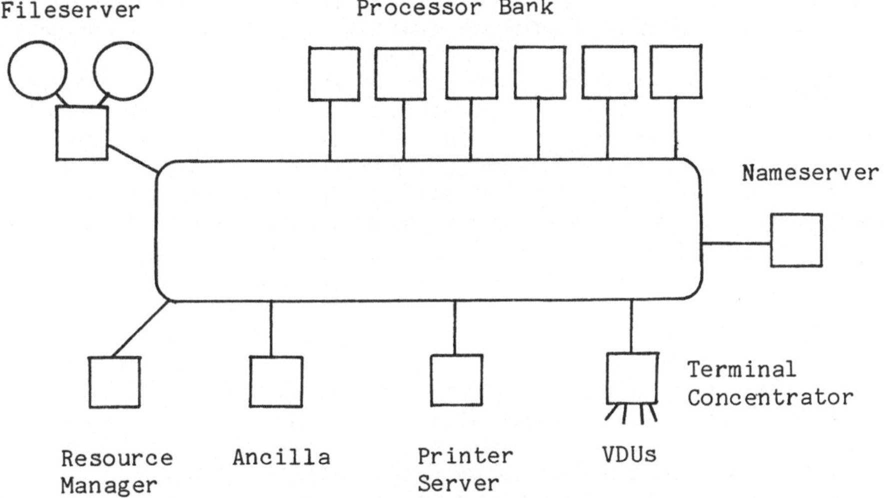

Figure 1 The Cambridge Distributed Computing System

The CDCS implements the concept of personal computers, not by the provision of a computer in every office, but by providing mechanisms whereby a user at a terminal within his office can either log onto one of the committed computer systems or, more importantly, gain control of a machine in the processor bank to run his choice of software exclusively [11,12,13].

The computers in the processor bank rely primarily on the servers on the network to provide them with essential services such as terminal connections, filing systems and printing etc. Each machine is provided equipped with a microprocessor-controlled intelligent ring interface. The interface provides the programmer with a channel to the ring and also facilitates remote bootstrapping and debugging.

To obtain a machine a client will indicate to the Resource Manager the type of machine required, the identity of the bootstrap image to load, and the total time for which the machine is required. The Resource Manager will check to see if a suitable machine is free and if so pass the names of the bootstrap image and the free machine to the Ancilla [14]. The Ancilla will then load the bootstrap image into the machine and start it executing. For every different type of machine in the processor bank there is an Ancilla service . This is because different machines have different loading properties and the object of the Ancilla services is to conceal these differences behind a uniform, simple interface.

Control is then passed back to the Resource Manager which connects the client to the appropriate machine. At the end of the session the user logs off and the machine is returned to the pool.

The CDCS processor bank and its associated services provides a powerful and convenient alternative to the personal computer model of distributed computing. The mechanisms support a framework for the allocation of computers in a manner similar to that of on-line sessions in a timesharing system. In addition the services required for the management of the processor bank can be duplicated to guard against hardware failures.

FUTURE DEVELOPMENTS

In the near future a chip version of the Ring will be available which will substantially reduce the cost of connecting devices to the Ring. In addition the chip version will allow the data part of the minipacket to be set to between 1 and 8 bytes. The minipacket has two extra control bits for general use.

Research is being carried out into the design and implementation of a 'Fast' Ring which will have a raw data rate of 100 Megabits/sec, high performance intelligent interfaces for connecting devices to the ring, and gateways for interconnecting Cambridge Rings.

Considerable research work is being carried out on the CDCS especially in the areas of protection, processor interaction, and the implementation of dynamic services. The CDCS also provides computing power for other researchers in the Laboratory.

A large scale project, called the Universe Project, to link up a number of Cambridge Rings on different sites throughout England by means of a satellite is well under way. One of the main aims of this project is to see how far the characteristics of a local area network can be preserved across the whole system.

REFERENCES

1] Kirstein, P.T.,"UCL Activities with the Cambridge Ring", Proceedings of IFIP WG 6.4- Local Area Networks Workshop, Zurich, August 1980.
2] Spratt, E.B.,"Developments of the Cambridge Ring at the University of Kent", Local Network and Distributed Office Systems, Online Publications, London May 1981, pp 503-518.
3] Sweetman, D, "A Distributed System Built with a Cambridge Ring", Local Network and Distributed Office Systems, Online Publications, London May 1981, pp 451-464.
4] Wilkes, M.V.,"Communication Using a Digital Ring", Proc. PACNET Conf. Sendai, Japan, 1975.
5] Wilkes, M.V. and Wheeler, D.J.,"The Cambridge Digital Communication Ring", Proceedings of The Local Area Communications Symposium, Boston, May 1979, pp 47-60.
6] Hopper, A. and Wheeler, D.J.,"Maintenance of Ring Communication Systems", IEEE Trans. on Communications, COMM-22, 1974.
7] Walker, R.,"Basic Block Protocol", Internal Documentation, Computer Laboratory, University of Cambridge.
8] Johnson, M.A.,"Ring Byte Stream Protocol Specification", Internal Documentation, Computer Laboratory, University of Cambridge, April 1980.
9] Dion, J.,"The Cambridge Fileserver", Operating Systems Review, Vol14 No 4, October 1980.
10] Wilkes, M.V. and Needham, R.M,"The Cambridge Model Distributed System", Operating Systems Review, Vol 14 No 1, January 1980, pp21-29.
11] Herbert, A.J.,"The User Interface to the Cambridge Model Distributed System", Proceedings of the 2nd International Conference on Distributed Computing Systems, Paris, April 1981, pp 503-508.
12] Needham, R.M.,"System Aspects of the Cambridge Ring", Proc. of Seventh Symposium on Operating Systems Principles, December 1979, pp 82-85.
13] Needham, R.M. and Herbert, A.J.,"The Cambridge Distributed Computing System", to be published by Addison Wesley, Spring 1982.
14] Shepherd, W.D.,"ANCILLA- A Server for the Cambridge Model Distributed System", to be published in Software Practice and Experience 1981.

DISTRIBUTED SYSTEM ACTIVITIES AT THE INSTITUTE OF CYBERNETICS OF THE UNIVERSITY OF MILAN

Marco Meli, GianPaolo Rossi

Institute of Cybernetics - University of Milan -
Via Viotti 5, 20133 MILAN (Italy)

ABSTRACT: a survey on LAN development and distributed processing at the University of Milan, with quoting to further projects is presented. The emphasis is on ERMES, a compilation-generated, low-cost LAN for distributed dedicated applications and on GCP, a derived version of Hoare's CSP, and their implementation as a distributed programming language for designing applications on the ERMES network.

Keywords: Local Area Network, Distributed System, GCP, ERMES.

The works here described are sponsored by: Honeywell Information Systems Italia, CP Project (GCP).

1. INTRODUCTION

The advent of the LSI technology has prepared the surroundings for a new approach in resolving the present computational problems. Nowadays microprocessors, together with high-bandwith, low error rate communication media now suggest to solve these problems in a distributed way instead of a centralised one. The Institute's main purpose has been to find out the means to use and how to solve some specific applications using a distributed system. Then it was decided to give particular importance to the environment for application's design instead of the flexibility at run time, in order to show that this kind of approach might lead to a low-cost, powerful system.In order to achieve this goal two projects were started:

K. G. Beauchamp (ed.), New Advances in Distributed Computer Systems, 239–243.

ERMES: a packet broadcast local area network developped as a joint venture with the Joint Research Center of the Commission of the European Communities.

GCP : an implementation of GCP as a concurrent application language by which the user can design its specific application taking full advantage of the distributed facilities.

2. ERMES LAN

The applications planned have been in general in the process control area: at JRC a deep interest was on fissile material control and measurement; now other applications are suggested, like monitoring of large industrial plants or embedded systems (eg weapon systems).

Even an application on office automation is in course of evaluation.

The ERMES LAN is a local area network based on a ring topology using optical fibers. The access method is of carrier-sense multiple access non persistent with collision detection (CSMA/CD). Our implementation provides that the requests from different stations are delayed by a random interval uniformly distributed between e given minimum and a variable value K:the value of K is determined by observing the number of collisions in a "window" of given width,so that such a value is (nearly) optimal with respect to the number of requests presently filed. More informations may be seen in |3|.

From a software point of view, ERMES may be seen as a world of distributed processes which communicate with each other by means of message exchange on virtual channels. The virtual channel is a bi-directional connection permanently established between two application tasks. All the necessary connections among the processes are established at compile time.

The Network Operating System implements a simple End-to-End connection among processes and the necessary flow control. It is based on a simple kernel of a multitasking Operating System called MICROS. The protocol implements at level two (NOS level) the station addressing and at level one the access protocol.

The application tasks (PLM/80 coded) are linked to the ERMES Library and to the configuration module which expands the operating system data structures.

More informations on ERMES may be found in |1| |2|.

3. GUARDED COMMUNICATING PROCESSES

A group of researchers at the University of Milan has developped, as part of their research on parallel programming, Guarded Communicating Processes (GCP), a modified version of Communicating Sequential Processes by Hoare. The main differences between CSP and GCP concern the communication primitives and the distributed termination convention. In CSP Hoare has proposed the synchronization of several sequential processes through the communciation among them only with input and output commands.
In GCP the communication primitives are input/output guarded processes which separate the condition that makes the communication possible from the action that performs the communication itself; and asymmetry among input guards, output guards, boolean condition are then avoided.

The other key difference from CSP is the reason of process termination: a sŷntactic distinction between Exoprocesses, (i.e. processes which terminated because of the termination of other processes communicating with them) and Endoprocesses (i.e. processes which terminate by reaching their final state) is stated, and a convention which enables the termination of process to cause the termination of Exoprocesses which communicate with him (distributed termination convention) is defined.

A more rigourous description of Hoare's CSP may be found in |6| about distributed termination convention in |7| |8|, on GCP in |4| |5|.

A restricted version of GCP has been implemented as a distributed programming language for ERMES; it consists of three objects: a preprocessor, a distributor and a new/higher level of the operating system.

The preprocessor generates the user sequential process modules and the configuration modules which will expand the system data strutures;

the distributor is an interactive station handler which allows the programmer to distribute the processes he wishes for each network station;

the new level of the operating system is the run time support of the application processes and is composed of four procedures, two which verify the availability of the two processes to communicate ; and the other two which perform the actual communication.

The GCP programming environment on ERMES widely uses the programming tools available on INTEL MDS 85 under ISI II operating system and the on purpose developped Distributed Dedicated Application Handler or DDAH, a Pascal program made of two sequential tasks, the preprocessor and the distributor.

The preprocessor receives as input a symbolic GCP coded application and produces a series of files containing the PLM/80 translation,and the virtual channel descriptors and other system tables.

The distributor allows the user to distribute the processes along the network by creating for each station to files, one containing the PLM/80 source of the application tasks and the other containing all the macroprototypes for the various data structures and the macrocalls to the system data structures.

The output source files are then compiled to obtain two object files to link with the operating system library.

4. CONCLUDING REMARKS AND FUTURE DEVELOPMENT

This first version of ERMES and GCP programming environment is characterized by many choices established at compile time. This choice, which is a very common choice in dedicated applications like for example process control, is in sake of semplicity of design and full testing. The following extensions will be on the implementation of an Exception Handler for interrupts and restart; the availability of more powerfull processes by means of dynamic virtual channel management, the implementation of whole GCP with nesting of concurrency and new form of communications (e.g.broadcasting) among processes is forecast.

About ERMES, further developments regard the full exploitment of the existing broadcast facility; also a down-line loader is in course of design.

5. BIBLIOGRAPHY

|1| Coppo N, De Grandi G., Rossi G.P. - The Operating System Architecture of ERMES - a Microcomputer Local Network, Datacom. Geneve, 17-19 june, 1980

|2| De Grandi G., Rossi G.P., - ERMES a Distributed System on Local Area - AICA'80

|3| Rossi G.P., Torelli M. - The Channel Access Protocol in ERMES Local Networks - Istituto di Cibernetica - Internal report - June 1981

|4| De Cindio F., De Michelis G., Simone C. - Guarded Communicating Processes - Proc. Fifth Honeywell International Software Conference, Minneapolis, 1981

|5| Bergamini G., De Cindio F., De Michelis G., Rossi G.P., Simone C. - Guarded Communicating Process as Distributed Programming Language on a Local Area - Submitted to International Symposium on Local Computer Networks - 19-21 April 1982 - Firenze

|6| Hoare C.A.R., - Communicating Sequential Processes - CACM 21,8 Aug. 1978, 666-677

|7| Francez N., - Distributed Termination - TOPLAS 2,1, Jan. 80 42-55

|8| Francez N., Hoare C.A.R., Lehman J.D., De Roever W.P. - Semantics on non-deterministic, concurrency and Communication - JCSS, 19, 3, Dec. 1979, 290-308.

AN INTRODUCTION TO A CALCULUS OF COMMUNICATING SYSTEMS

Matthew C.B. Hennessy

University of Edinburgh

ABSTRACT

A model for distributed computing systems, called CCS, is explained. It is based on the primitive notion of communication and comprises of a simple language for defining communicating agents and a theory for analysing their behaviour. The language is defined using a small set of combinators for defining new agents from existing agents. The theory is based on a set of axioms which can be used to prove the behavioural equivalence of agents.

1. In this paper we give an introduction to a model of distributed computing systems, called CCS (Calculus of Communicating Systems). The reader is referred to the references for more detailed expositions.

There are two important aspects of any model of distributed systems:

1. Description: it should be capable of describing a wide range of distributed systems and their associated behaviours.

2. Analysis: it should be possible to use the model to analyse, prove theorems about, and more generally predict the behaviour of the systems which can be modelled in it.

Thus it is not sufficient to have a modelling language. There must also be an associated theory. These present somewhat

K. G. Beauchamp (ed.), New Advances in Distributed Computer Systems, 245–262.

conflicting demands on the model: the former requires an *expressive* language whereas experience shows that to obtain a tractable theory the model should be as simple as possible.

In CCS simplicity is obtained by basing the model on a small set of primitive notions. One such notion is *communication*. There is only one kind of agent in the model, namely communicating agents. Descriptive power is obtained by defining a small set of combinators which can be used to build new agents from existing ones. The combinators are chosen not only for their expressiveness but also so that the relationship between the behaviour of the new agent and the behaviour of the agents from which it was built is mathematically tractable.

These combinators provide a language for computing agents. We can define an equivalence relation, $\sim$, over those agents, $P \sim Q$ meaning that P and Q have the same behaviour. Exactly what the "behaviour" of an agent should be is not very clear but we define $\sim$ in such a way that, (roughly speaking) whenever $P \sim Q$ then no computing agent under any circumstances would ever be able to detect any difference between P and Q. The theory then provides a method for *proving* the equivalence of computing agents.

This paper concentrates on explaining the basis of CCS. We refer the reader to [1],[7] for nontrivial examples of the use of CCS in modelling and analysing various systems.

2. It may be worthwhile to examine in detail an example of the type of model we have in mind: regular languages as a model for finite automata. This is a very successful model. It can describe all finite automata and it has associated with it a rich theory for analysing and comparing the behaviour of these machines. For example consider the two definitions

$$M_1 \Leftarrow a.(b.NIL + c.M_1) \qquad (1)$$
$$M_2 \Leftarrow a.b.NIL + a.c.M_2 \qquad (2)$$

They have associated with them the two finite automata

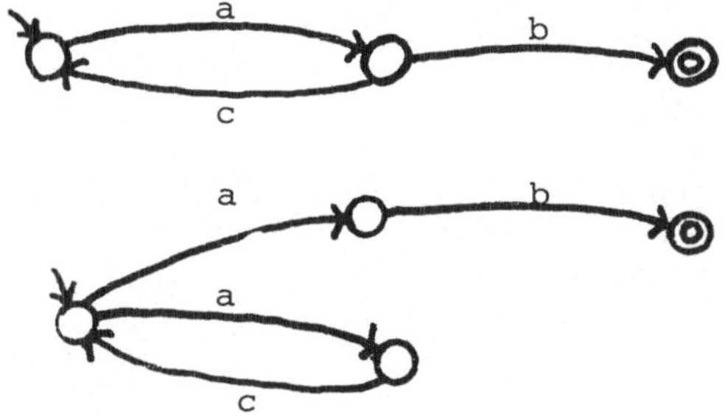

The initial and finite states are denoted as , , respectively. Each of these machines, considered as outputting devices, can output any sequence of symbols of the form acacac.... acab. We say that the language generated by the devices is (ac)*ab. If we define the behaviour of a device to be the language it generates (as is usual) then these two devices have the same behaviour and are therefore considered equivalent. Moreover there are well-known methods for manipulating the definitions (1), (2), which enable us to prove that they are equivalent.

Rather than considering a,b,c as symbols to be printed on an output tape they can equally well be considered as operations or actions. From this point of view M_1 defines a device, or agent which can perform the operation a, then either perform the operation b and halt or perform the operation c and revert back to its original state. In other words by a minor shift of interpretation regular expressions can be used as a language for defining rather simple computing agents. In this language there are four "combinators" or methods for constructing new agents from existing agents. We now define these precisely, together with a definition of how the behaviour of the new agent is determined by the behaviour of the agents from which it is constructed. This is accomplished by defining a relation $\xrightarrow{a}$ over agents for each operation a. Informally $P \xrightarrow{a} Q$ means that P can perform the operation a and thereby be transformed into Q. Thus for example M_1 above can give rise to the sequence

$$M_1 \xrightarrow{a} (b.NIL + c.M_1) \xrightarrow{c} M_1 \xrightarrow{a} (b.NIL + c.M_1) \xrightarrow{b} NIL$$

Thus it can be seen that these relations determine the behaviour of agents. In the sequel we assume some set of actions A.

Agent Construction 1: ACTION

If P is an agent and a is an action, then a.P is an agent.
Intuition: The new agent can perform the action a and then act as P.

Rule of Action: $a.P \xrightarrow{a} P$

Agent Construction 2: SUMMATION

If P and Q are agents then P + Q is an agent.
Intuition: The new agent P + Q can act as either the agent P or the agent Q. Whether it acts as P or as Q is determined non-deterministically.

Rules of Action:

If $P \xrightarrow{a} P'$ then $P + Q \xrightarrow{a} P'$
If $Q \xrightarrow{a} Q'$ then $P + Q \xrightarrow{a} Q'$

Agent Construction 3: INACTION

NIL is an agent.
Intuition: NIL has no possible actions.

Rules of Action: There are none.

Agent Construction 4: RECURSION

If $P_1, P_2, \ldots$ are agent identifiers and $D_1, D_2, \ldots$ are agents in which these identifiers appear then $P_1, P_2, \ldots$ are agents defined by

$P_1 \Leftarrow D_1$
$P_2 \Leftarrow D_2$
....
....

Intuition: The agents P_i are defined mutually recursively in terms of themselves. The possible actions of P_i are exactly the actions of D_i.

Rule of Action:

If $D_i \xrightarrow{a} Q$ then $P_i \xrightarrow{a} Q$

With these four combinators we can define every regular expression and the above definitions tell us how they can be interpreted as computing agents. The extensive theory of regular languages [6], can now be used to analyse and predict the behaviour of these agents.

Unfortunately this language for agents is virtually useless for our purposes. The agents which are definable exhibit no behaviour typical of distributed systems. They are in fact simple sequential machines. To model distributed systems additional combinators are required.

3. We introduce a new combinator with which we can combine two agents P,Q, to obtain the new agent P|Q. Intuitively P|Q is an agent which has two sub-agents P,Q, and these sub-agents can intercommunicate. Up to now agents could only perform actions but we can agree to define communication as the simultaneous occurrence of certain specified complementary actions. Usually the idea of communication involves one communicand passing a value to another. This value passing will be taken up in a later section but for the moment we deal only with synchronisation i.e. communication where no value is passed. This is often referred to in the literature as handshake. To define the complementary actions which will be used to model communication we add some structure to A, the set of actions. For any set B of actions let $\bar{B} = \{\bar{b} \mid b \in B\}$. The action $\bar{b}$ is said to be the complement of b. For convenience let $\bar{\bar{b}} = b$. From now on we assume that $A = L \cup \bar{L} \cup \{\tau\}$, for some set of actions L and some distinguished action τ. Then the definition of P|Q will include:

$$\text{if } P \xrightarrow{a} P',\ Q \xrightarrow{\bar{a}} Q',\ \text{then } P|Q \xrightarrow{\tau} P'|Q'$$

So the action τ is somewhat special since it denotes an internal action (communication). In the example above the communication between the sub-agents P,Q will be denoted as a τ-action of the composite agent P|Q. Because of this τ will enjoy properties peculiar to itself and therefore it is better to distinguish it from the other operations. In the sequel we let $OA = L \cup \bar{L}$. We now give the complete definition of the new combinator.

Agent Construction 5: Composition

If P and Q are agents, then P|Q is an agent.
Intuition: P and Q can communicate by synchronising their complementary actions.

Rules of Action

$$\text{If } P \xrightarrow{a} P' \text{ then } P|Q \xrightarrow{a} P'|Q$$
$$\text{If } Q \xrightarrow{a} Q' \text{ then } P|Q \xrightarrow{a} P|Q'$$
$$\text{If } P \xrightarrow{a} P',\ Q \xrightarrow{\bar{a}} Q',\ \text{then } P|Q \xrightarrow{\tau} P'|Q'$$

Example

If P_1, P_2 are defined by

$$P_1 \Leftarrow a.b.P_1$$
$$P_2 \Leftarrow c.\bar{b}.P_2$$

then the agent $P_1|P_2$ has the possible actions

$$P_1|P_2 \overset{a}{\rightarrow} b.P_1|P_2 \overset{c}{\rightarrow} b.P_1|\bar{b}.P_2 \overset{\tau}{\rightarrow} P_1|P_2 \ldots\ldots$$

and

$$P_1|P_2 \overset{c}{\rightarrow} P_1|\bar{b}.P_2 \overset{a}{\rightarrow} b.P_1|\bar{b}.P_2 \overset{\tau}{\rightarrow} P_1|P_2 \ldots\ldots$$

□

Having admitted the notion of synchronisation the intuitive meaning of the arrows $\overset{a}{\rightarrow}$ can be clarified. If a ε OA and $P \overset{a}{\rightarrow} P$ then an agent external to P can observe P performing a (and thereby being transformed into P'). A communication between P and Q therefore comes about by P observing Q and Q observing P. In this sense communication is <u>mutual observation</u>. For example the internal communication in the agent P|Q above is brought about by the mutual observation of its sub-agents P,Q. It should be emphasised that observation is not passive. One observes by communicating.

There are two more combinators used in CCS and we now proceed to explain these. Consider P_1 and P_2 defined in the previous example. The complementary actions $b,\bar{b}$ are used to synchronise the sequencing of the actions a,c. However this is not effective in all communicating environments. For example consider the agent $(P_1|P_2)|P_3$, i.e. an agent with three sub-agents P_1,P_2 and P_3, where P_3 is defined by

$$P_3 \Leftarrow b.P_3 + \bar{b}.P_3$$

Then this new agent has as a possible sequence of actions $\overset{a}{\rightarrow} \overset{\tau}{\rightarrow} \overset{a}{\rightarrow} \ldots\ldots$. In other words the presence of the agent P_3 disturbs the synchronisation between P_1 and P_2. An obvious solution to this problem is to introduce a new combinator which allows P_1 and P_2 to declare the actions $b,\bar{b}$ private to themselves.

<u>Agent Construction 6: Restriction</u>

If P is an agent and a ε L, then P/a is an agent.
<u>Intuition</u>: P/a has all of P's actions except its a and $\bar{a}$-actions.

<u>Rule of Action</u>

If $P \overset{b}{\rightarrow} P'$ and $b \notin \{a,\bar{a}\}$, then $P/a \overset{b}{\rightarrow} P'/a$.

Referring back to the example we can retain the required synchronisation between the actions a,c, by using the term $((P_1|P_2)/b)|P_3$. Now P_3 cannot interfere with the interaction between P_1 and P_2. This restriction combinator adds considerable power to the language in that it gives us the ability to internalise selected behaviour of agents.

Agent Construction 7: Relabelling

If P is an agent and S is a 1-1 mapping from A to A which respects complements and such that $S(\tau) = \tau$ then P[S] is an agent. Intuition: Each a action of P becomes an S(a) action of P[S].

Rule of Action: If $P \xrightarrow{a} P'$ then $P[S] \xrightarrow{S(a)} P[S]$.

This final combinator is very convenient for defining agents which have basically the same behaviour except that the actions of one are a renaming of the actions of another.

4. In this section we consider a simple example which may help to clarify the language. Consider the problem of ensuring a single reader and single writer access a memory in a disciplined manner. The reader and writer can be defined respectively by

$$R \Leftarrow r_1.r_2.R$$
$$W \Leftarrow w_1.w_2.W$$

The specification of the required composite system can be defined by

$$SPEC \Leftarrow \tau.r_1.r_2.SPEC + \tau.w_1.w_2.SPEC$$

To obtain the behaviour required by the specification from the reader and writer we impose a semaphore. A semaphore can be defined by

$$SEM \Leftarrow p.v.SEM$$

We now ensure that both the reader and the writer perform a p action before using the memory and a v action when they are finished with it. These new agents are defined by

$$R' \Leftarrow \bar{p}.r_1.r_2.\bar{v}.R'$$
$$W' \Leftarrow \bar{p}.w_1.w_2.\bar{v}.W'$$

Then the composite system,

$$((R'|W')|SEM)/p/v$$

will satisfy the specification SPEC. The restriction with respect to p,v ensures that the synchronisation is internal to the system. We are now faced with the problem of in what sense is the composite system equivalent to the specification SPEC. To explain the solution proposed in CCS we refer once more to finite automata. Two such devices are considered equivalent if they generate the same language and for this reason the two machines

M_1 and M_2, defined in §2, are equivalent. However in the more demanding world of communicating agents there is quite a serious difference between them. This is because in certain communicating environments M_2 causes a deadlock whereas M_1 does not. Consider the agent

$$P \Leftarrow \bar{a}.\bar{c}.P$$

Then $M_2|P$ can deadlock via the internal transition

$$M_2|P \xrightarrow{\tau} b.NIL|\bar{c}.P$$

This latter term is deadlocked since no communication is possible. However in the same environment M_1 does not cause a deadlock: $M_1|P$ can continue communicating indefinitely. Since deadlock is such an important property of systems we must conclude that M_1 and M_2 are behaviourally different even though they generate the same language.

One possible solution is to consider the <u>tree</u> of possible operations that an agent can perform. Then M_1 and M_2 will be different since the respective trees are

Referring once more to the reader and writer the specification SPEC gives rise to the tree:

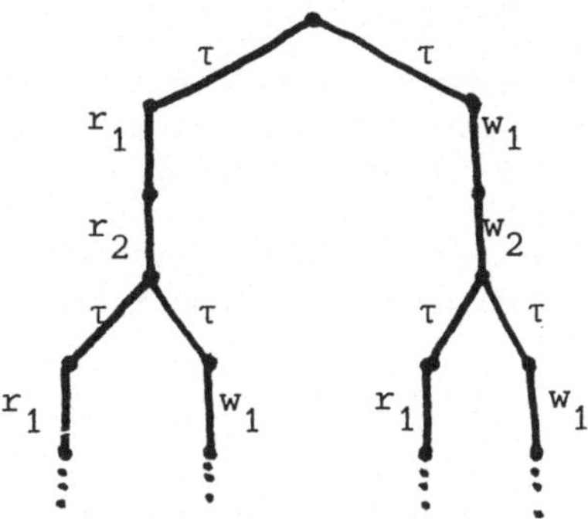

The composite system gives rise to the tree:

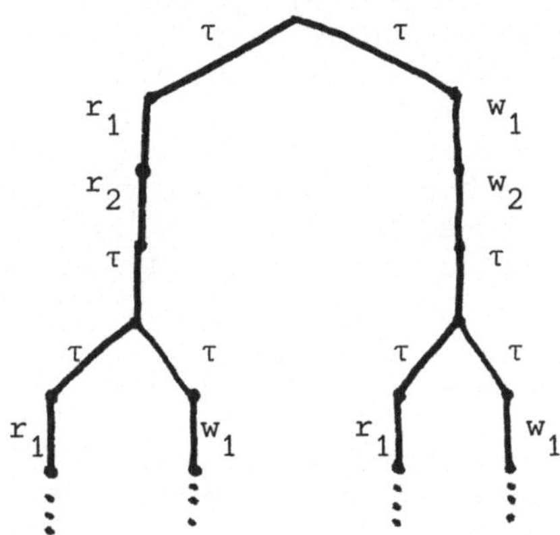

These two trees are quite similar, the only difference being that in the latter there is a profusion of τ actions. These however represent <u>internal</u> communications and therefore should be unobservable by external agents. This is essentially the approach taken in CCS but it is more convenient to present it in a slightly different fashion.

Define a new relation over agents, $\overset{a}{\Rightarrow}$, a ε OA, by:

$P \overset{a}{\Rightarrow} Q$ if $P \overset{\tau}{\rightarrow} \ldots P_n \overset{a}{\rightarrow} Q_1 \overset{\tau}{\rightarrow} \ldots \overset{\tau}{\rightarrow} Q$.

In the definition of observational equivalence we use these new relations instead of the relations $\overset{a}{\Rightarrow}$. The former capture the notion that τ-transitions are unobservable: if an observer demands an a action of P then P can compute internally both before and after performing a. The observer will be oblivious to these internal communications.
Let ~ denote the equivalence we are attempting to define, observational equivalence. Then we would expect ~ to satisfy the following property:

If $P \sim Q$ then for every $a \in OA$

i) $P \overset{a}{\Rightarrow} P'$ implies $Q \overset{a}{\Rightarrow} Q'$ for some Q' such that $P' \sim Q'$

ii) $Q \overset{a}{\Rightarrow} Q'$ implies $P \overset{a}{\Rightarrow} P'$ for some P' such that $P' \sim Q'$.

If we look upon agents as being in states then this means that whenever P changes state then Q can make a similar transition into an equivalent state, i.e. equivalence of states is preserved by the transitions.
There are many relations which satisfy this property. But we can define ~ to be the maximal relation which satisfies it. To see that such properties have maximal relations which satisfy them we refer the reader to [1],[3]. This equivalence has many nice properties but we wish to impose more stringent requirements on agents: we say that two agents P,Q are observationally congruent if either one can replace the other as a sub-agent without affecting the overall behaviour of the agent. Mathematically we write

$P \sim c\ Q$ if for every context $C[\ ]$, $C[P] \sim C[Q]$.

A context is merely an agent with a slot [ ], into which a sub-agent can be placed. For example in distinguishing M_1 from M_2 we used the context $[\ ]|P$.

5. In this section we give some properties of observational congruence. For example we can show that

$$(P_1|P_2)|P_3 \sim c\ P_1|(P_2|P_3)$$

This is quite an interesting property as it enables us to decompose agents in various different ways. Indeed if this property did not hold then the analytical power of the theory would be very limited. We also have that

$$P + \tau.P \sim c\ \tau.P$$

This can be represented diagrammatically as

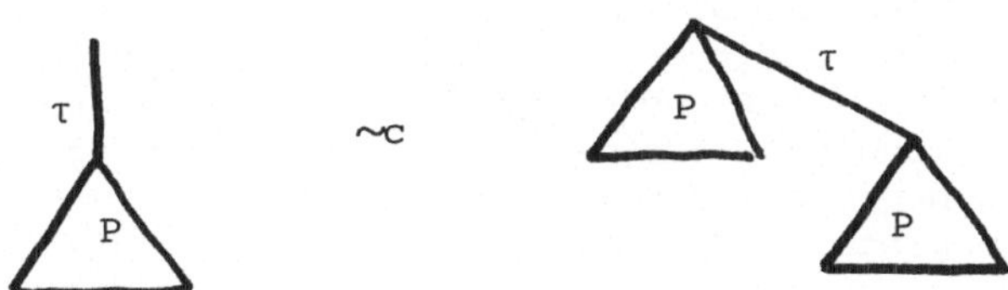

The reader may try his hand at designing a communicating observer which can distinguish between these two agents. The reason why no such observer can exist can be partially explained by considering the possible observable actions that each agent can make. If $P \overset{a}{\Rightarrow} P'$ we say that P' is an <u>a-derivative</u> of P. Then both τ.P and P + τ.P have exactly the same sets of derivatives no matter what action a we consider. Since this is the only information that an observer can discover no difference can be found. In a similar vein we have

and

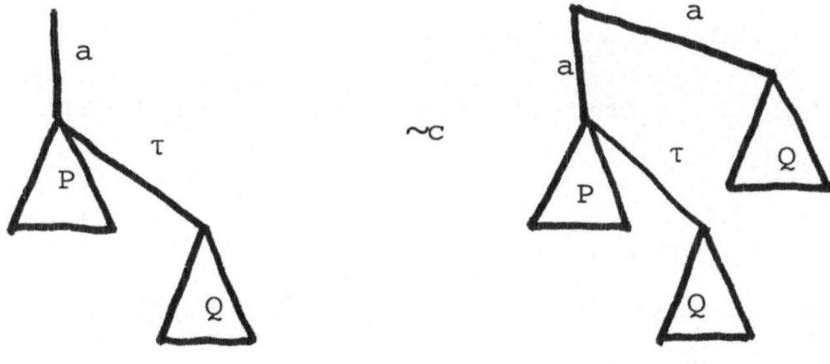

These identities form the basis of an axiomatic system for proving observational congruence. These axioms are presented in Table 1. In fact if P and Q are finite, i.e. do not involve recursive definitions then they are observationally congruent if and only if we can prove P = Q from these axioms. The axioms are said to be complete. The last axiom uses the notation $\sum_{i \in I} X_i$, which is merely an abbreviation for $X_1 + (X_2 + \ldots + X_n)\ldots)$. In deriving results it often happens that we apply the same sequence of axioms over and over again. It is useful to collect such sequences into a theorem which can be applied in one step. In some sense such theorems can be viewed as "macros". We give one example, the expansion theorem. A guarded sum is an agent of the form $\Sigma_I a_i.P_i$, each $a_i.P_i$ being a summand. We also use the notation P/B to denote $P/l_1/../l_n$, where $B = \{l_1,..l_n\}$.

Expansion Theorem

Let P denote $(P_1|\ldots|P_m)/B$, where each P_i is a guarded sum. Then $P = \Sigma\{a_i.((P_1|\ldots|P_i'|\ldots|P_m)/B \mid a_i.P_i'$ is a summand of $P_i, a_i, \bar{a}_i \notin B\}$

$+ \Sigma\{\tau.((P_1|\ldots|P_i'|\ldots|P_j'|..|P_n)/B \mid a.P_i'$ is a summand of $P_i, \bar{a}.P_j'$ is a summand of $P_j, j \neq i\}$

This can easily be derived by repeated application of the axiom (A14), using the associativity and commutativity of +, as stated in (A1),(A2).

We now give an instance of the use of the axioms and this expansion theorem, by proving that our solution to the reader/writer problem satisfies the specification. Let MM denote the composite system ((R'|W')|SEM)/p/v. We show that SPEC = MM. The definition of SPEC is of such a form (called bounded recursion in [1]) that there is a unique agent, with respect to ~c, which satisfies it. Therefore to show that MM = SPEC it is sufficient to show that MM satisfies the definition of SPEC, i.e.

$$MM = \tau.r_1.r_2.MM + \tau.w_1.w_2.MM$$

We do this using the axioms of table 1 and the expansion theorem.

$$MM = (\bar{p}.r_1.r_2.\bar{v}.R' \mid \bar{p}.w_1.w_2.\bar{v}.W' \mid p.v.SEM)/p,v$$

using the recursion definitions of R',V',SEM

$$= \tau.((r_1.r_2.\bar{v}.R' \mid \bar{p}.w_1.w_2.\bar{v}.W' \mid v.SEM)/p,v)$$
$$+ \tau.((\bar{p}.r_1.r_2.\bar{v}.R' \mid w_1.w_2.\bar{v}.W' \mid v.SEM)/p,v)$$

using the expansion theorem

A1 $X + (Y + Z) = (X + Y) + Z$

A2 $X + Y = Y + X$

A3 $X + X = X$

A4 $X + NIL = X$

A5 $X + \tau.X = \tau.X$

A6 $a.\tau.X = a.X$

A7 $a.(X + \tau.Y) = a.(X + \tau.Y) + a.Y$

A8 $NIL/a = NIL$

A9 $(X + Y)/a = X/a + Y/a$

A10 $b.X/a = NIL$ if a is b or $\bar{b}$

$= b.(X/a)$ otherwise

A11 $NIL[S] = NIL$

A12 $(X + Y)[S] = X[S] + Y[S]$

A13 $(a.X)[S] = S(a).(X[S])$

A14 Let X,Y denote $\sum_I a_i.X_i$, $\sum_J b_j.Y_j$ respectively.

$$X|Y = \sum_I a_i.(X_i|Y) + \sum_J b_j.(X|Y_j) + \sum_{\bar{a}_i = b_j} \tau.(X_i|Y_j)$$

TABLE 1

$= \tau.r_1.r_2.((\bar{v}.R' \mid \bar{p}.w_1.w_2.\bar{v}.W' \mid v.SEM)/p,v)$
$+ \tau.w_1.w_2.((\bar{p}.r_1.r_2.\bar{v}.R' \mid \bar{v}.W' \mid v.SEM)/p,v)$

using the expansion theorem four times

$= \tau.r_1.r_2.\tau.((R' \mid \bar{p}.w_1.w_2.\bar{v}.W' \mid SEM)/p,v)$
$+ \tau.w_1.w_2.\tau((\bar{p}.r_1.r_2.\bar{v}.R' \mid W' \mid SEM)/p,v)$

using the expansion theorem

$= \tau.r_1.r_2.\tau.MM + \tau.w_1.w_2.\tau.MM$

using the recursive definitions of R',W'

$= \tau.r_1.r_2.MM + \tau.w_1.w_2.MM$

using A6.

It follows that MM = SPEC.

It should be emphasised that each use of the expansion theorem can be replaced by a suitable number of applications of the axiom (A14) and therefore the proof relies only on the axioms of table 1. Such proofs are tedious but are quite mechanical and some form of machine-assistance could easily be implemented. An alternate specification for the reader/writer problem is given by

$$SPEC' \Leftarrow r_1.r_2.SPEC' + w_1.w_2.SPEC'$$

This is different than SPEC. The latter states that a reader can be serviced, a writer can be serviced and also that the memory can decide of its own volition which to service. This last property is missing from SPEC' but it appears in SPEC in the form of the possible internal actions τ. Since there is a difference between SPEC and SPEC' then the reader is invited to find a communicating context which differentiates between them.

We conclude this section with some remarks on τ-actions. If they are unobservable why should they appear in the theory at all? Even though they are unobservable they do indirectly affect the observable behaviour of agents. For example in agent $(a.NIL + b.NIL \mid \bar{a}.NIL)/a$ the purely internal communication to a deadlocked state, via the actions $a,\bar{a}$, can prevent an observer from demanding the b action. Without some notion in the theory for internal communication this possibility of deadlock cannot be represented.

6. In this final section we turn our attention briefly to the type of communicating agents which pass values between each other. We arbitrarily decide that a.P is an agent which <u>receives</u> values whereas $\bar{a}$.P <u>sends</u> values. It is also convenient to change

somewhat the informal interpretations of the actions. Since there are values flowing between agents we can think of every a ε L as a <u>channel</u>, $\overline{a}$.P being now an agent which sends a value along the channel a and a.P one which receives a value from this channel. The exact nature of the values involved and the operations on them will not be specified. Instead a set of expressions and variables will be used informally and the reader will be expected to make sense of them as we go along. To denote the values involved in communications we change the syntax slightly. The agent construction 1 is replaced by a new three part agent construction.

<u>Agent Construction 1a: INPUT</u>

If P is an agent, a ε L, x a variable, then ax.P is an agent.
<u>Intuition</u>: The agent ax.P can input a value along the channel a and bind it to x in P.

<u>Rule of Action</u>: ax.P $\xrightarrow{av}$ P[v/x]

<u>Agent Construction 1b: OUTPUT</u>

If P is an agent, a ε L and e an expression, then $\overline{a}$e.P is an agent.
<u>Intuition</u>: The agent $\overline{a}$e.P can output the value of e along the channel a and continue as P.

<u>Rules of Action</u>: $\overline{a}$e.P $\xrightarrow{\overline{a}v}$ P if e has no free variables and v is the value of e.

<u>Agent Construction 1c: CONDITIONAL</u>

If P,P' are agents, b a boolean expression, then

IF b THEN P ELSE P' is an agent.

<u>Intuition</u>: The agent IF b THEN P ELSE P' acts either as P or as P', depending on whether b is true or false.

<u>Rule of Action</u>:

i) P $\xrightarrow{av}$ Q implies IF b THEN P ELSE P' $\xrightarrow{av}$ Q provided b has no free variables and the value of b is true.

ii) P' $\xrightarrow{av}$ Q implies IF b THEN P ELSE P' $\xrightarrow{av}$ Q provided b has no free variables and the value of b is false.

A simple example of an agent is

SW <= ax.IF even(x) THEN $\overline{b}$x.SW ELSE $\overline{c}$x.SW

SW is an agent which receives a value along the channel a and either sends it along channels b or c, depending on whether or not it is even.

Note that the transitions are now of the form $\xrightarrow{av}$ for some value v. Complementary transitions are simply pairs of the form $\xrightarrow{av}$, $\xrightarrow{\bar{a}v}$, for some v. With this notation the remaining combinators still make sense. Instead of dwelling on the theory of this new set of combinators we give an example of their use in designing protocols. These examples are taken from [5].

A uni-directional channel for transmitting values can be specified by an agent such as

$$C \Leftarrow inCx.\overline{outC}x.C$$

A medium which corrupts the value transmitted can be modelled as

$$M_1 \Leftarrow inMx.\overline{outM}\ corrupt(x).M_1$$

where corrupt is some function over values which models the degradation of the value during transmission. If we are lucky enough to be able to correct the corrupted values, using a function called correction, then it is easy to design a protocol to use with M_1. This is in the form of a sender and a receiver:

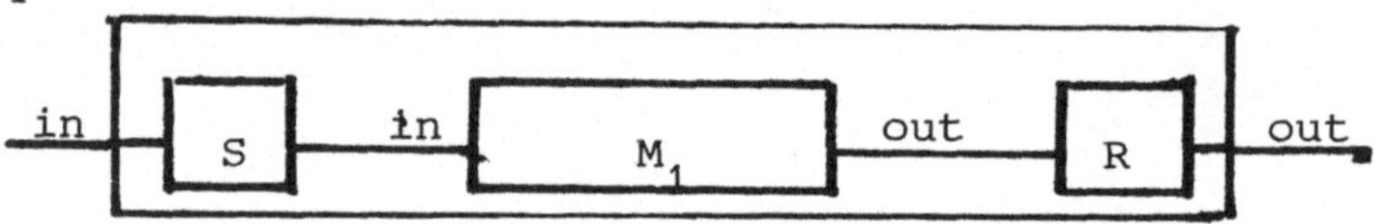

These are defined by

$$S_1 \Leftarrow inCx.\overline{inM}x.S_1$$

$$R_1 \Leftarrow outMy.\overline{outC}correction(y).R_1$$

Provided that correction(corrupt(x)) = x we can deduce that

$$C \sim c\ (S_1|M_1|R_1)/inM,outM$$

In fact the axioms of Table 1 can be modified to take into consideration value-passing and then we can derive

$$C = (S_1|M_1|R_1)/inM,outM$$

using these modified axioms and a form of induction. A medium which can do irreparable damage to the transmitted value may be defined by

$$M_2 \Leftarrow \text{inMx}.(\overline{\text{outMx}}.M_2 + \overline{\text{outM}}.\text{error}.M_2)$$

where 'error' is some special value.
To use such a medium the receiver must be able to send an acknowledgement to the sender saying whether the value received is irreparable or not. To transmit this acknowledgement we use a medium similar to M_2, defined by.

$$M_2' = M_2[S]$$

where S maps inM,outM to inM',outM', respectively and leaves other actions untouched. The protocol can be visualised as

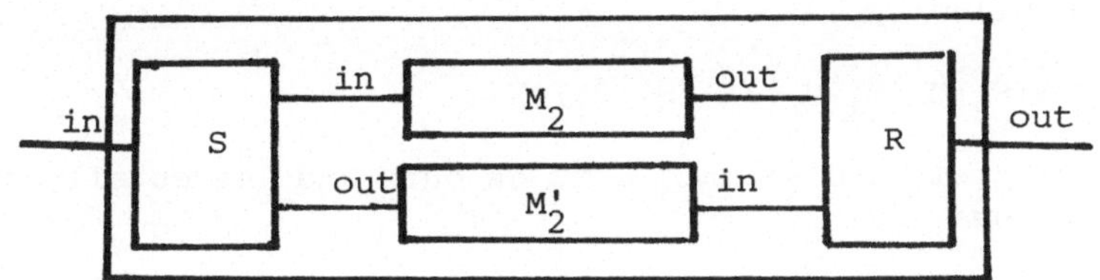

The new sender and receivers are defined by

$$S_2 \Leftarrow \text{inCx}.S'$$

$$S' \Leftarrow \overline{\text{inMx}}.\text{outM'y}.\ \text{IF } y = \text{error THEN } S' \text{ ELSE } S_2$$

$$R_2 \Leftarrow \text{outMx}.R'$$

$$R' \Leftarrow \overline{\text{inM'}}\text{x}.\ \text{IF } x = \text{error THEN } R' \text{ ELSE } \overline{\text{outCx}}.R_2$$

A new transmission medium can now be defined by

$$(S_2 | M_2 | M_2' | R_2)/X$$

where X = {inM,outM,inM',outM'}

This simple protocol is however not sufficient to give us a channel with the same characteristics as C. In fact

$$C \not\approx (S_2 | M_2 | M_2' | R_2)/X$$

For a description of a more complicated protocol which uses this medium the reader is referred to [5].

Finally we give one more example to show how the calculus can model hierarchical protocols. Suppose we have a medium which can both damage values irreparably and corrupt them. These can be modelled by

$$M_3 \Leftarrow inMx.(\overline{outM}\ corrupt(x).M_3 + \overline{outM}\ error.M_3)$$

A new protocol, built from the previous two, visualised as

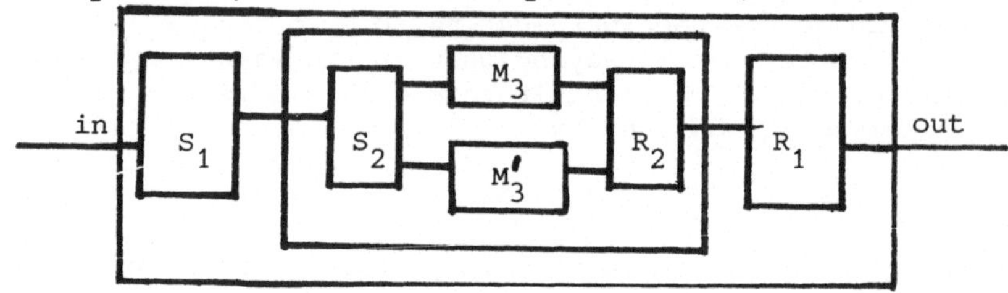

can be modelled by

$$(S_1 | ((S_2 | M_3 | M_3' | R_2)/X)[S] | R_1)/X$$

where S relabels inC as inM, outC as outM and leaves all other actions alone.

ACKNOWLEDGEMENTS

The author is supported by the S.R.C. under grant no. GR/A/75125.

REFERENCES

LNCS n stands for Lecture Notes in Computer Science, Volume n, Springer-Verlag.

[1] Milner, R., *A Calculus of Communicating Systems*, LNCS 92, 1980.

[2] Milner, R., *An Introduction to a Calculus of Communicating Systems*, IUCC Conference, Exeter University, 1980.

[3] Hennessy, M., Milner, R., *On Observing Nondeterminism and Concurrency*, LNCS 85, 1980.

[4] Hennessy, M., Plotkin, G., *A Term Model for CCS*, LNCS 88, 1980.

[5] Zhou Chao Chen, Hoare, C.A.R., *Partial Correctness of Communication Protocols*, Oxford University, 1981.

[6] Hopcroft, J., Ullman, J., *Formal Languages and their Relation to Automata*, Addison Wesley, 1969.

[7] Milne, G., *A Mathematical Model for Concurrent Computation*, Ph.D. thesis, Computer Science Dept., University of Edinburgh, 1978.

MODELLING OF DISTRIBUTED COMPUTING SYSTEMS

P.J.B. King

Computing Laboratory,
University of Newcastle upon Tyne,
Newcastle upon Tyne, NE1 7RU, U.K.

We outline several recent studies of distributed computing systems by probabilistic models. The studies are used to attempt to answer two questions. What is the effect of different scheduling strategies on the performance of distributed systems, and what are the relative advantages and disadvantages of using a single processor or multiprocessor to service a given workload. The possibility of a conservation law for multiprocessors, like that for single processors, is discussed.

This paper surveys some studies of distributed computing systems by probabilistic modelling methods. Two broad categories of problem are considered: (a) the effect of different scheduling strategies on the performance of distributed systems, and (b) the relative advantages and disadvantages of using single-and multi-processor systems to service a given workload.

First, a model for evaluating the trade off between distributed and centralised computing is examined, taking into account the communication costs. The model is applied to the problem of optimising the workload routing in an existing star-like network. The optimal allocation of computer power at the various sites in the network is also considered. Both these problems can be treated assuming that sites co-operate to optimise the overall performance of the system or assuming that each site acts selfishly and tries to optimise the performance of locally originated jobs.

The second problem area concerns multi-processor systems in which the demand consists of different job classes. One has to decide

K. G. Beauchamp (ed.), New Advances in Distributed Computer Systems, 263–277.

how to assign processors to job classes so as to provide them with appropriate levels of service. A family of scheduling strategies is proposed and investigated; these strategies allow both extreme and moderate discrimination between job classes. The strictly pre-emptive priority disciplines are members of the family. The possibility of a conservation law for multi-processors is discussed.

Finally, the effect of breakdowns on the performance of single and multi-processor systems is studied. The optimal number of processors under different conditions is calculated.

1. THE TRADE-OFF BETWEEN DISTRIBUTED AND CENTRALIZED COMPUTING

As the price of hardware decreases, so the availability of minis, micros and intelligent terminals increases and greater flexibility is provided to organizations which use computers. All their needs can be met in-house with a local machine; they may export all their workload to a remote main frame provided by a bureau; part of their workload may be processed locally and the remainder exported; these are just a few of the options available and it is difficult to decide which will of the many possibilities be the "best" solution.

Centralised computing brings with it economies of scale, since one large processor outperforms a group of small processors with the same total capacity; it has the disadvantage of increased communication costs, and added congestion at the central processor. In favour of distribution, one may wish to balance the workload over existing machines.

The communication costs involved take the form of expenditure on equipment, communication lines etc. and transmission delays. We model the costs as being proportional to the number of jobs transferred and also the volume of data transferred. This will mean that the best jobs to transfer are those that do a large amount of computing, and relatively small amounts of input and output.

First, we consider the allocation of workload in an existing star network. Each site decides what proportion of its workload to process locally, and what proportion to send to the central site, which is shared between all the local sites. Our second problem considers the allocation of computing power among the sites, subject to a fixed budget, as well as workload routing. The routing and allocation problems cannot be separated. Both problems can be considered either selfishly, with a single site attempting to optimise the performance of its own locally originated jobs, or assuming cooperation between sites, where optimising the performance of the network as a whole is the goal. We start by considering the selfish optimisation, and later consider the changes needed to perform a global optimisation. All these problems are fully discussed in [7].

1.1 Workload allocation for given processor speeds

As seen by one of the local stations, the star network appears as in Figure 1. The jobs submitted at the local station have two attributes as far as our model is concerned; a length X in instructions and a class R. Both are assumed to be positive real numbers. The decision whether or not to process a job locally is taken purely on the class to which it belongs, based on a threshold value α. If $R \leq \alpha$ it is processed locally, otherwise it is sent to the central station for processing. Given the distribution of jobs by class and the distribution of lengths of jobs dependent on class, we can calculate the total workload submitted locally, L, and the fraction which is processed locally, $l(\alpha)$. The workload from other local stations submitted to the central station, which for the time being we assume to be given, is denoted by L_0. Our first problem is to choose the threshold α to minimise the costs of locally originated jobs. Our cost function, which we seek to minimise, must include terms to account for the processing costs, both locally and centrally, and the communication costs. We use a function $D(\alpha) = cN(\alpha) + c_0N_0(\alpha) + T(\alpha)$, where $N(\alpha)$ and $N_0(\alpha)$ are the expected numbers of jobs being processed locally and centrally when the threshold is α. $T(\alpha)$ represents the communication costs, and is of the form

$$T(\alpha) = b_1Y(\alpha) + b_2 [L - l(\alpha)]$$

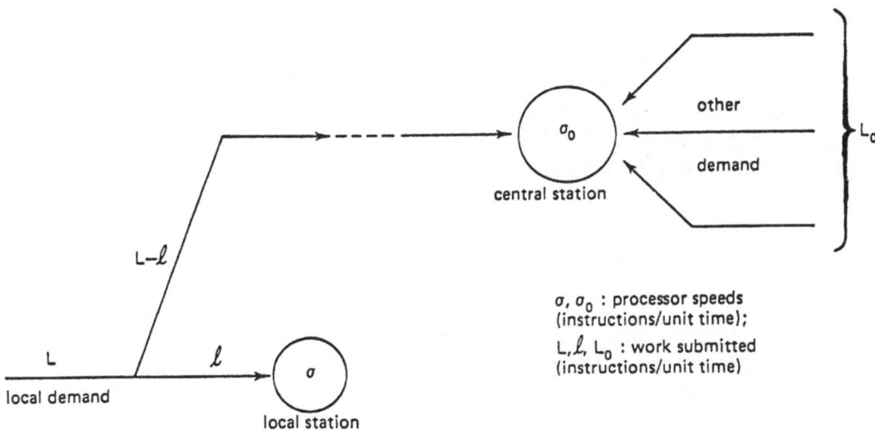

Figure 1

where $Y(\alpha)$ is the number of jobs transferred in unit time, and $L-l(\alpha)$ the workload transferred. c, c_0, b_1, b_2 are non-negative constants. By assuming that arrivals form a Poisson stream at the local stations, and that all processors use the processor sharing discipline a closed form expression for $D(\alpha)$ can be found. Using this closed form, various conditions can be established under which the optimal policy is to process all jobs locally $(\alpha=\infty)$ or process them all centrally $(\alpha=0)$. If these conditions do not hold an α^* exists which is the threshold value that minimises the cost. This assumes that the speeds of both processors, the local and the central, are adequate for any routing chosen. If this is not the case, for example, if L, the locally generated load, is greater than the speed of the local processor, then at least some of the workload will have to be sent to the central processor. This will impose an upper limit α_{max} on the class of jobs to be executed locally. A lower limit may also be imposed if the central processor is not fast enough to cope with all the workload submitted locally and what it receives from other stations.

1.2 Workload and processor speed allocation

So far, we have assumed that the network already exists; the speeds of the local processor, σ, and the central processor, σ_0 were considered as fixed. If, however, we are at the stage of planning a new network, or modifying an existing network, the budget available can be spent in several ways. All can be spent on local processing power; all can be added to the central funds; or some can be spent locally and some contributed to the central station's cost. The speeds of the two processors depend on the allocation of money. We now have a two dimensional optimisation problem; what is the best combination of money to spend locally, and workload to process locally.

We assume that the speed of the processor is proportional to a certain power of the amount of money invested. Assuming that the local station's manager has an amount M, of which he invests m locally and contributes the rest to the central funds, and that the other stations contribute a total of M_0 to the central fund, the speeds of the processors will be

$$\sigma(m) = a\, m^{\gamma}$$
$$\sigma_0(m) = a\,(M-m+M_0)^{\gamma}$$

Economy of scale in purchasing computer power implies that $\gamma > 1$; Grosch's Law indicates $\gamma \simeq 2$. We now have a cost function $D(\alpha, m)$ similar to the previous one, but depending on two parameters. Only some combinations of α and m values may give speeds of processor and threshold values which are feasible, and allow the processors to deal with their workloads. One could find the optimal pair as follows. For each α find the optimal allocation of money m; this gives a curve $\psi(\alpha)$. Then find the minimum value of $D(\alpha, \psi(\alpha))$. This is a non trivial task, since there may be non-global minimal and either one or two feasible regions to search.

When $\gamma = 1$ special cases arise and some progress can be made analytically. If, furthermore, $b_1 = 0$ the optimal allocation of funds is either $m = 0$ or $m = M$. However, if $b_1 > 0$ the optimal pair (α^*, m^*) can be in the interior of the feasible region.

Figure 2 shows an example constructed to show this. In this case, the function $l(\alpha)$ can be explicitly inverted, and the optimisation is done in terms of $D(l,m)$. The function $\psi(l)$ is seen on the plane $D = 0$, and the curve $D(l, \psi(l))$ above it. Here $D(l, \psi(l)$ has only a single minimum, and it is a global one, but this is not always so.

1.3 Global optimisation

When we wish to minimise the costs of the whole Network, basically the same methods are used, except that L_o can be replaced by a sum of the centrally submitted load of all the other processors, the

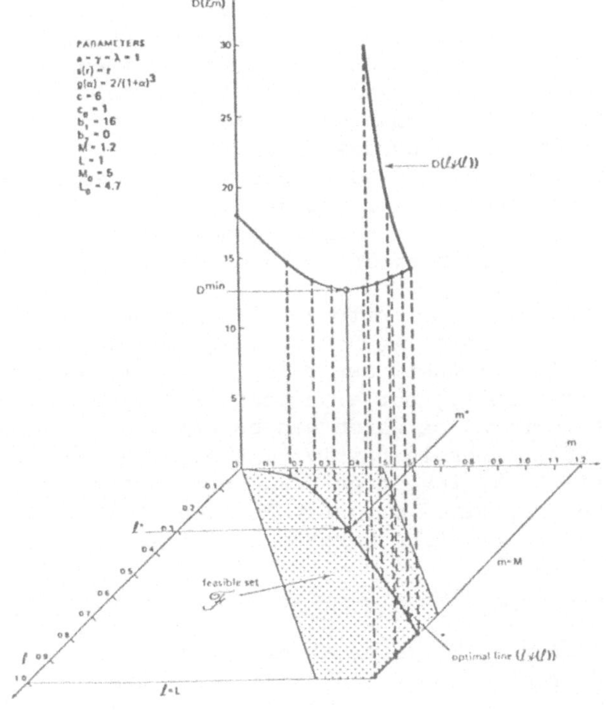

Figure 2

threshold α becomes a vector of thresholds, and so on. There are now several interrelated decision problems, one for each local site. They can be conveniently formulated as a dynamic programming problem. At stage j, local station j decides how much of its load to process locally, depending on the optimal decisions already taken by processors 1,...j-1 and assuming that processors j+1 .. k will make optimal decisions too.

1.4 Summary

In conclusion, one can say that these problems, although conceptually simple, are very difficult to solve. It seems likely, in practice, that the optimal allocation of money is either all locally or all centrally, since it needs a little care to construct an example where this is not the case.

2. MIXED PRIORITY SCHEDULING

In computer systems the demand often falls into several distinct job classes. For example, one might wish to favour terminal users of a system in order to give them a good response to requests. Alternatively, some users might be paying more than others, in the expectation of better service. Scheduling algorithms give a means of providing appropriate levels of service. Rather than provide ever more complex algorithms applicable to uni-processors, the increasing availability of multi-processor systems admits another alternative. Give different classes of jobs priority on different processors. To reduce the response time of a particular class of jobs, one will increase the number of processors allocated to that class. An example can be seen in Figure 3. The system has 4 identical processors; 3 give priority to class 1 jobs; one gives priority to class 2.

Clearly, the pre-emptive priority disciplines belong to this family of disciplines which we have called "mixed-priority". They are interesting in their own right, not only as members of the mixed priority family, but also because they probably give the extremes of performance which can be attained. Analysis of this family of strategies is fairly complex, even for 2 classes of job. We shall first investigate the pre-emptive priority case, then the more general mixed priority case and finally discuss the problems involved in treating more than 2 classes of customer.

2.1 Multi-processor systems with pre-emptive priorities

We consider a system of N identical processors offering service to jobs of two different classes. Class 1 jobs have pre-emptive priority over those of class 2, which resume from the point of interruption when a processor becomes free. A full analysis can

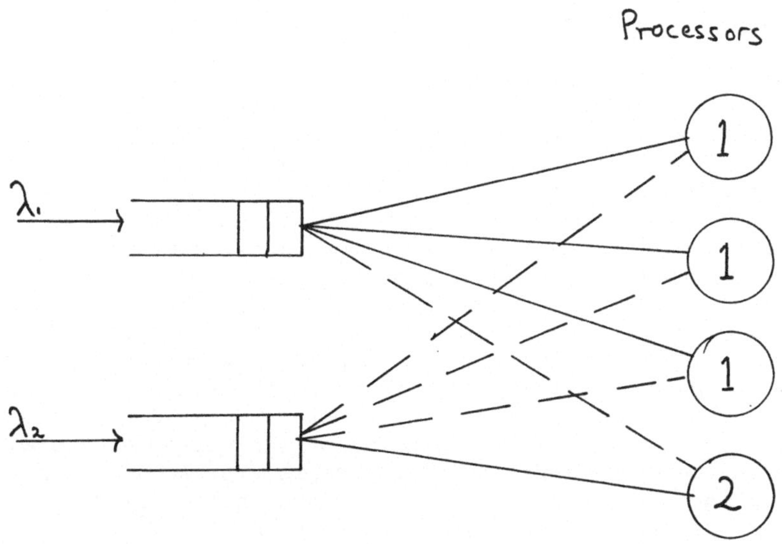

Figure 3

be found in [6]. We shall give an outline. By assuming Poissan arrivals, and exponentially distributed lengths of jobs, the systems is Markovian, and its state can be totally described by a pair of numbers of jobs present from each class.

The state space divides into 3 regions; one where the transition rates between states depend on the numbers of jobs of both classes; one where they depend only on the number of class 1 jobs; and one where they are independent of the number of jobs. Manipulation of generating functions in the regions where the rates are state independent or dependent on only one variable can give equations for the generating functions in terms of the unknown probabilities of the states in the region where the rates are state dependent. Using these equations, the balance equations from that region, and the normalising equation, the unknown probabilities can be found and all the usual performance measures calculated.

Figures 4 and 5 show the results obtained for two example sets of systems. In both cases, the total traffic intensity has been kept constant. The extreme points of the lines represent the vectors of response time pairs (W_1, W_2) which are achieved by giving pre-emptive priority to class 1 and class 2 respectively. Corresponding pairs of vectors have been joined by a line in the figures. The existence of a conservation law would imply that whatever scheduling discipline was applied, the performance vector would lie on the line.

2.2 Mixed priority scheduling for 2 classes of job

We assume a set of N identical processors, of which k are allocated to give pre-emptive priority to class 1 jobs. As before arrivals from a Poisson stream, and have exponentially distributed lengths. Again the system is Markovian and the state can be described by a pair of integers representing the number of jobs in each class. The state space now divides into 4 regions; one where transition rates are state dependent; one where they depend on the number of class 1 jobs; one where they depend on the number of class 2 jobs; and one where they are state independent. This falls within a family of models with "limited state dependency", which are analysed in [2]. We give a brief outline here. Once again generating functions for the regions in which the transition rates on state independent or only dependent on one of the class variables are used. The two variable generating function in the state independent region is expressed in terms of the generating functions in the two semi infinite regions, and the unknown probabilities in the state dependent region. Using a classical solution to a problem in complex variable theory, the Riemann-Hilbert problem, the coefficients of a set of linear equations can be calculated, which when added to the balance equations from the state independent region enable all the standard measures of performance to be calculated. Figure 6 shows the performance vectors obtainable for a particular set of systems, again chosen so that the total traffic intensity of the system remains constant. It can be seen that for any number of processors, the vectors given by different choices of k, lie in a straight line; the extreme points in the line representing the

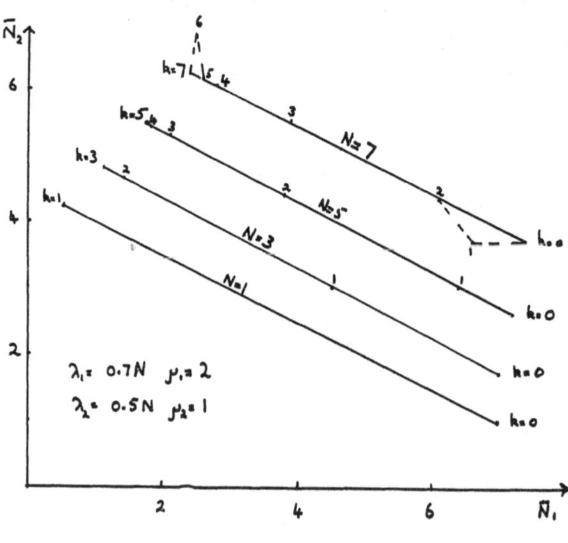

Figure 6

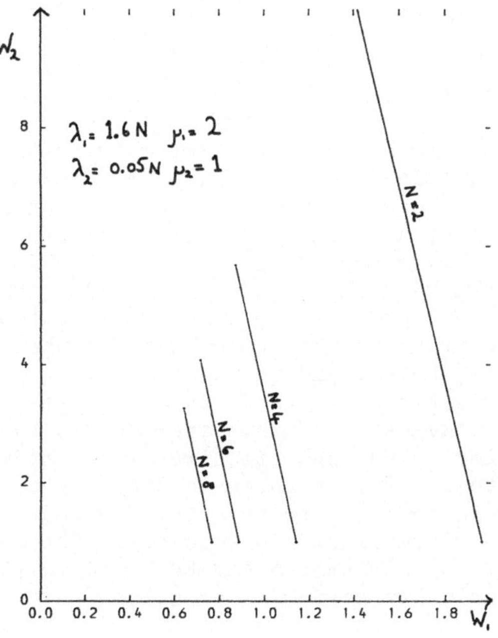

Figure 4

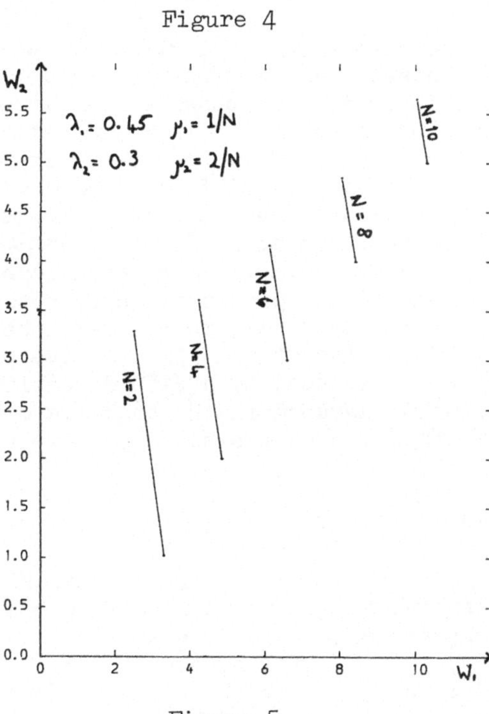

Figure 5

pre-emptive priority disciplines. This lends some support to the conjecture that a conservation law exists.

2.3 Other methods of solution

It is possible to attack the problem of solving these systems directly using the balance equations. Since these are an infinite number of these, the system must be truncated - effectively making it finite, although very large. There are then several ways to solve this large set of equations. Bounds can be put on the errors incurred by such truncation methods. Unfortunately, it is not possible to predict the size of truncation needed to give a particular error bound; one must solve the equations, calculate the error bound, and if it is too large take a larger truncation and repeat the process. This can clearly be expensive in CPU time. The advantage of this approach is that it is conceptually very simple, and only requires minor modifications in order to solve other types of infinite state space Markov processes, whereas the methods outlined in the previous section were specific to a small class of problems. An account of the methods used and some of the problems encountered can be found in [3]. It uses the pre-emptive priority system as an example.

2.4 Extension to more than 2 job classes

The methods that were used in 2.1 and 2.2 above are specific to the case of 2 job classes. If we wish to consider systems with more than 2 classes there are two possibilities. We can use the 'brute force' method outlined in 2.3 or we can attempt to approximate many classes by 2. Consider the pre-emptive priority system with R classes, from the point of view of class r jobs ($3 \leq r \leq R$). Since they pre-empt all jobs in classes r+1...R, we can safely ignore these classes. Jobs of classes 1... r-1 all pre-empt class r, so we can form an aggregate class by lumping these together. In the single processor case, nothing is changed by this process. For multi-processors however, there are two sources of error in this process; the pre-emptive priority within class 1 .. r-1 has been replaced by a FIFO discipline; and the exponential distribution assumed in the analysis of the two class multi-processor system has been replaced by a hyperexponential distribution.

2.5 Conservation law

In uni-processor systems there is the well known Kleinrock conservation law, which states that no matter what scheduling discipline is employed (so long as no idle periods are introduced while there is work to do) $\sum \rho_i W_i$ will be constant ρ_i is the traffic intensity of class i jobs, W_i is their mean response time, and the sum is taken over all classes. The existence of such a conservation law was used to show that all achievable performance vectors lie within a polytope whose vertices correspond to the vectors achieved by pre-emptive priority disciplines [1]. If such a conservation law holds for multi-processor systems, it might be possible to prove a similar result, and hence establish the feasibility of achieving desired performance from given configurations of processors.

Also the effect on other classes performance caused by decreasing (or increasing) the response time of one class, by whatever means, can be calculated. All the numerical evidence that we have from solving many systems, strongly supports the existence of such a conservation law, at least approximately.

3. MULTI-PROCESSOR SYSTEMS SUBJECT TO BREAKDOWNS

It is well known that in the absence of breakdowns, a single processor with rate of service μ gives a better average response than a system of N processors each with rate of service μ/N. This can be proved under certain assumptions; intuitively, the single processor system uses its full capacity so long as only work remains whereas the multi-processor only provides the same rate of service when there are more than N different jobs in the system.

However when the processors are subject to breakdowns, a counter argument in favour of multiple processors can be made. A single breakdown reduces the rate of service of a uni-processor to zero, whereas in a multi-processor system useful work can continue, albeit at a reduced rate. The natural question now arises, "What is the optimal number of processors in a multiple processor system subject to breakdowns?" Clearly, this will depend on the criteria selected for optimality and on the pattern and rate of breakdowns.

Our basic model is one of an N processor system with Poisson arrivals at note 1 and where each processor gives service at rate μ/N. We have considered two models of breakdowns. In the first, processors are alternately operative and undergoing repair. The second considers repairs as special 'jobs' which have pre-emptive priority over ordinary jobs. The service time of these special 'jobs' is the time to repair a processor. In this model it is possible for more than N breakdowns to exist. Since our results show that the models are qualitatively similar, we only present examples of the first model here, taken from [4].

The model has already been analysed, assuming exponential distributions for breakdown and repair times [5, 8] but these analyses have not been used to evaluate different systems subject to breakdown. We use the mean response time of jobs as our criterion of performance, and Figure 7 shows the response times obtainable for different numbers of processors, and different rates of breakdown. The curves are labelled with p, the probability that a processor is operative. This is a typical situation. For small breakdown rates (large p) the response time increase monotonically, a single processor is optimal. For a higher breakdown rate, the response time curve has a decreasing portion followed by an increasing tail. As the breakdown rate increases, so does the optimal number of processors.

The dependence of the optimal number of processors N* on the breakdown rate, ξ, is shown in figure 8 for different values of μ and the repair rate, η. Note that ξ is bounded above to ensure that the system is non saturated. Two vertical lines have been drawn on each graph. They represent the values of ξ which correspond to 90% and 95% of saturation. It seems that unless the system is close to saturation, the optimum number of processors is quite small.

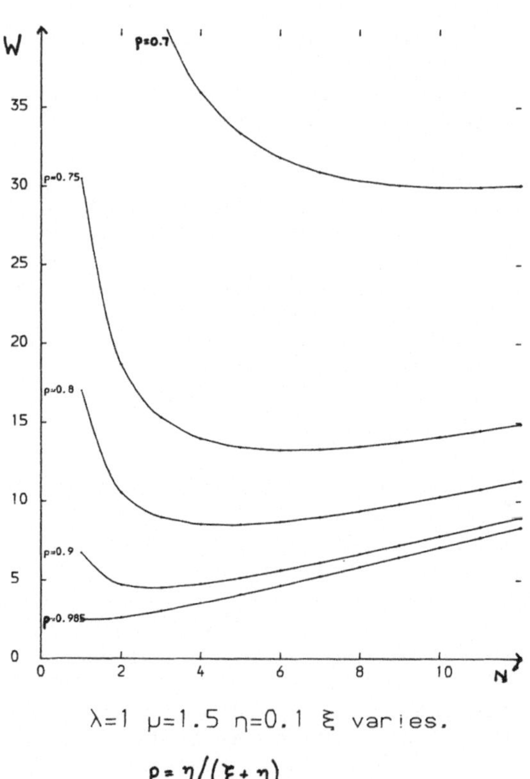

Figure 7
Response time curves

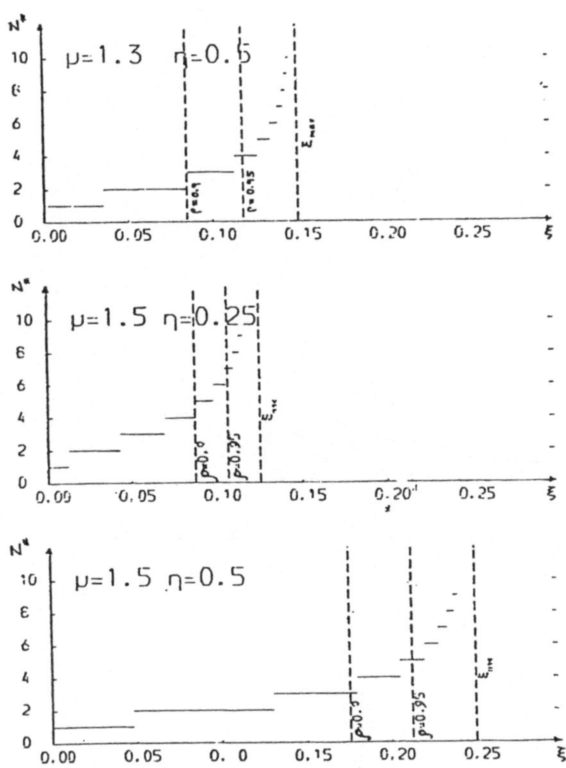

Figure 8
Optimal number of processors as function of ξ

Figure 9 shows the effect of lengthening the mean operative period and mean repair times as well, keeping the overall probability of an operative processor constant (at p = 0.83). The general trend is as in figure 7 except that increasing the breakdown rate (and the repair rate) now decreases the response time. One might argue that smaller processors are more reliable than large ones, letting the breakdown rate decrease N. This gives the graph in Figure 10; the optimum number of processors is more pronounced.

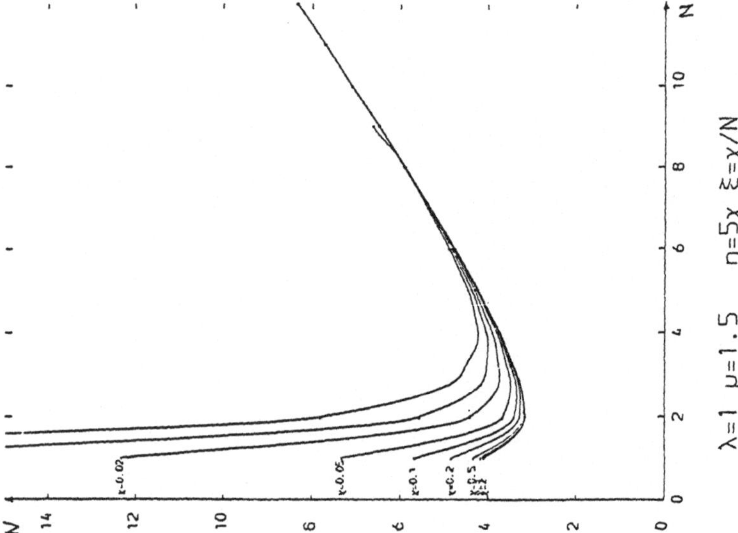

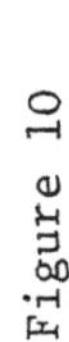
Figure 10

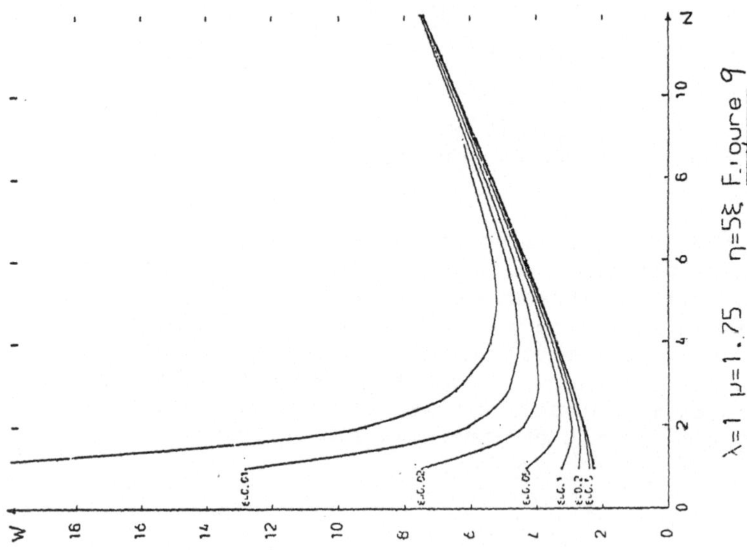

Figure 9

REFERENCES

[1] Coffman, E.G., and Mitrani I., Selecting a scheduling rule that meets pre-specified response time demands. Proceedings 5th Symposium on Operating System Principles, Austin (1975).

[2] Fayolle, G., King, P.J.B., and Mitrani, I., The Solution of Certain Two-dimensional Markov Models. To appear in Journal of Applied Probability Earlier version: Performance Evaluation Review, 9,2 (1980) pp. 283-289.

[3] King, P.J.B., and Mitrani, I., Numerical Methods for Infinite Markov Processes, Performance Evaluation Review, 9, 2 (1980), pp. 277-282.

[4] King, P.J.B., and Mitrani, I., The Effect of Breakdowns on the Performance of Multi-Processor Systems, submitted to Performance '81 conference.

[5] Mitrani, I. and Avi Itzhak, B., A Many Server Queue with Service Interruptions, Operations Research 16, pp. 628-638 (1968).

[6] Mitrani, I. and King, P.J.B., Multiprocessor Systems with Pre-emptive Priorities, Performance Evaluation 1, 2 (1981).

[7] Mitrani, I. and Sevcik, K., Evaluating the Trade-off Between Centralised and Distributed Computing. Proceedings, 1st Int. Conf. on Distributed Computing, Huntsville (1979).

[8] Neuts, M.F. and Lucantoni, D.M., A Markovian Queue with N servers subject to Breakdowns and Repairs. Management Science 25, pp. 849-861 (1979).

DATA SECURITY IN DISTRIBUTED COMPUTING SYSTEMS

Wyn L Price

National Physical Laboratory, Teddington, UK

ABSTRACT

The needs for data security in distributed computing systems are considered in general terms, followed by a summary of the development of encryption techniques from the classical to the most modern systems available for general use. A brief account is given of the historical development of the Data Encryption Standard (DES). Lastly the location of encryption devices in a distributed network is considered.

INTRODUCTION

We shall assume for the purpose of our discussions that a distributed computing system consists of a communication subnetwork to which are connected various host computers and user terminals. The security of data which is processed within and carried by such a system depends upon several factors. The chief of these are the control of access to services and files (including the identification of authorised users) and the protection of data from disclosure or alteration by means which do not necessarily have regard for the access rules of the system.

The very nature of a distributed computing system makes more severe the problem of providing adequate security. In the days when computer installations were confined to particular sites it was moderately simple to provide sufficient security by physical means; access could be restricted, rooms could be locked and tapes could be filed in secure cupboards. With the advent of access via local networks from distributed terminals,

K. G. Beauchamp (ed.), New Advances in Distributed Computer Systems, 279–292.

the problem of assuring adequate security began to emerge. First there was the problem of identifying authorised users at terminals and second there was the problem of preventing unauthorised access via components of the local network. Systems were developed in which passwords were used for user identification; these restricted the use of terminals to those duly authorised. Little attention was paid at this time to making the local network secure against intrusion, perhaps because the rewards for the intruder were not necessarily very great in comparison with the technical difficulty of successful intrusion by tampering with the network. In passing one might at this point observe that private networks consisting of lines leased from some PTT are, if anything, less secure than networks in which many users share the same facilities. This is because leased lines are easily identifiable at exchanges, thereby easing the task of the intruder; traffic using shared facilities is less easily identified. With the development of widespread distributed networks the nature of the problem has been transformed. Opportunities for intrusion have become far more numerous, with traffic from many users sharing network lines and nodes, though the selection of relevant traffic may present a technical problem. However, some communication systems almost go out of their way to make things easy for the intruder; an example is the system of in-band telephone signalling. 'Interesting' traffic can be selected by monitoring the numbers called.

The value of information carried by distributed computer networks must be enormous. Amongst the first organisations to rely upon this new technology were the banking institutions. A large proportion of their transactions is now carried by one network or another; in the international field the SWIFT network is very significant. The effect of this development is beginning to be felt directly by the ordinary consumer. Automatic teller machines have been with us for several years and are gradually spreading to new sites; the most satisfactory method of operation (from a security point of view) of such machines involves direct communication with a central computer; off-line machines are not necessarily as secure. It will not be long before point of sale terminals become widespread; customers will present machine readable cards at shop checkouts and the cost of their purchases will be debited directly to their bank accounts. This operation must also have access to a central computer system, preferably on-line. Several experimental systems have already been tried out.

Other commercial and industrial organisations also rely heavily upon computer networks for information exchange between organisational units. Much of this information is highly confidential and of great value, for example, the geological survey results of the major oil companies; many stock exchange transactions are also conveyed in this manner.

Privacy of personal information is now a major issue, with many countries introducing legislation for protection of this kind of data. Personal data is now held by a very large range of organisations, ranging from government departments to credit rating agencies.

From these few, but important, examples it should be plain that careful attention to security is now an essential feature of the design of any distributed computing system. The rewards for the successful intruder are now potentially very great. The technology of distributed computing systems, with shared facilities such as nodes, favours the technically skilled intruder unless specific measures are taken to frustrate illegal access. Clearly network users will be reluctant to entrust their valuable data to networks unless they have assurance of the security of the communication system.

It is very difficult to estimate the degree to which criminals are at present intruding upon computer networks. Studies of computer crime are being undertaken both at the UK National Computing Centre and the Stanford Research Institute (eg 1). Reported crime seems to relate more to abuse of access privileges by authorised users than to a highly technological attack upon the communication system. Unfortunately detection of computer crime is not easy and it is possible, even likely, that a large proportion goes undetected, particularly that which may involve a high technology attack. The activities of 'phone-freakers' in recent years is evidence that some people rise to the challenge of breaking a technologically advanced system.

Our main concern in this and the related sessions will be the way in which data encryption techniques can be of help in achieving the requirement for security within distributed computing systems, paying particular attention to the security of the communication sub-network. It is whilst being carried by or temporarily stored within this sub-network that data is most at risk. Communication links may be comparatively easy to tap, either passively (just reading data in transit) or actively (altering data that is being read). Security of network node computers requires careful attention.

It is in any case extremely difficult to design an operating system in a distributed context which has provably secure access rules. However interesting and difficult this problem may be, we shall not be concerned to any significant degree with operating system design.

CLASSICAL ENCRYPTION TECHNIQUES

Encryption of communication has been used at least since the early Egyptian and Greek civilisations, and possibly even earlier than that. Some of the encryption methods used in early times, such as the Caesar cipher (allegedly invented and used by Julius Caesar) are now seen to have been laughably insecure. However, encryption methods have acquired increasing sophistication (but not necessarily strength) as the centuries have gone by; techniques of cryptanalysis (decryption by some third party) have also advanced.

It is necessary to distinguish between ciphers and codes; in ciphers the text tokens which constitute a message are manipulated and altered in some way according to prescribed rules whose action may be controlled by an encipherment key; on the other hand codes are concerned with words or groups of words, for which arbitrary substitution of other words or letter groups is made. Codes have seen much use in commercial applications, partly because they usually lead to compression of messages and therefore reduce transmission costs. However, they depend upon look-up tables, which may be very extensive, and are not really suitable for use in computer networks. No key is involved in encoding, unless one regards the whole tabulation of code equivalents (commonly called a 'code-book') as a key in itself. To change a code means the replacement of one look-up table by another. Combinations of codes and ciphers have been successfully used; a message is first encoded and then encrypted, the technique being known as superencipherment.

The basic components of encryption algorithms are substitution and transposition. Substitution means replacement of text tokens (eg letters) by others according to some rule; the Caesar cipher was based on a substitution of letters by others found three places later in the alphabet. The substitution rule may be much more complicated; polyalphabetic ciphers use different substitution tables for successive text tokens; alternatively the rule may be constructed according to some complex procedural algorithm. Transposition means the reordering of text tokens, which may also be done in a complex way. The behaviour of both substitution and transposition algorithms may be influenced by reference to an encryption key. In many applications in the past ciphers have been used which consist only of either substitution or transposition; however it is generally recognised that more secure ciphers can be constructed by using combinations of these methods. In some cases these combinations may look very complex, but apparent complexity is no guarantee of security; for example repeated monoalphabetic substitution confers no additional strength.

Design of encryption algorithms requires rigorous critical attention.

Until the first world war ciphers were usually applied by 'manual' methods, sometimes assisted by simple mechanical devices. By the end of that war various cipher machines had been invented and were coming into use. Early cipher machines were either entirely mechanical, electromechanical or pneumatic. An extremely important development at this time was that of Vernam, who conceived what has since been developed into the 'one-time pad', a technique which is still in use for high-level security. We shall see that Vernam's idea, which involved combining the message with another data stream by an 'exclusive-OR' operation (figure 1), forms part of the method of operation of some modern encryption methods. In the figure the first exclusive-OR operation produces ciphertext from plaintext, whilst the second, using the same additive data stream, restores the plaintext. The additive data stream constitutes the encryption key and should ideally be completely random; a well chosen pseudo-random stream may be acceptable.

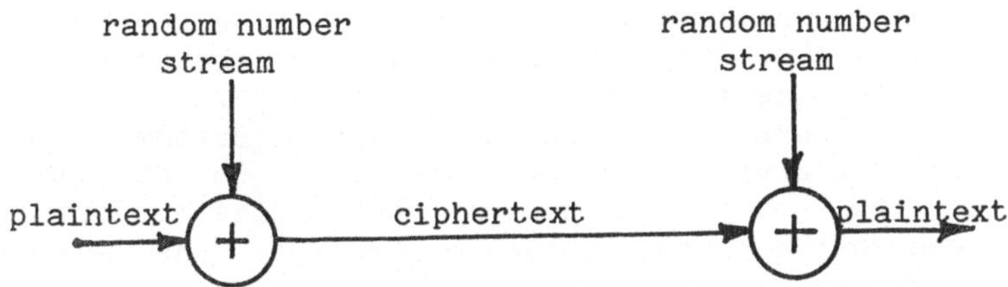

Figure 1. Principle of the Vernam cipher

It is now well-known that a major impetus in the development of digital computers, first electro-mechanical and then electronic, during the second world war, came from the need to cryptanalyse the traffic carried by ciphers of the previous generation, developed between the wars. It is beyond question that the successes obtained against the latter were facilitated by the advent of the new computing machinery. However, with the passage of time, the new machines themselves permitted the development of better encryption techniques, thereby somewhat restoring the balance in favour of the cryptographer (the system designer) rather than the cryptanalyst (the potential intruder). As a generalisation computing complexity can be said to be on the side of the cryptographer rather than the cryptanalyst. The latest encryption techniques in the public domain (we are not concerned here with modern military or diplomatic ciphers) depend on application of microprocessors and special-purpose LSI devices.

Generally speaking, classical ciphers need the same key for the twin processes of encryption and decryption. Communicating users must therefore be in possession of exactly the same key if they are to communicate securely using ciphers of this type. Establishment of key pairs at the two users, possibly located at remote sites, presents an organisational task which must itself be carried out securely if the system is to provide the required degree of security. Assistance from the communication sub-network, in the form of some enhancement of the communication protocol, may be necessary for the proper control of this process. We shall meet in a later session the concept of cryptosystems which do not require the same key at source and destination of a protected data flow. The existence of such cryptosystems may change radically the nature of the key management task.

The task of an intruder seeking to circumvent the security provided by an encryption system is in general two-fold. He needs to discover the nature of the encryption algorithm and the key that is in use in association with the algorithm. This involves a process of cryptanalysis. It is the aim of the cryptographer to make the process of cryptanalysis as difficult as possible. Where a standard encryption algorithm is in use, and therefore the cryptanalyst need not strive to discover the details of the encryption process, then, clearly, the way in which the encryption proceeds under control of the encryption key must be suitably complex.

From the point of view of the cryptanalyst, the task of breaking a cipher is made easiest if a 'chosen plaintext' attack is possible. In such an attack the cryptanalyst selects a plaintext to suit his method of attack; 'all zeros' or 'all ones' may be appropriate. It is assumed that the cryptanalyst is able to introduce this text into the encryption device and to produce the corresponding ciphertext. Next best to chosen plaintext is 'known plaintext' where the cryptanalyst is in possession of pairs of plaintext and corresponding ciphertext. If the character set of the plaintext is known, then some progress in cryptanalysis may be made even if the actual plaintext is not known; this is because text expressed in most character sets has some redundancy, for example, parity bits in ASCII code. The process in this case would involve decrypting the ciphertext with a succession of keys, noting which keys produce correct parity in the output. By taking further pairs of known plaintext and ciphertext, the field can be narrowed down to the correct key.

For those who are interested, an excellent history of cryptography and cryptanalysis may be found in the book 'The Codebreakers', by David Kahn (2).

DATA SECURITY IN A DISTRIBUTED NETWORK

Within a private network the requirement of providing security of data by encryption can readily be met with proprietary encryption devices obtainable from any of a wide range of manufacturers. Devices available are generally designed for point-to-point installation, for example line-bracketting as in figure 2. An element of intelligence may be built in, so that control characters or message headers are not necessarily encrypted; alternatively the encryption devices may encrypt all bit patterns that are offered to them. Encryption is under the control of an algorithm built in to the device and influenced by a user-supplied key. The range of keys is usually very large, 40 decimal digits being not untypical. The details of the algorithm in devices of this kind are usually kept secret by the manufacturer (the design being regarded as valuable commercial property), so that the user relies upon the reputation of the manufacturer for the strength of the encryption system.

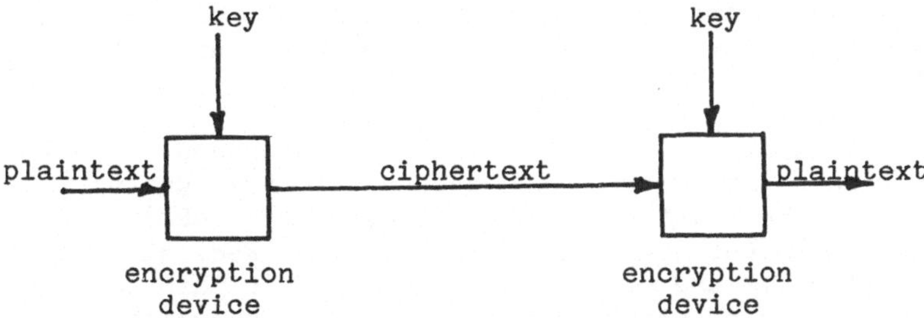

Figure 2. Line-bracketting encryption

Another important characteristic of commercial devices of this kind is that like can only communicate with like. Devices must either be used in pairs, as in the line-bracketting context, or in groups, as in a multi-drop configuration. The main point is that it is not possible to communicate between devices of different design. If users from many different organisations, connected to the same communication subnetwork, wish to exchange encrypted messages, then they must choose to use matching encryption equipment. This could mean choosing a particular proprietary device and rejecting all others. No attempt has been made to reach any such agreement and it is highly doubtful whether it could ever be reached.

The secret nature of proprietary devices has further repercussions. Devices may need to be installed in unsupervised, insecure locations where they may be subject to attack by intruders seeking to discover the details of the encryption process. If the security of the system depends in part upon the secrecy of the algorithm, then discovery of the algorithm by some intruder reduces the level of security. If the key handling procedure is well designed, then the remaining level of security may still be more than enough for the requirements of the user. On the other hand, any reduction in system security may be unacceptable. To restore security in such circumstances would require the withdrawal of all compromised units and their replacement with devices of different design. On a large scale basis this could be a horrendous task, both in terms of effort and of expense.

For these and other reasons interest has been aroused in the last decade in the possibility of establishing standard encryption algorithms which are acceptable to manufacturers and users alike. The details of the encryption algorithm of devices conforming to the standard would, of course, be public. The security of such a system depends firstly on the complexity of the text transformation under the control of the key and secondly on the security with which the keys are stored and transmitted.

The initiative for developing standard encryption algorithms came from within the domain of the United States Government. The National Bureau of Standards (NBS) is charged under the Brooks Act with responsibility for Federal standards for the effective use of computer systems. In this context NBS initiated a study programme for computer data security in 1971. Strictly speaking the NBS remit is restricted to the field of computer systems, with data communication standards coming within the aegis of the National Communication System; where security is concerned there is clearly an overlap of interest, because the same encryption algorithm may be used for both requirements. We shall see later how this has affected the development of standards.

The first aim of NBS was to find and develop a suitable encryption algorithm and in 1973 an invitation was issued to inventors to submit details of available cryptographic systems which could be used to protect data in transit and in storage. Design requirements included

a) the system should be economically viable
b) the details of the algorithm should be made public

c) the algorithm should be expressible in special purpose LSI hardware.

Hardware implementation was required because of the need to avoid placing an additional heavy load upon system hardware. The order code of conventional digital computers is not ideally suited for implementation of cryptographic algorithms. Furthermore hardware implementations are capable of more satisfactory validation procedures than are software implementations and they are reckoned to be less vulnerable to tampering.

The 1973 invitation to submit designs produced no useful result. Many ideas were submitted that needed further development, which was not what was wanted. Therefore the invitation was re-issued in 1974. This time a further range of designs was submitted, including one from IBM. The latter was based on an IBM design, code-named Lucifer, for an encryption algorithm; this had originally been developed in connection with a design for an automatic teller machine. The details of the Lucifer algorithm had been published in various papers (eg 3) and therefore the open publication requirement was met.

The IBM design was successful in the 1974 competition and the process of setting up a Federal Data Encryption Standard (DES) began. It was issued in 1975 as a proposed standard and finally published as a Federal Information Processing Standard in January 1977 (4), becoming effective six months later in July 1977. From that date US Federal Agencies were required by law to comply with the standard for the purposes of data protection or to show good reason for not doing so.

We shall examine the Data Encryption Standard in some detail in the next session. For the purpose of this introduction let it suffice to say that the algorithm deals with data in blocks of 64 bits, producing 64 bit blocks of ciphertext from 64 bit blocks of plaintext; the exact behaviour of the algorithm is governed by a 64 bit encryption key, of which only 56 bits are actually active (the remainder indicating parity). Therefore the total number of different keys which are possible is 2**56 or about 7.2x10**16. The operation of the algorithm is a complex blend of substitution and transposition. The change from encryption to decryption mode, upon application of the appropriate command, is achieved very elegantly within the algorithm.

For reasons which will be made plain in the next session it is not good practice to segment data into blocks of 64 bits and encrypt these in succession. Rules for methods of using the encryption standard need development. Just as the encryption algorithm itself is the subject of a standard, so the rules for methods of use must also be expressed in standard terms if users are to communicate correctly.

The DES is meant to be used, in US Government communications and data processing applications, in conjunction with existing Federal Information Processing Standards (FIPS) and Federal Telecommunications Standards (FTS), but additional standards are needed, and are being prepared, for electrical, mechanical and functional aspects of stand-alone, add-on communications security equipment using the DES. For an effective security system, standards are required for incorporating DES devices in terminals and communications processors.

It must be stressed that the DES is not intended for military or diplomatic applications; the encryption techniques used in these domains are not made public.

THE LOCATION OF ENCRYPTION DEVICES

From the point of view of the security of the data the ideal locations for the encryption function are at the source from which the data is flowing and the sink to which it is going. If it is possible to achieve this arrangement, then the data does not exist in plaintext form anywhere along its route through the communication medium, and thereby its exposure to possible intrusion is minimised. Clearly the header information which accompanies messages in transit must be readable by intermediate elements of the communication sub-network, otherwise network traffic management would be impossible. Therefore encryption devices used in this mode (figure 3), commonly called end-to-end, must be capable of distinguishing between message headers and message text, delimited by some standard flags. This calls for a level of intelligence in the encryption device which is higher than that required to control encryption on a direct point-to-point line. Another important factor in the design of an end-to-end encryption system is the requirement that identical keys shall be made available at both ends of the communication channel (unless an encryption system is being used which does not need identical keys at either end). In a very large network where any user may wish to communicate with any of a large selection of other users, this may not be easy to achieve.

Furthermore, end-to-end encryption, with message headers in clear, is open to a traffic analysis attack, in which the intruder seeks to acquire information from an analysis of the amount of traffic flowing from particular sources and to particular destinations. Frustration of an attack of this kind

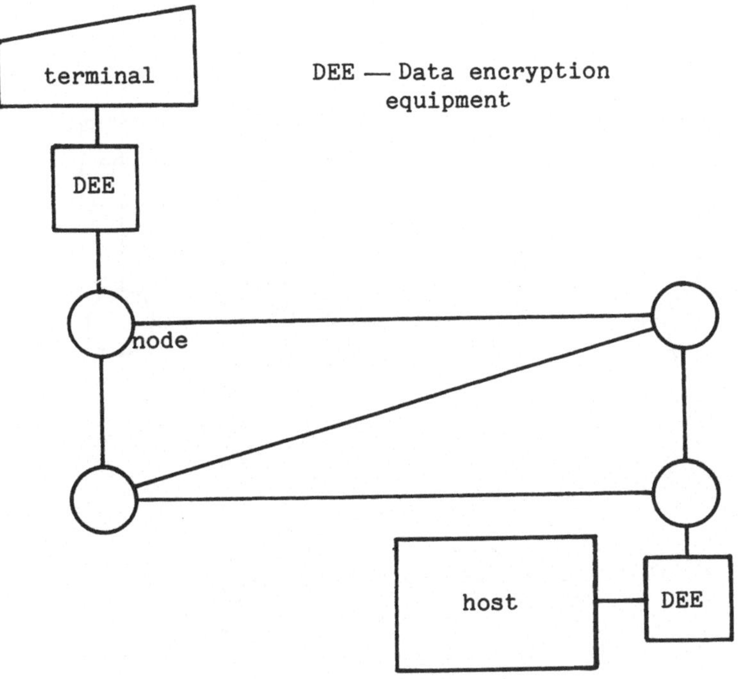

Figure 3. End-to-end encryption in a network

must depend upon encryption being applied at other levels in the sub-network hierarchy. We have already met the concept of point-to-point encryption on a direct link. This may be introduced on lines between network nodes (line encryption, figure 4) when key management may be relatively simple, involving the exchange of keys between adjacent nodes. For this function of key exchange, keys are encrypted under master keys used only for the purpose of secure key transfer. Line encryption by itself cannot be recommended as sufficiently secure for safe sub-network communication. The reason for this is that messages protected by line encryption are re-encrypted on every line traversed, and therefore exist in plaintext at all nodes traversed. The physical security of nodes may not be sufficiently assured and, in any case, an accidental malfunction in a node may misdirect a message, thereby laying the latter open to disclosure. A bug installed in the node may achieve this by deliberate intent.

A higher level of security is achievable if node-by-node encryption (figure 5) is used. In this mode of operation a part of the node storage is set aside for handling of plaintext messages; this part of the node is made as secure as possible

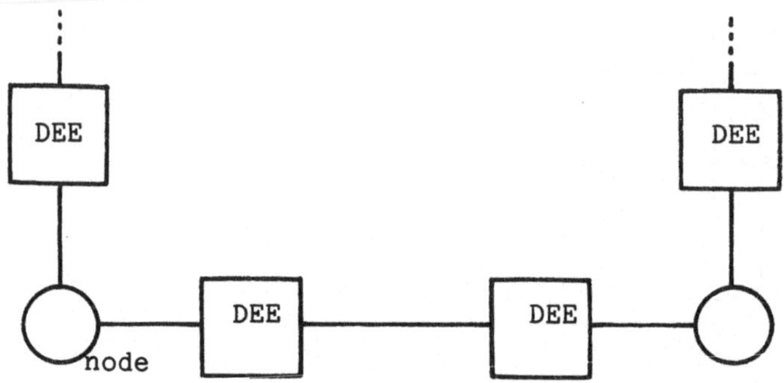

Figure 4. Link encryption in a network

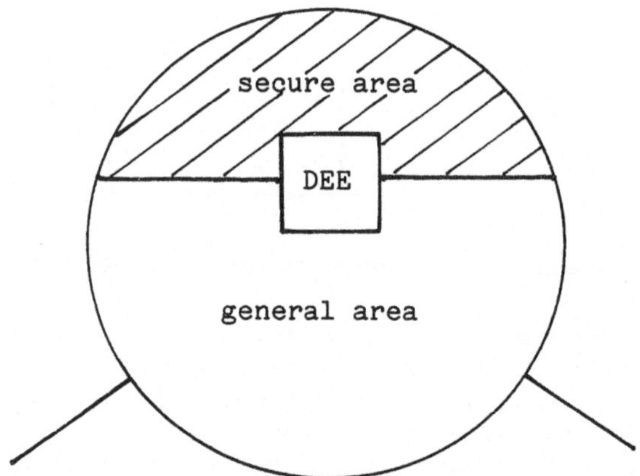

Figure 5. Node-by-node encryption

by appropriate system design. Messages arrive in the node with their headers in clear, but with the data encrypted under the control of a key unique to the link just traversed. The node reads the cleartext header, decides (using its routing algorithm) on the next link to be traversed, and then passes the message text to the secure module to be decrypted under the key of the link just traversed and then encrypted under the key of the link next to be traversed. The encrypted message emerges from the secure area and is reunited with its corresponding header. The message therefore is only in plaintext form within the secure module and is protected against disclosure by this means.

For many applications, in order to achieve adequate security, it may be necessary to apply encryption in several ways within the subnetwork and associated communication processes, combining end-to-end encryption with node-by-node encryption for example. Multiple encryption of this kind must be part of an integrated system design and can be fitted into network architectures such as that of Open Systems Interconnection.

Where assistance from the network is required for key management, this may be entrusted to a key distribution centre (KDC) located at a network host. The function of the KDC will be to respond to requests for encryption keys for secure communication between a calling and called user. Communication of encryption keys between KDC and users must be protected by encryption under master keys used specifically for this purpose. Communication between users and KDC must also be protected by integrity checks which prevent a sophisticated attack in which an intruder masquerades as a KDC and imposes keys upon unsuspecting users, thereby gaining access to their protected communication. Various studies of the KDC function have been carried out (5,6).

A somewhat parallel function within a network is provided by a register of public keys where a public key cryptosystem is in operation. Indeed the presence of a public key cryptosystem may ease the task of distribution of keys for such ciphers as the DES.

A further centralised function that may be required within a network is that of arbiter of message signatures produced with an application of public key cryptosystems. We shall describe this function in some detail in a later session.

REFERENCES

1. Parker, D.B. Crime by Computer. Charles Scribner's Sons, New York, 1976.

2. Kahn, D. The Codebreakers. Macmillan, New York, 1967.

3. Feistel, H. "Cryptography and computer privacy." Scientific American, 228, 5, 1973, pp. 15-23.

4. National Bureau of Standards. "Data encryption standard." NBS FIPS Publication 46, January 15 1977.

5. Heinrich, F. "The network security center: a system level approach to computer network security." NBS Special

Publication 500-21, Vol 2, January 1978.

6. Price, W.L. & Davies, D.W. "Issues in the design of a key distribution centre." NPL Report DNACS 43/81, April 1981.

THE DATA ENCRYPTION STANDARD AND ITS MODES OF USE

Wyn L Price

National Physical Laboratory, Teddington, UK

ABSTRACT

An account of the algorithm of the Data Encryption Standard is given; it is shown how the same device can conveniently carry out encryption and decryption. Various criticisms of the cryptographic strength of the algorithm are considered. Four modes of use of the DES are described and recommendations are made regarding the context in which each may be used. Finally reference is made to the need for a well designed key management system.

THE DATA ENCRYPTION STANDARD

The standard, born from a desire for some degree of uniformity in techniques for encrypting data within computer systems and data communication networks, was conceived by IBM and developed by the US National Bureau of Standards in conjunction with IBM. Its status as a standard is that of a Federal Information Processing Standard (1), which has mandatory application within the domain of the US Government.

The standard algorithm is designed to accept blocks of 64 bits of plaintext and to convert these to 64 bits of ciphertext (there is therefore no data expansion) under the control of a 64 bit key (only 56 bits actually influence the encryption). The encryption process makes use of a complex blend of substitution and transposition; in substitution, groups of bits are exchanged for other bit patterns, using look-up tables; in

K. G. Beauchamp (ed.), New Advances in Distributed Computer Systems, 293–309.

transposition, individual bits are moved to other parts of the field.

For US Government purposes it is required that the algorithm shall be expressed either in special purpose LSI hardware or in a microprocessor implementation; software implementation is not acceptable because this is less easily validated and may be more easily altered. However, we shall see that other national and international standards bodies have been considering the Data Encryption Standard (DES) and that some of these may find a software implementation acceptable.

The logical structure of the DES is shown in figure 1. This contains various registers and functions. Registers L and R, each of 32 bits capacity, form part of the main loop of the algorithm; register TEMP (32 bits) is not usually shown in flow diagrams of the standard, but can be seen to be logically necessary for correct operation. Registers C and D are shifting registers of 28 bits each which are concerned in the cycle key generation. Functions IP, IP-1 and P are permutations in which bit positions are changed within the relevant field; P is a particularly irregular permutation function. E is an expansion permutation which accepts 32 bit inputs and produces 48 bit outputs, replicating half the input bits in the process. PC1 and PC2 are permuted choice functions. PC1 takes the 64 bit encryption key as input, omits the 8 parity bits, and rearranges the order of the remaining 56 bits. (These 56 bits are divided into two fields of 28 bits each to be loaded respectively into registers C and D.) PC2 accepts 56 bits as input and produces a rearranged selection of these as a 48 bit output. The 'boxes' labelled S1 to S8 are look-up tables; these each produce a 4 bit output corresponding to a 6 bit input ('S' could stand for 'substitution'). The tables represented by each of the S boxes are all different.

The exact details of all the permutations and of the S boxes are tabulated in and form part of the description of the standard algorithm.

A plaintext block of 64 bits to be encrypted is entered into the input register, whereupon it passes through the initial permutation IP and is split into two fields of 32 bits each which are respectively entered into registers L and R. There follow 16 cycles of operations during which the permutations and substitutions take place.

The first events in the cycle are the transfer of the current contents of register R to register L and the expansion of the contents of register R by the expansion permutation E. The result of permutation E is combined in an exclusive-OR operation with the current cycle key (this will be defined

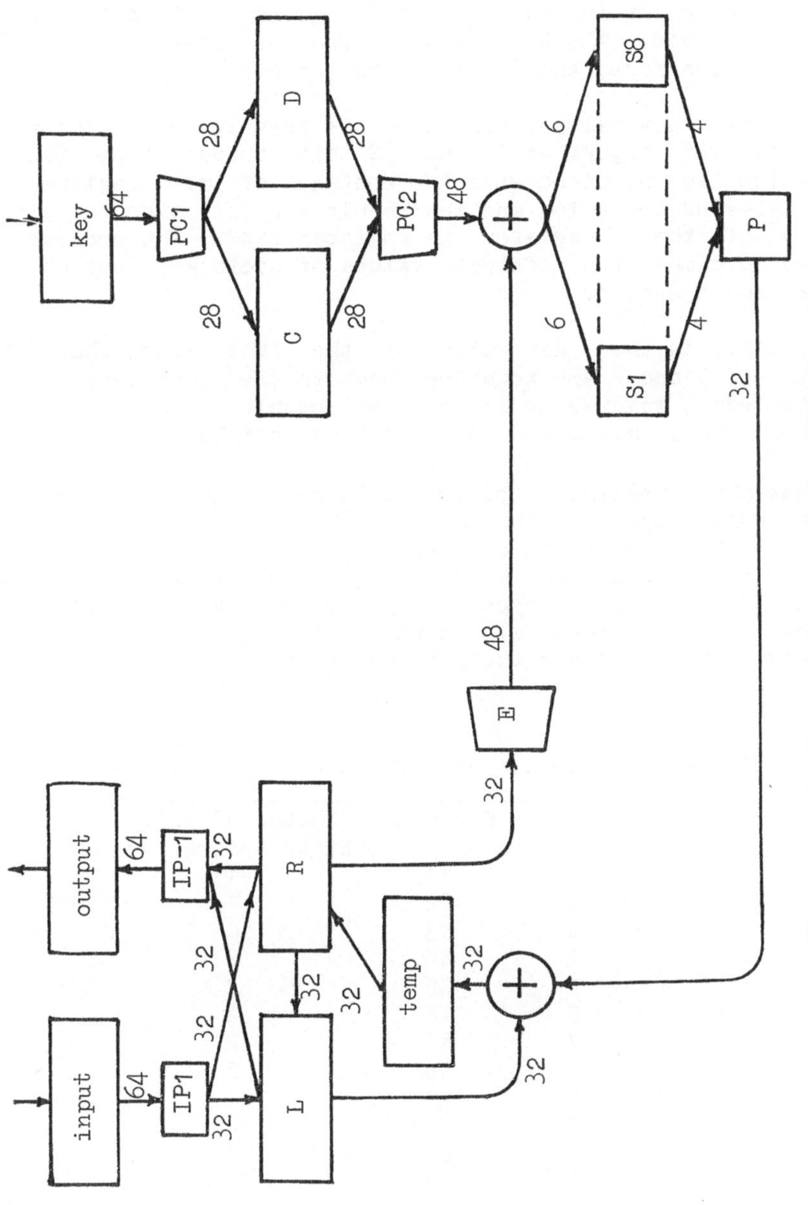

Figure 1. Logical structure of the DES algorithm

shortly). At this stage the data field is 48 bits wide. These 48 bits are split into 8 groups of 6 bits each; each group is then input to an S box and a conversion takes place to the corresponding entry in the S box table, giving a 4 bit output from each S box. The 8 outputs are brought together into a 32 bit field which forms the input to the P permutation.

We have now almost completed the description of the main loop of the DES algorithm. The 32 bit output from the P permutation is combined with the contents of the L register in an exclusive-OR operation and the result replaces the original contents of the R register via register TEMP. The process is repeated 16 times with different values of cycle key (output of PC2) on each occasion.

Finally, as the last step of the full algorithm, the contents of L and R are together input to the final permutation IP-1 (the exact inverse of IP) and the result is the 64 bit ciphertext block, appearing in the output register.

That which remains to be explained is the generation of the 16 successive cycle keys from the 64 bit encryption key. We have seen how 28 bits of the encryption key are sent respectively to registers C and D. The input to PC2 from C and D is varied between encryption cycles by cyclic shifts of registers C and D. These cyclic shifts are of either 1 or 2 bit positions. Table 1 shows the sequence of shift operations over the 16 cycles.

Table 1. Schedule of left shifts in encipherment

Cycle	K Block	Number of left shifts before PC2
1	K1	1
2	K2	1
3	K3	2
4	K4	2
5	K5	2
6	K6	2
7	K7	2
8	K8	2
9	K9	1
10	K10	2
11	K11	2
12	K12	2
13	K13	2
14	K14	2
15	K15	2
16	K16	1

An elegant feature of the algorithm is the way in which it can be switched simply between encryption and decryption. For encryption the cyclic shifts of registers C and D take place in one direction; for decryption the cyclic shifts take place in the other direction. Table 2 shows the order of cyclic shifts for decryption. Apart from the order of presentation of cycle keys at the output from PC2 the operation of the algorithm is identical for encryption and decryption.

Table 2. Schedule of right shifts in decipherment

Cycle	K Block	Number of right shifts before PC2
1	K1	0
2	K2	1
3	K3	2
4	K4	2
5	K5	2
6	K6	2
7	K7	2
8	K8	2
9	K9	1
10	K10	2
11	K11	2
12	K12	2
13	K13	2
14	K14	2
15	K15	2
16	K16	1

It is quite easily possible to see how presentation of the cycle keys in the reverse order allows the algorithm to change from encryption to decryption. Consider the 'ladder' diagram of figure 2. In this figure keys K1, K2, etc, are each 48 bit cycle keys, not to be confused with the 64 bit encryption key. In the figure, for the sake of compactness, we show only four cycles of encryption followed by four cycles of decryption. R and L stand for the respective contents of the R and L registers; the function F stands for the combination of the expansion permutation E, the exclusive-OR of the result with the cycle key (Kn), the substitution in the S-boxes and the effect of the permutation P. The symmetry of the diagram depends on the nature of the exclusive-OR operation; if, in this way, the same quantity is successively combined twice with a parameter, then the result is to yield the original value of the parameter. Note that in both encipherment and decipherment modes the R and L states alternate regularly in the diagram, except for the last stage; this represents the fact that an essential feature of

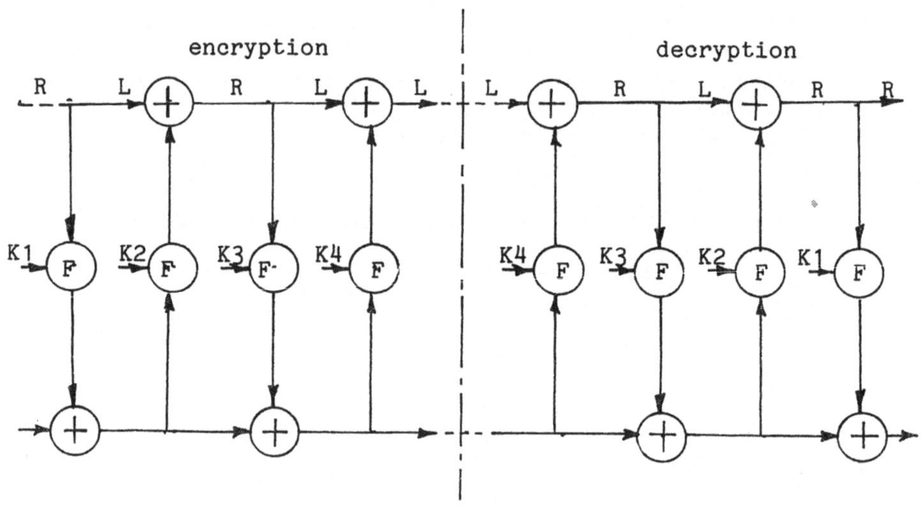

Figure 2. Ladder diagram showing symmetry of DES encryption and decryption

the operation of the algorithm, after the last cycle, is one final exchange of the contents of R and L.

Now a requirement of an acceptable cipher function is that any change to a bit of the input block should cause random changes of output bits; it should not be possible to relate particular changes of input to the way the output changes. If this were not so then the algorithm would be too easily invertible by a cryptanalyst. A careful study of the influence of input changes on the output has been made by Meyer and Matyas; an account of this work is expected to appear in their forthcoming book (2). It is asserted that after 5 cycles of encryption a complete mixing of bits has taken place, obscuring any effect that input patterns may have on the output pattern. Since the algorithm contains 16 cycles it is believed that inversion of the algorithm without knowledge of the encryption key is effectively impossible. On this account the DES algorithm is believed to have sufficient strength.

Attempts (3) to demonstrate weaknesses in the choice of the S-box tables, and thereby to define an algorithmic method of cryptanalysis, have not been successful. The design principles of these tables have not been published and there can be seen certain unexplained structure, but no method has been devised to exploit this.

Serious criticisms of the strength of the DES algorithm have been based on the size of the key domain. Since the effective key length is 56 binary bits, the number of different possible keys is about 7.2x10**16. Hellman has put forward suggestions (4) that, on certain assumptions, it might be possible to build a cryptanalysing machine which, given matched pairs of plaintext and ciphertext, would search for the corresponding key by trying all possible keys; this process is commonly called exhaustive search.

Hellman's main assumption is that a DES chip could be built that would carry out the algorithm in 1 microsecond; the function of the chip would be to test a key against a known ciphertext-plaintext pair. Input and output would be minimal because the chip would be programmed to try the next key in succession until success was achieved. By assembling one million such chips (the feasibility of which is another assumption), all working in parallel on different parts of the key domain, Hellman asserts that it should be possible to identify the correct key within half a day on the average. There is no doubt that such a cryptanalysis machine would be extremely expensive and would consume an immense amount of power; furthermore it might be difficult to maintain. The fastest DES chip known today carries out the algorithm in just under five microseconds; but a factor of five is not really significant in this context. It is possible that advances in technology may make feasible the building of such a cryptanalysis machine, though these advances are quite likely to take place rather slowly. A fair estimate of the future useful life of the DES algorithm is until 1990. After that it may be necessary to seek to define a replacement algorithm.

Though, in general, the DES algorithm generates pseudo-random bits in its output, it is necessary to remember certain regularities in its behaviour; in particular it is advisable to avoid certain key values. For example if the effective encryption key bits are either all zeros or all ones, then all the sixteen cycle keys will have identical values. The result is that two successive encryption operations with either of these keys results in restoration of the original plaintext without any decryption operation. The same is true for those keys which fill the C register with ones and the D register with zeros, or vice versa. Keys of this class are known as weak keys and should be avoided.

Davies has identified (5) six further key pairs which he

has called 'semi-weak'; their behaviour is such that the effect on text of encryption with one key of each pair is exactly equivalent to decryption with the other key of the pair. This is due to the symmetrical character of the pattern of shifts used in encryption and decryption within the C and D registers. It seems advisable in practice to avoid using keys in this category.

We give in table 3 a listing of all the weak and semi-weak keys in hexadecimal form. It is worth noting that at least one manufacturer of equipment incorporating the DES algorithm has built in tests to detect and reject the weak keys should these be chosen accidentally by any process of key choice.

Table 3. Weak and semi-weak DES keys (hexadecimal notation)

Weak							
01	01	01	01	01	01	01	01
1F	1F	1F	1F	1F	1F	1F	1F
E0	E0	E0	E0	E0	E0	E0	E0
FE	FE	FE	FE	FE	FE	FE	FE
Semi-weak pairs							
{01	FE	01	FE	01	FE	01	FE
{FE	01	FE	01	FE	01	FE	01
{1F	E0	1F	E0	0E	F1	0E	F1
{E0	1F	E0	1F	F1	0E	F1	0E
{01	E0	01	E0	01	F1	01	F1
{E0	01	E0	01	F1	01	F1	01
{1F	FE	1F	FE	0E	FE	0E	FE
{FE	1F	FE	1F	FE	0E	FE	0E
{01	1F	01	1F	01	0E	01	0E
{1F	01	1F	01	0E	01	0E	01
{E0	FE	E0	FE	F1	FE	F1	FE
{FE	E0	FE	E0	FE	F1	FE	F1

Choice of keys for use in the DES should be made in such a way as to make them unpredictable to any potential intruder. Ideally they should be completely random, chosen by some process such as coin tossing. If complete randomness is not possible, then as good a random number generator as possible should be used.

MODES OF USE OF THE DES

Electronic Code Book Mode

We have already seen that the DES is essentially a block cipher, acting on blocks of text of 64 bits. A long message may be segmented into blocks which can be encrypted successively by repeated operation of the algorithm with the same key, producing an encrypted message of the same length as the original. Unfortunately this technique has at least two serious drawbacks.

If, within the plaintext, two 64 bit blocks occurring at different locations within the message are identical, then the two ciphertext blocks produced to correspond to these will also be identical. The result is that, though an intruder may not be able to decipher the blocks themselves and obtain the corresponding plaintext, the fact of repetition of plaintext will be plainly obvious. This may betray material information to an intruder carrying out a passive attack.

The second, and more serious, objection concerns an active attack in which the intruder seeks to change the message without alerting the intended recipient. If the DES is used in the simple mode just outlined, then there is nothing to prevent the intruder from altering the encrypted message, deleting selected blocks, rearranging their order or inserting blocks recorded from other transmissions which had been made under the protection of the same encryption key.

Because of its similarity to techniques of encoding messages, as distinct from enciphering them, mentioned in our first session on data security, this mode of use of the DES algorithm has been given the name 'Electronic Code Book' or ECB. For every 64 bit block of plaintext there is a defined 64 bit block of ciphertext determined by the key currently in use. The analogy with a code book is obvious.

It is recommended that ECB mode should be restricted to certain very specific purposes and should not be used for general message encryption. The purposes for which it is acceptable relate to exchange of encryption keys, or certain other encryption parameters of restricted length, between communicating entities implementing the DES. Because of the random nature of encryption keys, it is safe to encrypt these for transmission in ECB mode. An intruder is unlikely to gain any advantage because of the use of ECB mode for this purpose. The superior level key used to encrypt the key to be transmitted is sometimes termed a master key; its use will be strictly restricted to the protection of transmitted keys.

Now if ECB mode is unacceptable for the protection of long messages, then some other means must be devised for secure use of the DES algorithm. What is required is to make elements of the ciphertext depend not only on the corresponding elements of the plaintext but also on all the preceding elements of the plaintext. If this can be achieved, then identical repeated blocks of plaintext will be encrypted differently when they recur. Furthermore, it will no longer be possible for the intruder to interfere in an undetectable manner with the ciphertext by deletion, substitution or addition of ciphertext blocks.

Cipher Block Chaining

We shall now describe three techniques which have been devised to meet this need, beginning with the technique which operates in block mode. This is known as 'Cipher Block Chaining' or CBC and is illustrated in figure 3. Prior to encryption of a plaintext block, the latter is combined bit by bit in an exclusive-OR operation with the previous ciphertext block. At the receiver the process is reflected by first decrypting the received block and then carrying out an exclusive-OR operation on the decrypted block with the previous ciphertext block as the other input to this operation; the result of this is to yield the original plaintext. The symmetry of figure 3 about its centre line illustrates the symmetry of the operation.

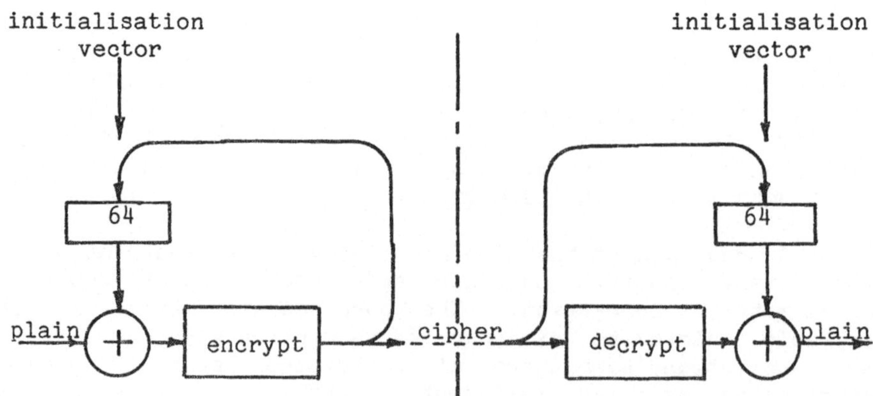

Figure 3. Cipher block chaining

Clearly the first plaintext block must be treated in a different way from later blocks as there is for this no previous ciphertext block. The solution to this is to decide what to load into the two 64 bit registers of figure 3 at the start of the operation. If the two registers contain unrelated random patterns, then the first plaintext block arriving at the receiver will be garbled; after that the system would self-synchronise and would perform correctly. The preferred method is to arrange that the contents of the two 64 bit registers should be identical at the start of transmission, each containing the same 'initialisation vector' or IV. The IV must be exchanged between the communicating entities in a secure fashion; for this purpose ECB encryption mode may be used.

In formatting plaintext messages for encryption it is highly desirable to introduce variability into the beginning of each message. This is because structured messages often begin with a standard preamble. Variability can be introduced very easily by placing a serial number right at the beginning of the message. When CBC encryption is in use the effect of the serial number is propagated right through the message when encrypted; therefore plaintext messages which are otherwise identical except for their serial numbers will appear totally different when encrypted.

Because of the dependency of received plaintext blocks on the previous ciphertext, it is pertinent to discuss the effect of transmission errors. If an error occurs during the transmission of a particular encrypted block, then clearly the corresponding plaintext produced at the receiver will be completely wrong. The effect is communicated also to the next block, but only in respect of the particular bits affected by the transmission error. After this received plaintext blocks will be correct unless a further transmission error occurs. An intruder who deliberately interferes with a transmitted block cannot predict what effect this will have on the corresponding plaintext, but can choose exactly the effect he wishes to have upon the following block. Therefore, if a transmission error is detected, for example by a cyclic redundancy check, then the whole received chain should be rejected and retransmission called for. The communication protocol should include provision for this.

The CBC mode of operation makes efficient use of the DES, because each 64 bit block of plaintext is processed by one operation of the algorithm, exactly the same as if ECB mode operation were being used. The only overheads are the message serial number at the beginning of each message and the manipulation detection check placed at the end of each. The transmission of the IV accounts for very little overhead, because the same IV can be used for many messages protected by DES in CBC mode. Note that the plaintext message is often called a 'chain' and the serial number is denoted a 'Chain Identifier' or CID.

One problem remains to be discussed in the CBC context, that of messages which do not segment into an integral number of 64 bit blocks. What should be done with the final segment which is less that 64 bits? One suggestion has been to pad out the final block with random bits. A more elaborate arrangement is shown in figure 4. Quite simply the principle is to take the final complete ciphertext block, re-encrypt this with the encryption key, then take as many bits from the result as are necessary and combine these bit-by-bit in an exclusive-OR operation with the partial final plaintext block. In the diagram P1 and P2 are complete plaintext blocks, while P3 represents the final incomplete plaintext block. C3 is the final transmitted ciphertext block formed by an exclusive-OR of P3 with m bits of the result of a second encryption of C2. An intruder can, by tampering with the transmission of this block, affect selected bits of the received plaintext. However, the redundancy of the chain should make this detectable and the recommendation is that any error should cause rejection of the whole chain.

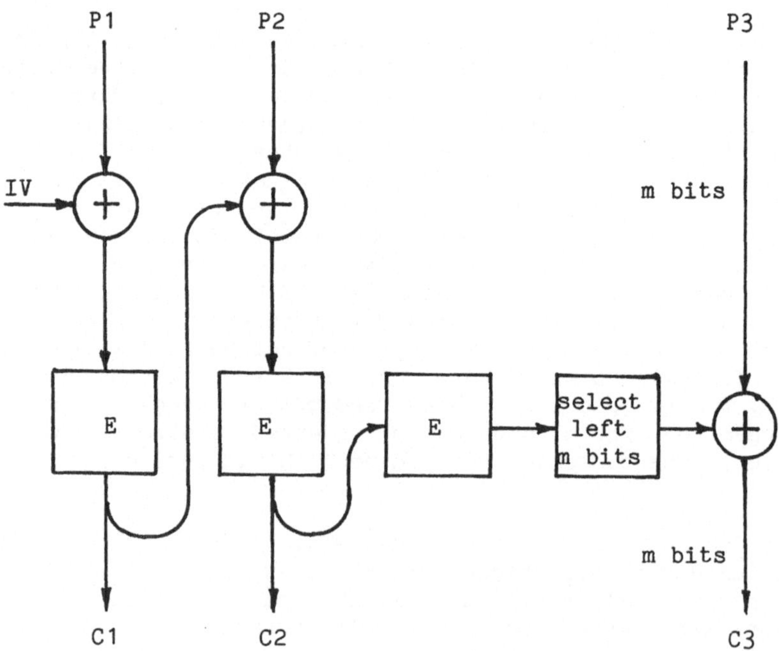

Figure 4. Final segment in CBC

CBC mode operation is evidently well suited where the data is presented in the form of fairly large units as in procedures such as the HDLC link protocol or the X25 interface. However there are applications where the data becomes available in much smaller units, sometimes even bit by bit, and must be transmitted without delay. For these applications some form of 'stream encryption' must be devised.

Cipher Feedback

The preferred mode of using the DES for encryption of a character stream is 'Cipher Feedback' or CFB; this is illustrated in figure 5. Note that here the encryption operation appears not in the direct path of the data, but in the feedback path. Note also that the DES device is placed in encrypt mode at both sender and receiver. Effectively the DES algorithm is being used to generate a pseudo-random bit stream which is being combined with the data stream in an exclusive-OR operation. The same additive bit stream is generated at both sender and receiver, so the result of the second exclusive-OR is to restore the plaintext.

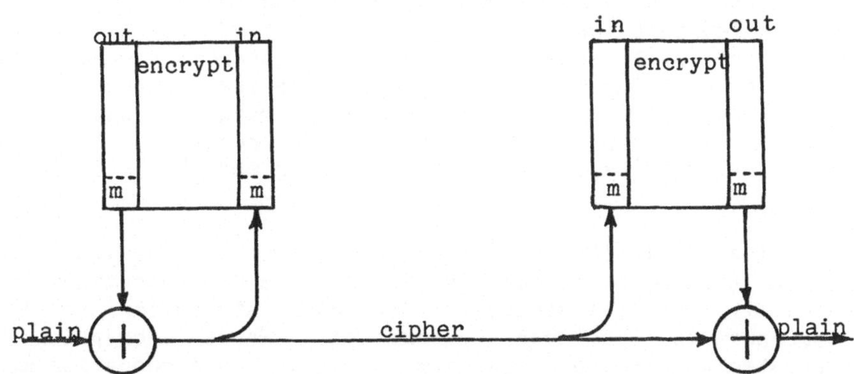

Figure 5. Cipher feedback

The size of the data 'character' may be anything from 1 to 64 bits, but the recommended values are 1 and 8. The size of the additive character taken from the output of the DES device is chosen to correspond to the plaintext data character size; the remaining bits of the DES output are discarded. The character produced by the exclusive-OR is transmitted and also fed into the input register of the DES device, which operates as a shift register; as each character is processed, the corresponding number of bits is lost from the other end of the shift register. For every plaintext character processed, one full operation of the DES algorithm is required; therefore the method is not as efficient in operation as is the CBC mode of use.

As in CBC mode an initialising vector is required; this is entered into each of the input registers of the communicating DES devices at the beginning of transmission. If this is not done, then a number of characters will be received in garbled form at the beginning of transmission, but the system will then self-synchronise. The IV can be exchanged between communicating devices before transmission; it is not necessary to encrypt this, because the DES operation stands between it and the data to be encrypted; so, the key being secret, an intruder gains no advantage from knowledge of the IV.

Transmission errors affect the correct receipt of messages in much the same way as was observed with the CBC mode. Any errors are fed into the input register of the DES device at the receiver and will cause the additive streams at sender and receiver to be different until the error has been shifted out of the input register. If the data character size is 8 bits, then 9 incorrect characters of plaintext will be received, the one directly affected by the transmission error and the 8 following ones.

Despite the inefficient use made of the DES device, CFB mode is the most frequently found mode of operation in commercially available devices. The main attraction is its transparency to the data transmitted. Where encryption is incorporated within a protocol structure, CBC mode operation may be more appropriate. We shall examine the incorporation of DES encryption in a protocol structure in the third lecture on data security.

Output Feedback

The remaining method of stream encipherment is superficially similar to CFB. Output feedback or OFB uses DES devices in a feedback mode as in CFB; the devices at sender and receiver are both placed in encryption mode. A flow diagram of the process is given in figure 6. The difference lies in the nature of the data fed into the DES input registers. In OFB the input consists of the bits selected from the DES output for exclusive-OR addition to the plaintext character. The effect of this is that the transmitted characters do not enter the DES input registers (the plaintext has no effect on the output stream from the DES device) and that transmission errors therefore do not affect any characters other than those directly involved; there is no error propagation.

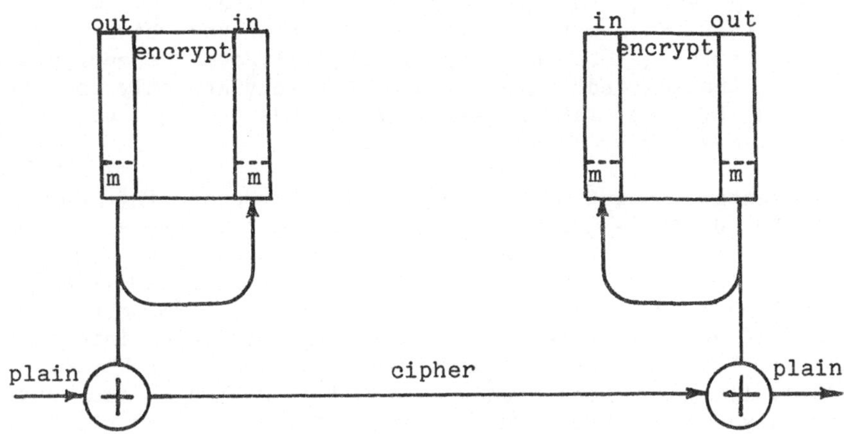

Figure 6. Output feedback

The OFB mode has been devised particularly with such applications as satellite transmission in mind. Here the bit error rate in transmission may be sufficiently high as to make multiplication of the error rate by a further factor of 8 or 10 due to the operation of CFB encryption unacceptable to the user.

The method has certain disadvantages, particularly that the current ciphertext depends only on the current plaintext and the current state of the pseudo-random stream drawn from the DES output; it does not depend on the previous plaintext in any way and therefore the valuable property of message chaining has been lost. It is particularly important with OFB not to use the same initialising vector with the same key for different messages. If this is done, then the pseudo-random streams produced on each occasion will be identical and a simple logical operation will enable the effect of the additive stream to be eliminated; cryptanalysis can then proceed purely on the basis of a 'likely word' attack on the two messages.

KEY MANAGEMENT FOR THE DES

It is of obvious importance that communicating DES devices shall contain the same encryption key. This may be achieved quite simply where the devices are connected in the same physical line. Each device will hold a master key which is used for secure exchange of other keys which will form the encryption keys for transmitted data. The master keys are only used for encryption of the latter keys, sometimes known as data encrypting keys (DEK) or session keys.

The establishment of master keys in communicating devices cannot be done securely by key exchange protected by DES encryption. This is because no superior level key would be available for their protection. Therefore DES master keys must be established by a physical visit by somebody entrusted with the task of safe key transport. Keys are entered into some commercial devices by using dials, thumbwheels or keyboards; the 'travelling key man' could carry a list with him and enter master keys into each device visited according to his list. Because the list may be lost or stolen, or even betrayed by the man, this is not the most secure method of master key transport. A better method is to use a key dispensing module, such as some manufacturers supply; this module contains a set of keys which it will issue, one at a time, to each DES device into which it is plugged. Once the key is issued by the key dispenser, it is deleted from the dispenser. For additional security one manufacturer uses an optical link from the dispenser to the cryptographic device. Design and operation of key dispensers requires careful study; Davies has examined (6) some of the factors involved.

Encryption key management may be made a task of the computer or communication system which is being protected, in which case appropriate enhancement of protocols and provision of special functions will be necessary. We shall examine these in some detail in the next session.

REFERENCES

1. National Bureau of Standards. "Data Encryption Standard." Federal Information Processing Standard 46, January 15, 1977.

2. Meyer, C.H. & Matyas, S.M. Cryptography - a New Dimension in Computer Data Security. (To be published by John Wiley & Sons, New York.)

3. Hellman, M., Merkle, R., Schroeppel, R., Washington, L., Diffie, W., Pohlig, S. & Schweitzer, P. "Results of an initial attempt to cryptanalyse the NBS data encryption standard." Information Systems Laboratory, Stanford University, Report SEL 76-042, Sept. 1976, revised Nov. 1976.

4. Diffie, W. & Hellman, M.E. "Exhaustive cryptanalysis of the NBS data encryption standard." Computer, 10, 6, 1977, pp. 74 - 84.

5. Davies, D.W. "Some regular properties of the 'Data encryption standard' algorithm." NPL Internal Note.

6. Davies, D.W. "The security of key transport methods." NPL Internal Note.

STANDARDISATION AND IMPLEMENTATION OF DATA ENCRYPTION

Wyn L Price

National Physical Laboratory, Teddington, UK.

ABSTRACT

Details are given of the current state of development of standards for DES applications, both in national and international contexts. Standards are needed both for modes of operation and for secure device environment. The incorporation of an encryption function as an enhancement of various communication protocols is discussed. A range of implementations of the DES is described. Lastly, the subject of key management is developed further.

INTRODUCTION

We have seen that the internal design of the DES algorithm is moderately complex. The effect of the basic algorithm on text is relatively simple in operational terms, a 64 bit block of plaintext is transformed to a 64 bit block of ciphertext under the influence of a key which is effectively 56 bits in length.

However, in order to attain adequate security it is necessary to use the basic algorithm in somewhat complex ways. To communicate successfully, devices implementing the DES must obey the same procedural rules concerning data formats, field sizes, methods of handling keys and initialisation vectors, etc. Now it is possible to create these rules on an ad hoc basis for particular applications, but it is far better to develop agreed standards to which any DES user will conform.

K. G. Beauchamp (ed.), New Advances in Distributed Computer Systems, 311–325.

The DES became effective as a Federal Information Processing Standard in July 1977, publication of FIPS 46 (1) having taken place in the previous January. This established the algorithm as a standard in the domain of the United States Government, but gave it no status outside this domain. Work on a standard encryption algorithm for general public application in the United States began within ANSI committees early in 1978. The current ANSI committees with responsibilities in this area are X3T1 and X3S3.8. Publication of an ANSI standard for data encryption is expected shortly at the time of writing (April 1981). This will be identical in every functional particular with the DES expressed as in FIPS 46; the title of the standard will, however, be 'Data Encryption Algorithm', to comply with conventional nomenclature, and the description of the algorithm has been re-written.

In the international domain Technical Committee 97 of ISO has set up a special working group (Working Group 1 or WG1) to consider standardisation of encryption procedures. This is a novel arrangement for an ISO committee. WG1 held its first meeting in January 1981 and is expected to meet again later in 1981. As regards a data encryption algorithm, WG1 has recommended the adoption of the US DES as the basis for an international standard, but calling it 'Data Encryption Algorithm 1', to make the point that it may not be the only algorithm in this area that may require standardisation. The text of the DEA1 draft standard is at an advanced stage of preparation.

Input to WG1 has come from various member organisations of ISO; much of the original impetus in its formation came from the British Standards Institution, which set up committee DPS/21, later to be renamed OIS/21, to consider data encryption matters.

MODES OF OPERATION STANDARDS

The chief modes of operation devised for safe use of the DES have been outlined in the previous session on data security and it has already been said that they require definition as standard procedures. As might be expected the first efforts to produce such standard definitions came from within US Government agencies. This work began at about the same time as the publication of FIPS 46 and was shared between the National Communications System and the National Bureau of Standards, respectively responsible for Federal Standards in the domains of data communication and data processing; the formal responsibility lay with the former organisation.

Federal Telecommunications Standard (FTS) 1026 (on interoperability and security requirements for use of the Data Encryption Standard in data communication systems) has been in preparation over a number of years, has passed through a number of drafts and is still (April 1981) not available in published form. The objectives of FTS 1026 are to achieve interoperability and the required level of security; in respect of the latter, prevention of plaintext disclosure, detection of fraudulent insertion, detection of fraudulent deletion, detection of fraudulent modification and detection of replay are required. A recent draft of this standard includes a description of ECB, CFB and CBC modes of operation, together with detailed definitions of plaintext formats and service message formats used for setting cryptographic modes and exchanging cryptographic parameters. A description is given of a method of using the DES as the basis for data authentication. Though the emphasis of FTS 1026 is upon secure communication over a direct line, reference is also made to the place of data encryption procedures within the structure of standard communication network architectural models.

The National Bureau of Standards is independently publishing a Federal Information Processing Standard, FIPS 81 (2) of 1981, which gives an operational description of the modes of operation, including that of OFB, excluded from the May 1980 draft of FTS 1026; authentication-only mode is relegated to an appendix, and no details are given either of plaintext data formats or of service message formats. The document is simply a straight-forward statement of the main modes of operation and on this account is easy to follow. Its existence somewhat preempts the publication of FTS 1026. It is understood that the latter, when published, will not repeat the material contained in FIPS 81, but will use this as a reference document.

The National Bureau of Standards has also in preparation a publication entitled 'Guidelines for Implementing and Using the NBS Data Encryption Standard', a volume in its 'Guideline' series. This will provide an excellent general discourse on the DES and its modes of use, together with subjects of wider scope such as key management.

Activity on standardisation of the modes of operation has also been going on within ANSI and BSI. OIS/21 has considered drafts of modes of operation standards which will be passed on to WG1 of ISO/TC97. One such draft elegantly expressed the modes of operation in terms of formal statements in the high level language ADA.

GENERAL SECURITY REQUIREMENTS FOR THE DES

Not only are communicating devices implementing the DES required to conform to standards regarding compatibility, but it is also necessary to make them proof against various physical attacks intended to interfere with their operation or to cause disclosure of unprotected plaintext or of encryption keys. Work has been proceeding in the United States under the aegis of the National Security Agency to produce a Federal Standard covering this aspect of the use of the DES. This standard, FTS 1027, has, like FTS 1026, been under preparation for a number of years, but it is understood that in this case publication may be expected soon.

The aim of FTS 1027 is to define manufacturing procedures and usage techniques that will prevent inadvertent transmission of plaintext, prevent the theft of equipment with encryption keys installed, prevent disclosure or unauthorised modification of encryption keys, delete encryption keys if tampering is detected, prevent transmission of encryption keys and disable the equipment if malfunction is detected. Attention is given to physical security, including electrical screening, locks and methods of key entry. Furthermore procedures are defined by which the equipment carries out integrity checks upon the performance of the DES algorithm.

This aspect of DES application has not yet received attention from the BSI committee nor, as far as is known, from the ANSI committees. ISO/TC97/WG1 has not yet considered the matter. Nevertheless these standards bodies may eventually be called upon to prepare security standards for the design of equipment incorporating the DES.

ENCRYPTION AND DATA COMMUNICATION PROTOCOLS

Already considerable attention has been given by various organisations to the inclusion of data encryption as an option within various data communication protocols. It is convenient in the present discussion to consider these within the structure of the ISO Open Systems Interconnection (OSI) architectural model, which is illustrated in figure 1. This structure is intended more as a guide than as an authoritative taxonomy of functions. Even so it is acquiring considerable authority as a medium for describing communication protocols.

The actual definition of OSI (3) makes reference to the possibility of incorporating data encryption only at level 6, the presentation level. Nevertheless data encryption may in principle be installed at any of the levels, or at more than one

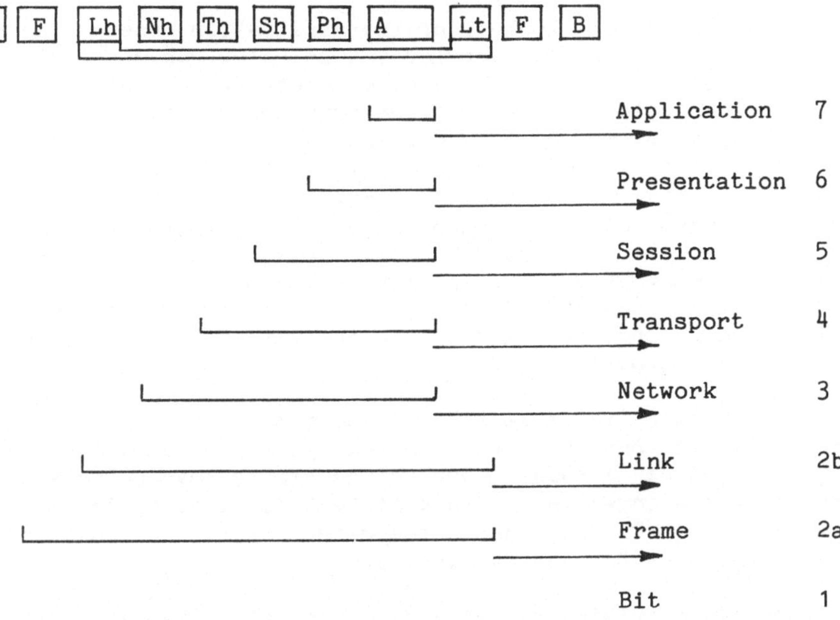

Figure 1. Extent of encryption within OSI architecture

of them. The standard protocols which figure at the various levels of OSI are in various stages of development. Levels 1 to 3, physical, link and network, are already well defined, and level 4, transport, protocols are in an advanced state of preparation. Much remains to be done at the higher levels, 5 to 7, session, presentation and application. The incomplete state of protocol definition hinders consideration of the encryption function at some levels. A preliminary discussion of this subject may be found in Price (4).

A protocol hierarchy which is more completely defined is that of the Teletex service, the subject of several CCITT recommendations. These include a full definition of the document level, corresponding to OSI level 6, and the session level, corresponding to OSI level 5. Proposals have been put forward (5) for enhancement of the Teletex protocol recommendations with encryption capability and it seems likely that work will commence shortly to put these ideas into practice.

It is a matter of some debate whether encryption should be provided within the communication subnetwork by the network authority or within the domain of the higher level protocols under the more direct control of the application user. Should the function be provided at level 4 or below, or at level 5 or above? Provision of encryption by the network authority implies that the encryption keys and the encryption function would be under the control of the authority and that encrypted data could be read in plaintext by the authority should it so desire. Indeed the regulations which govern the operation of data communications under some authorities require that the authority shall be able to read all the communications carried. This is especially relevant when trans-national border data flow is considered. However, though interesting, these are semi-political considerations which are not strictly relevant to our present discussion.

That which is relevant is the fact that encryption at a high level in the hierarchy effectively protects the data contained in a message, but of necessity leaves the message headers of the lower levels in plaintext form, readable by any eavesdropper. Figure 1 shows the extent of protection afforded by encryption at the various levels.

Clearly provision of encryption at all levels would be expensive and might impose substantial overheads on the protocols at some levels. Adequate protection is attainable by installing encryption at two or, at the most, three levels. Encryption at level 2 and at level 6 seems a reasonable proposal.

We examined in the first lecture on data security the merits of various ways of installing encryption in the subnetwork, at the lower levels of the OSI architecture. Proposals from the ANSI committee X3T1 on encryption at levels 1 and 2 are in an advanced state of preparation. TC97/WG1 has set up a working party to examine this area.

We now turn to consider encryption at level 4 and above. The aim of OSI level 4, providing the transport service, is to allow transparent transfer of data in a reliable and cost-effective manner between the session entities located in the layer above, relieving these entities of any concern for the detailed way in which this is carried out. It is at the transport service level that end-to-end call control is available on a fully developed basis. Therefore this level is a natural candidate for the location of an encryption facility.

Encryption provided by the transport service will be applied in response to a request from the session level for an encrypted call. Encryption may also be imposed because the distant session entity will only accept a call on condition that this is encrypted.

The transport service peer protocol provides for a negotiation phase in which desired call parameters are offered to a distant location; the latter may either indicate acceptance of the call as offered, it may reject the call or it may offer to accept the call provided alternative call parameters are substituted. A negotiation phase is a very important part of the protocol of any transport service encryption enhancement (or indeed for encryption at any level other than the lowest). The potential communicating parties must at least achieve agreement on:

a) the encryption algorithm to be used
b) the mode in which the algorithm is to be used
c) the encryption keys
d) the initialisation parameters.

Exchange of encryption parameters may begin during the call negotiation phase, but it is unlikely that agreement on the encryption parameters can be completed during normal call negotiation, because encryption is an enhancement at level 4; the encryption negotiation protocol does not form part of the connection protocol. Encryption negotiation can only be completed after a call has been set up, at which stage the call is unencrypted. If the encryption negotiations are successfully concluded, then the call will be switched into an encrypted mode. If encryption negotiation fails, then the call must be closed down without delay. It is not recommended that a call shall be switched successively in and out of encryption mode several times during its lifetime.

It is permissible, without loss of security, to negotiate the particular encryption algorithm and the mode of its use in a fashion which can be overheard by potential intruders. It is particularly important not to expose encryption keys and, in CBC mode, the initialisation parameters in this way. When the DES is being used, as distinct from public key cryptography (which we shall meet later), this demands that a master key shall exist at each transport entity; exchange of secret parameters takes place under the protection of the master key. Alternatively lists of secret parameters may be prepared beforehand and installed by a physical visit at each communicating location; selection of parameters may then take place openly during the negotiation dialogue without loss of security. Public key cryptosystems present an important alternative to these methods of parameter management.

Once an encrypted call has been set up, the integrity of the transferred data must be assured. Cryptographic data chaining must be employed; at the transport level a suitable form is Cipher Block Chaining which we have already encountered. Successive plaintext chains must be linked together by a system of chain identification numbering which forms part of the ciphertext chain. The transport service protocol does not guarantee delivery of data, but it does guarantee that data delivered will be in the correct sequence as it is offered to the session entity. If a malfunction of the call happens, for example if a Reset occurs, then some data may be lost. Loss of data should be reported to the session entity on whose behalf the transport connection is operating; this entity should decide on the point from which retransmission should commence. Entire retransmission of the data of an encrypted call should be avoided on efficiency grounds; retransmission of the same data with different cryptographic parameters must be avoided on the grounds of cryptographic security.

When the stream of encrypted data is ended, the call must be closed down in an organised fashion. It is important to ensure that the Disconnect message is sent only after the delivery of data to the destination (including the complete cryptographic integrity check, in case some retransmission is needed). Otherwise data may be overtaken by the Disconnect message and lost, making it impossible to complete the check correctly. Once it is clear that the data transfer is satisfactorily complete, the call should be closed down in the normal way, but the encryption enhancements at both source and destination should overwrite any current data encryption keys and initialisation parameters with random data. Nothing should be left that would enable an opponent at some later time to make sense of encrypted data recorded in transmission somewhere within the subnetwork.

The lowest architectural level that will be under user control is the session layer, level 5. If encryption is installed here the user has considerable assurance that his data is well protected. However, a session layer may work on behalf of different users; separation within the layer must be assured. Encryption at level 6, presentation, provides better separation between users, but, since this layer is concerned with data formatting and has a heavy work load, it may be considered that encryption might add too much to this load. Encryption at layer 7, application, would seem to conflict with the requirement that layer 6 be able to carry out formatting functions since these would need access to the plaintext data. Nevertheless there are bound to be cases where the required level of security can only be provided at level 7 or within the user processes themselves; in such instances level 6 cannot carry out formatting functions in the normal way. Opinion within the standards bodies leans at present towards providing encryption at level 6.

IMPLEMENTATIONS OF THE DES ALGORITHM

During the period since the announcement of the standard algorithm in 1977 at least twelve manufacturing organisations have produced devices which incorporate the DES. All but one of the devices which we shall mention are of American origin and are subject to export control under the US Munitions Act. They range from single chips, through fast multiple chip sets, circuit boards and armoured line bracketting boxes, to elaborate devices which are associated with sophisticated key management systems. Validation of the various devices as containing true implementations of the DES algorithm is offered as a service by the National Bureau of Standards. The techniques used for validation have been described by Gait (6). NBS has so far carried out about 15 device validations.

Table 1. DES chips and chip sets

Special purpose LSI	Maximum Speed	Modes Available
Advanced Micro Devices		
Am Z8068	> 1 Mbyte/sec.	ECB, CFB & CBC
Burroughs		
MC884	208 Kbytes/sec.	ECB & CFB
Fairchild		
9414	1.6 Mbytes/sec.	ECB, CFB & CBC
Motorola		
MGD68NE	50 Kbytes/sec.	Not known
Western Digital		
WD2001E/F	167 Kbytes/sec.	ECB
WD2002A/B		ECB
WD2003		ECB & CFB
Microprocessor		
Intel		
8294	80 bytes/sec.	ECB
Texas Instruments		
9940	600 bytes/sec.	ECB
American Microsystems		
S6894	1.45 Kbytes/sec.	ECB

Table 1 gives a list of some single chips and chip sets, both special purpose LSI and microprocessor based. All of the devices offer ECB mode, most of them offer CFB and one or two provide CBC. We have no information at present on the availability of devices with OFB mode. The algorithm time of operation ranges from 100 milliseconds for one of the microprocessors (intended for low bandwidth operation with cash terminals) to just under 5 microseconds for an LSI four chip set (offering an effective data rate of 1.6 Mbytes/second and therefore suitable for encryption of bulk data on a fast transmission line).

Devices such as those listed in Table 1 are intended for installation in OEM equipment. Physical protection is not built in to the devices, which must be used in secure installations.

Table 2. DES circuit boards

	Maximum Speed	Modes Available
Motorola		
DES1100DSM		CFB
MGD6800DSM	2.5 Kbytes/sec.	ECB & CFB
MGD8080DSM	2.5 Kbytes/sec. (both in CFB)	ECB & CFB
Rockwell-Collins		
CR-300	5.6 Kbytes/sec. (ECB mode)	ECB & CFB

Two manufacturers offer printed circuit boards (listed in Table 2) with DES capability, which provide more elaborate functions, particularly with regard to key management. They are designed to be compatible with various microprocesor systems. The Motorola devices afford two encryption keys, effectively corresponding to the data encryption and key encryption categories which we have already met. The Rockwell-Collins devices allow for as many as 32 different data encryption keys stored in RAM on the encryptor card, either in plaintext or encrypted under a master key. To prevent access by an intruder to RAM-stored keys, the whole unit would need added physical protection. All these devices offer ECB and CFB modes of operation; their operating speeds range from about 2.5 Kbytes/second to 5.6 Kbytes/second.

Several manufacturers have produced armoured encryption boxes incorporating the DES. These are intended for installation in circumstances where the physical security of the device cannot be assured by other means. A list of several such boxes is given in Table 3. Most of the devices are suitable for line-bracketting or point-to-point installation; some are in addition able to handle multi-drop communication with as many as 32 terminals and even more complex configurations.

Table 3. DES armoured boxes

	Maximum Speed	Modes Available
Burroughs		
TA3200	2.4 Kbytes/sec.	CFB
Comcrypt		
TX100	6 Kbytes/sec.	ECB, CFB & CBC
IBM		
3845/46	2.4 Kbytes/sec.	CFB
Motorola		
DES3100NSM	1.2 Kbytes/sec.	ECB & CFB
DES4100NSM	7 Kbytes/sec. (ECB mode)	ECB & CFB
Racal-Milgo		
Datacryptor	1.2 Kbytes/sec.	CFB
Rockwell-Collins		
CR-200(220)	1.2 Kbytes/sec.	ECB, CFB & CBC

It is fundamental to the design of all armoured encryption boxes that it should be virtually impossible to interfere in an undetected manner with the correct operation of the unit. If cryptographic key setting is available on an integral keyboard, this is either located behind a locked panel or is protected in some other way by physical locks. Often a cryptographic key can be loaded via a special key issuing module to be plugged into the box under the control of a physical key. Attempts to break into the box result in the destruction of any cryptographic keys loaded within the device; interference with the communication connections has the same effect. An internal auxiliary power supply is usually provided which sustains the key store if external power is interrupted.

A complete cryptographic system has been designed by IBM around the DES algorithm; this is an integral part of the IBM Systems Network Architecture and is available within the operating systems of IBM 360 and 370 computers. it is designed to protect data when in transit between host and terminals and when in storage within hosts. The DES cryptographic facility is provided within host computers either through the Programmed Cryptographic Program Product or through a piece of hardware, the 3848 Cryptographic Unit, which needs special software, the Cryptographic Unit Support Program Product, to control it. At terminals connected to 360/370 hosts, the encryption capability is provided by special hardware. The maximum data transfer rate is 1.5 Mbytes/second, the mode of encryption being CBC.

A novel feature of the 3848 is the way in which keys are handled. As in many devices two key levels are operated, one for data encryption and one for key encryption. The master key which is used to encrypt the data keys is effectively twice the length of a normal DES key in order to give enhanced protection to the data keys stored for use within the device. This is effected by splitting the master key into two parts; if these are designated respectively Ka and Kb, then the encryption of a data key requires the following sequence of operations-

1) Encrypt under Ka
2) Decrypt under Kb
3) Encrypt under Ka

In addition to the system surrounding the 3848 device, IBM have an elaborate key management scheme which has been described in detail by Ehrsam et al (7). This provides for a whole range of master key variants used to protect communication keys and file storage keys. A set of functions available to users provides for translation of encrypted data from one key to another; this is required, for example, if data has to be sent to another host computer. Precautions are taken to prevent misuse of the cryptographic functions by any user or by an intruder.

The IBM key management scheme is built around the concept of a host computer which controls a set of connected terminals and is able to communicate with other host computers of peer status. This is not applicable to communication networks in general, which may contain less well ordered structures. We shall conclude our consideration of standardisation and implementation of data security with a brief look at methods of key management in general.

METHODS OF KEY MANAGEMENT

We have underlined several times the need for communicating cryptographic devices to hold the same key and, possibly, other cryptographic parameters in common as well. We have seen that devices used in a line bracketting mode may exchange keys if each device is primed with a master key by a physical visit from a security officer. We have just mentioned the elaborate IBM key management scheme. In the more general networking context we have indicated the need for peer protocol entities to be able to negotiate cryptographic parameters.

The enhancements that must be made to communication protocols will permit secure exchange of cryptographic parameters only if the protocol transactions are proof against tampering or disclosure to third parties. In a small network this could be achieved by equipping each communicating entity with a master key specific to each of the other entities with which communication will be required. Since master keys would be equally valid in either direction between entities, for a network of N communicating entities N(N-1)/2 different master keys would be sufficient. Each entity would need to store (N-1) master keys.

For a very large network with many communicating entities the number of possible master keys could become unmanageably large. In such circumstances another solution must be found. This can be by the provision of a key distribution centre (KDC) which participates in the key management task. Each communicating entity requires a master key for communication with the KDC. To establish a data encryption key for communicating with a particular correspondent the KDC can be used in several ways. A survey of these has been made by Price and Davies (8). Generally speaking it is necessary when DES encryption is being used that the users shall trust the KDC to respect their secrets. Therefore this kind of system may be more acceptable where all the users belong to one organisation. If a public key cryptosystem is in use, then the same degree of trust may not need to repose in the KDC, though it is vitally necessary to prevent an intruder masquerading as the key distribution centre and taking over its role.

We shall look at one DES-based technique which involves the KDC. This is illustrated in figure 2, where user A wishes to communicate securely with user B. User A sends to the KDC a

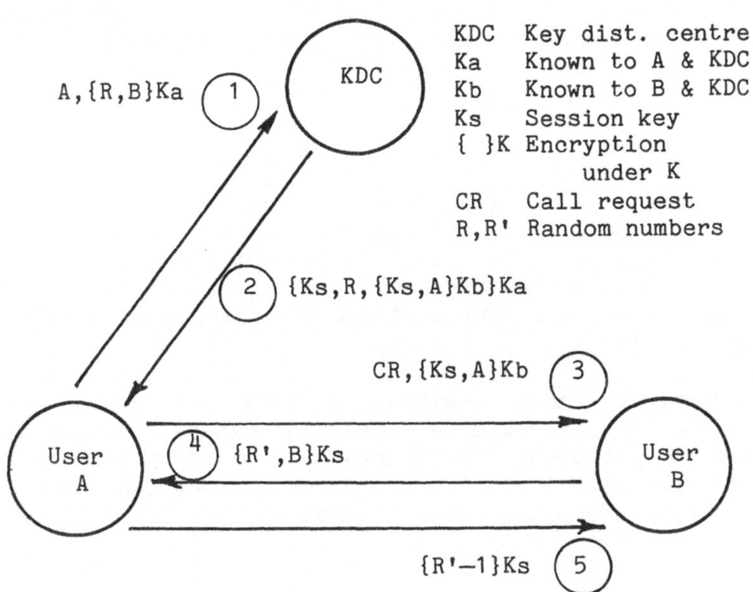

Figure 2. KDC with DES encryption

request for a session key for secure communication with user B. In his request he identifies himself (in plaintext); he also chooses a random number R and concatenates this with B's identity and enciphers this joint field with Ka, the key he holds in common with the KDC. The KDC in reply sends back the chosen session key Ks plus A's random number R, all enciphered under Ka. Also within the same enciphered field he includes a second copy of Ks plus A's identity enciphered under Kb, which is the KDC-user key of B; this part of the communication is therefore under double encipherment. A deciphers the whole under Ka, checks the validity of R (thereby confirming that the communication is currently valid and not a replay by an intruder of previous KDC reply) and installs Ks for message encryption to B. A sends a call request to B, including the copy of Ks plus A's identity enciphered under Kb. B deciphers this and thus obtains Ks. To make certain that the call request is a genuine live request and not a replay by an intruder, B chooses a random number R' and sends this, enciphered under Ks, to A in his call response. Finally A modifies R' in a standard way (eg R'-1), enciphers the result under Ks and sends it to B. At the end of this process A and B both possess the same encryption key, Ks, and are sure that they are talking to each other in a live call.

Where initialisation vectors are required to be exchanged, they can be handled and protected in similar ways.

Communication between user entities and the KDC should be sustained by the normal communication protocols supported by the network and its users.

CONCLUSIONS

Considerable progress has been made in defining standards for data encryption in the communication context. This is particularly true at the lower architectural levels. Much remains to be done at the higher levels, though a start has been made in this direction. The earliest results may be expected in the Teletex field.

Manufacturers have designed a range of devices which implement the DES algorithm in ways which can be incorporated in communication protocols. The emphasis has chiefly been on units which can be built in to OEM equipment or that can stand alone to protect point-to-point communication.

REFERENCES

1. National Bureau of Standards. "Data Encryption Standard."

Federal Information Processing Standard 46, January 15, 1977.

2. National Bureau of Standards. "DES Modes of Operation." Federal Information Processing Standard 81, expected mid-1981.

3. International Standards Organisation. Open Systems Interconnection. Document ISO/TC97/SC16 N227, October 1979.

4. Price, W.L. "Encryption in computer networks and message systems." Proc. IFIP TC-6 Intl. Symposium on Computer Message Systems, Ottawa, April 1981.

5. Davies, D.W. "Enhancement of Teletex procedures to incorporate encipherment and signature." NPL Report DNACS 42/81, April 1981.

6. Gait, J. "Validating the correctness of implementations of the NBS Data Encryption Standard." National Bureau of Standards, Special Publication 500-20, November 1977.

7. Ehrsam, S.M., Matyas, S.M., Meyer, C.H. & Tuchman, W.L. "A cryptographic key management scheme for implementing the Data Encryption Standard." IBM Systems Journal, 17, 2, 1978, pp. 106-125.

8. Price, W.L. & Davies, D.W. "Issues in the design of a key distribution centre." NPL Report DNACS 43/81, April 1981.

PUBLIC KEY CRYPTOSYSTEMS, AUTHENTICATION AND SIGNATURES

Wyn L Price

National Physical Laboratory, Teddington, UK.

ABSTRACT

The concept of public key cryptosystems is introduced; brief reference is made to the 'knapsack' system, whilst the so-called RSA system is treated at greater length. The need for message authentication leads naturally to a consideration of the role of public key cryptosystems as generators of digital signatures. Finally we see how the invention of public key systems has affected the task of key distribution within networks.

INTRODUCTION

The encryption algorithms which we have met so far in our discussion of data security have been of the type which employ one secret key for encryption and decryption. Entities which require to communicate in secret under encryption via such a system must each possess the same encryption key. As we have seen, this requirement presents additional organisational tasks for the supporting system.

On this and other accounts the concept of a 'public key cryptosystem' must be attractive. The principle is that one key is used for data encryption and another, different, key is used for decryption. The encryption and decryption algorithms which use these keys may be identical or different in structure, depending on which basic principle is being used for the encryption. In operation, widespread publication is given to the encryption key, whilst the decryption key is kept a secret

K. G. Beauchamp (ed.), New Advances in Distributed Computer Systems, 327–340.

by the receiver of encrypted messages. Clearly, there must be some kind of algorithmic relationship between the encryption and decryption keys. For adequate security it must be totally infeasible to derive a secret decryption key from a published encryption key. The encryption and decryption algorithm(s) are public knowledge.

The concept of public key cryptosystems first appeared in print in 1976, with the publication of the important paper 'New directions in cryptography' by Diffie and Hellman (1). At about the same time, thinking on similar lines was being developed by Merkle, though his results were not published until 1978 (2). Neither of these papers proposed a practicable public key cryptosystem, though there was evidence that such could be found without too much difficulty.

We show in figure 1 the principle of a public key cryptosystem; two keys are specified, ke for encryption, public knowledge, and kd, for decryption, secret. As we have already implied, ke and kd must be algorithmically related. In the figure we have represented this property by the generation of ke and kd from a common starting seed, ks; functions F and G respectively generate ke and kd from ks. Unlike encryption with the DES it is not possible to pass information in both directions protected by the same encryption key; each receiver of data has to generate his own key pair. It is particularly important that function F which produces the public key ke shall be 'one-way' - it must not be possible to invert function F and

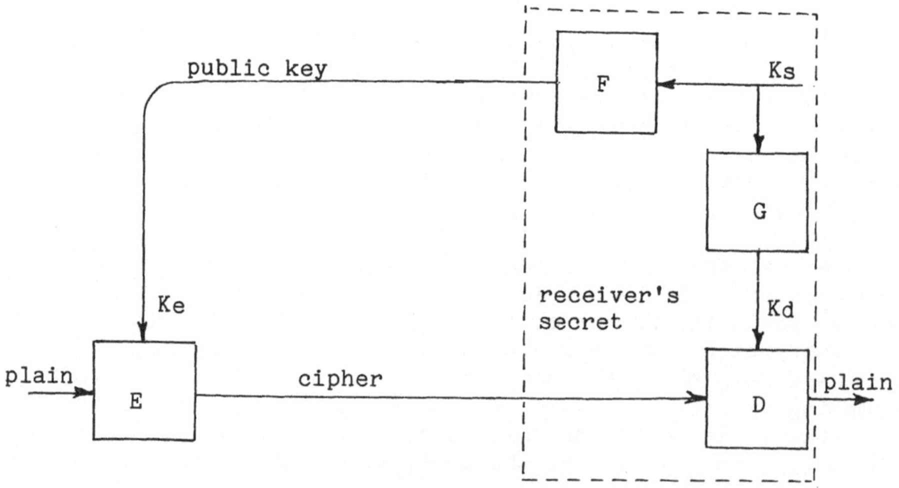

Figure 1. The principle of a public key cryptosystem

thereby derive ks, the starting seed, from which the intruder could easily derive kd. In order that inversion of the function may not be achieved by simply calculating all values of ke for all inputs to F and sorting these to order, the range of values of ke must be made suitably large.

1978 saw the publication of two separate and quite different practicable ideas for public key cryptosystems. These came respectively from Merkle and Hellman (3) and from Rivest, Shamir and Adleman (4).

THE TRAP-DOOR KNAPSACK

Merkle and Hellman's idea was to use the 'knapsack' problem, constructing a special kind of knapsack which could easily be solved by the possessor of secret information, but which would otherwise defy solution within a viable time. Knapsacks which could be solved easily by those in possession of secret information were called 'trap-door' knapsacks by Diffie and Hellman. Trap-door knapsacks do not have the property of commutation, that is to say it is not possible meaningfully to carry out encryption and decryption in either order; we shall see as our discussion develops that this property has significance when digital signatures are sought.

Some criticism has been levelled at trap-door knapsack systems; Schroeppel and Shamir (5) have published details of a time- memory trade-off which might yield solutions more rapidly if sufficient storage capacity is available. Shamir and Zippel (6) have described a variant on the basic knapsack idea, which they believe to give greater security. Shamir has published a method (7) of adapting knapsack cryptosystems for obtaining digital signatures.

The computational load of the knapsack algorithm is not particularly arduous, so that it can be carried out efficiently by standard computing equipment. However, the quantity of key material required for adequate strength is formidable. If several keys have to be held for communication with a range of users, the amount of storage could easily be prohibitive.

THE RSA PUBLIC KEY CRYPTOSYSTEM

We shall not say any more about the knapsack public key system, because we believe that it is more appropriate to examine the system of Rivest et al, known from the authors' initials as the 'RSA' system, in greater detail. We shall see that the RSA system has properties which make it particularly suitable for generating digital signatures. The RSA system is based on certain results in number theory which are discussed at some length in Davies et al (8). We have not the space in the present context to describe the basic principles in extenso.

The operation of the RSA algorithm depends upon a process of finite exponentiation. The plaintext is raised to a chosen power in modular arithmetic, with a defined modulus; to obtain the plaintext from the ciphertext a similar operation is carried out with a different exponent.

The modulus used in the finite exponentiation is a composite number formed as the product of two large prime numbers. It is critical to the security of the encryption method that an intruder shall not be able to factorize the modulus and discover the component prime numbers. This implies that 'large' should mean about 200 - 250 binary bits size for each prime.

The exponents used respectively for encryption and decryption are related algorithmically to the prime numbers which together constitute the modulus. If we call the two primes p and q, and the two exponents e and d (for encryption and decryption), then, if e is chosen to have no factors in common with p-1 and q-1, d will be given by the expression

$$de = 1 \text{ modulo } [\text{LCM } (p-1), (q-1)].$$

The operation of the encryption and decryption steps of the algorithm may be illustrated by the equations

$$y = x^e \text{ modulo } m \quad \text{and} \quad x = y^d \text{ modulo } m,$$

where x and y are respectively the plaintext and ciphertext.

The encryption exponent e and the modulus m are made fully public, whilst the decryption exponent d remains the secret of the owner of the key pair. It is because of this property that systems of this kind are known as public key cryptosystems. Senders of encrypted data to the owner of the key pair make use of the appropriate public key; only the intended recipient should be able to carry out decryption.

Clearly, if the ciphertext y is to be secure against decryption by taking the discrete logarithm, then the latter operation must be made infeasible. This can be done by ensuring that both p-1 and q-1 shall have large factors.

The strength of the RSA system depends on the difficulty of

obtaining the prime factors p and q by factorising the modulus m. The best available method of carrying out this operation is of the same order of complexity as the discrete logarithm problem; the number of elementary operations required for this purpose is given by the expression

$$\exp(\mathrm{sqrt}[\ln(m).\ln\ln(m)]).$$

The estimate of size of p and q required for necessary strength is based upon this formulation.

A cryptanalytic attack (9) on the RSA system which does not aim at discovering the decryption key but seeks to find the plaintext corresponding to a given ciphertext is based on the principle of repeated exponentiation with the encryption exponent. If such an operation is carried out, then eventually a result will be obtained which is identical with the original ciphertext. The result of the immediately previous operation will then represent the plaintext from which the ciphertext was generated by the exponentiation operation. The plaintext is thus revealed by a process of iteration without discovery of the decryption exponent.

Frustration of this attack depends upon a further constraint on the choice of p and q. If the large factors of p-1 and q-1 referred to earlier as requirements for strength are respectively u and v, then the further constraint on choice of p and q is that u-1 and v-1 should also have large factors. The attack by iteration will then be impracticable because the number of necessary successive exponentiation operations becomes infeasibly large.

A serious disadvantage of the RSA encryption system is the computational load placed on the computer in which the system is implemented. The computational operations normally available in most computers cannot be combined to perform the RSA calculations efficiently. A small machine might take say 10 minutes to carry out encryption or decryption. This is not likely to be acceptable in a real-life situation. Large machines might carry out the function in seconds.

Serious application of the RSA public key cryptosystem depends upon the availability of special purpose hardware designed to carry out the required operations efficiently. Hardware of this type has been designed at MIT and an LSI chip is at present undergoing testing; the design allows for encipherment of 512 bit numbers in about half a second. General availability of this hardware is not yet predictable.

A further problem facing users of the RSA system is that of

selecting suitable primes for use in the calculation. Finding suitable primes can be a time-consuming process unless powerful hardware is available; this could be embarrassing for the small user. One solution is to purchase prime pairs from a supplier; indeed offers of primes for sale have already been made. This is not likely to be acceptable where high security is sought; the primes would be known to third parties, who could easily calculate the encryption and decryption key pairs. A better solution is to have the special purpose chip programmed to calculate pairs of primes. The MIT chip has been designed with this aim in mind and can find prime pairs in less than 30 seconds. It would be an easy extension of the design to provide that the encryption key can be read from the chip, whilst the reading of the decryption key is prevented.

MESSAGE AUTHENTICATION

Straightforward message authentication can be achieved in several ways. The aim is to show that a received message has not been altered in any way during transmission. Encryption of a message with the DES can provide authentication, provided that the component parts of the message are properly chained together in the kind of way we have already outlined in earlier sessions. The receiver of a meaningful message that has been protected in transit by encryption has a good degree of assurance that the message has come from a source where the same encryption key is available as at the receiver. This is normally enough to satisfy the receiver of the message's provenance. The modes of operation standard for the DES includes a section on an authentication-only mode. This is used where a message may be sent in cleartext with an appended authentication field which is generated by a process involving the DES algorithm.

However, since the receiver, like the sender, possesses the means for enciphering as well as deciphering text, the receipt of such an enciphered text, or a plaintext with DES-generated authentication field, cannot be used as proof in any contractual dispute. The enciphered message or the DES-based authenticator could just as well have been created by the receiver as by the sender. Some device is necessary where a 'signature' is generated using information only available to the sender, but verifiable by the receiver or by an arbitrating third party. It is in this context that public key cryptosystems become particularly significant.

GENERATION OF DIGITAL SIGNATURES

It is expected that it will become increasingly important

to be able to prove the origin of messages received over a data communication network. If contracts are to be exchanged by such means, then it must be possible to show that they came from the right source and have not been falsified either by some third party or by their alleged recipient.

Attempts have been made to adapt single key cryptosystems, such as the DES, to allow secure digital signatures to be generated. Unfortunately such methods as have been published depend either upon the existence of expensive special purpose systems (10) or upon the availability of large quantities of special keys used for signature generation (11).

Figure 2 shows how some public key cryptosystems can be used to generate digital signatures. If the twin operations of encryption and decryption can be carried out in either order, regarded as arbitrary transformations, then a text to be signed may be 'decrypted' first, using the secret key information of the sender; the transformed text is transmitted as a signed document. The receiver of such a document takes the public key of the alleged sender and carries out an 'encryption' operation. If the result is meaningful, measured by the presence of some redundancy or other agreed material in the text, then the signature of the sender is established; the signed document could have come from no other source.

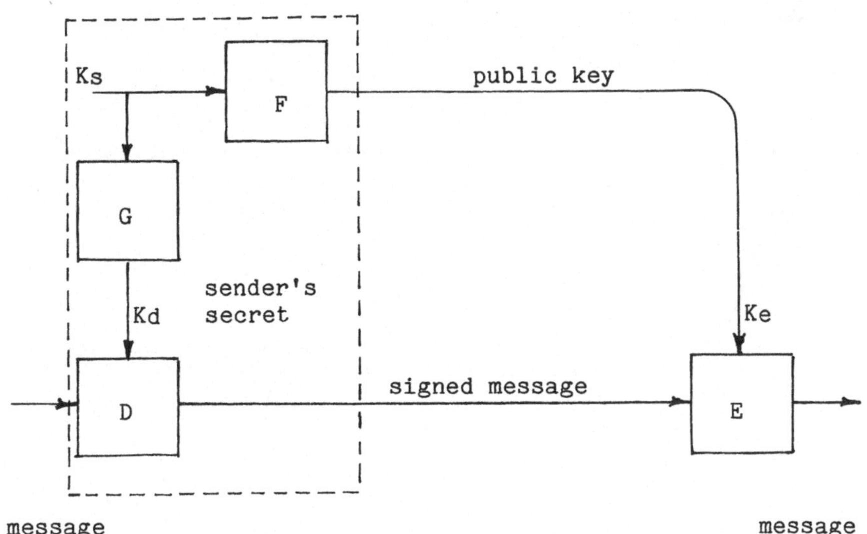

Figure 2. Message signature with a PKC

The integrity of such a signature system depends upon the completely reliable availability of the public key of each participating user. There must be a public key register of unimpeachable integrity available on the network for consultation by all users. We shall return to this point when we later discuss the question of key management in a network.

In the context of signature achieved by means of public key cryptosystems there remains a further important consideration, that of efficiency. We have already observed that the RSA public key cryptosystem is not computationally efficient unless special hardware is provided. In the absence of special RSA hardware it seems unlikely that RSA generated signatures would be viable if based on direct signature of entire messages. However, we can show that it is not necessary to apply the RSA algorithm to the entire message in order to obtain a signature.

If it were possible to generate a digest or summary of the message in which each message bit was reflected in a complex way in the summary, then signature of the summary would be a satisfactory alternative to direct signature of the whole message. Figure 3 illustrates this point. The sender of the

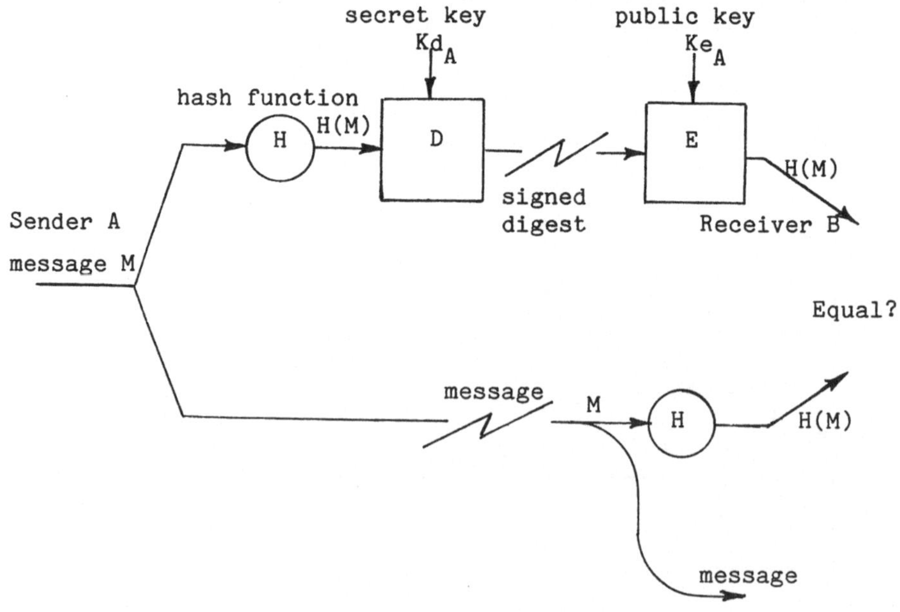

Figure 3. Signature of a message digest

signed message calculates the digest and signs it with his secret RSA key; he then sends the cleartext message accompanied by the signed digest. The receiver re-calculates the digest from the received plaintext message and also applies the public key of the sender to the received signed digest. If the results of these two operations are identical, then the receiver has verified the signature.

The design of the digest generation must be such that it is not possible for compensating changes to be made to the message which yield a deliberately altered message with the same digest. Methods of secure digest generation have been discussed in Davies and Price (12). We give here a method which is based on use of the DES algorithm since this can be carried out rapidly in appropriate hardware.

An apparently suitable digest may be generated by segmenting the message to be summarised into successive 56 bit blocks, choosing an initialisation vector and encrypting this repeatedly with the successive blocks of the message used as encryption keys. This is shown in figure 4. Note that for checking the signature the receiver must be in possession of an authenticated copy of the initialisation vector. The digest plus the initialisation vector together make up 128 bits which is a lot less than the field size usually used for RSA encipherment. Therefore the 128 bits may be replicated as many times as are necessary to more nearly fill the appropriate field.

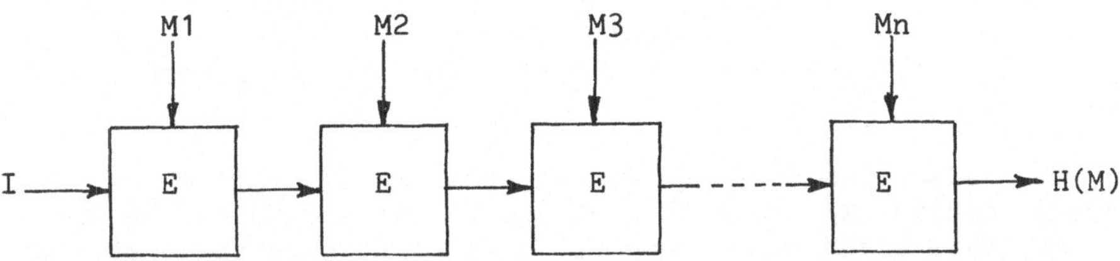

Figure 4. Calculation of message digest with DES cipher

Unfortunately a digest generated in this way is open to a form of attack known as the 'birthday method'; this was pointed out by Ralph Merkle and is discussed at some length by Davies and Price (12). Because the DES decryption operation is available to a potential attacker who wishes to generate false messages bearing an apparently valid signature, it is possible to take the digest and work backwards through the message, finding out intermediate values in the digest calculation using the message segments as keys in the decryption process which this involves. In a similar way the attacker can start from the initialisation vector and work forwards with the encryption operation finding out the intermediate values. The attacker selects the midpoint of the process and tries out message variations of the two halves of the message until a combination is found that chances to yield the same mid-value intermediate result. By sufficient trials it is likely that a message will be found that is different from the original but which produces the same digest generated by the method outlined. This message can then be substituted for the original and presented with the signed digest as an apparently correctly signed transmission.

An attack of this kind can be satisfactorily countered by carrying out twice the series of encipherments to form the digest. Each segment of the message is therefore involved twice in different parts of the digest generation and a birthday method attack is frustrated.

The question of arbitration of digital signatures is important. If public keys are available at some central source for consultation by any user, then the user who receives a signed message can check to his own satisfaction that the received signature is valid. Proving this validity to some third party involves more organisation. Indeed some public arbitration service may well be necessary. It is conceivable that this may be organised in connection with the register of public keys. What is needed is some enhancement to the communication protocol that allows a message signature to be submitted to the arbitrator; the latter responds in a way that is designed to give authoritative validity to a correct signature, with unambiguous attribution to the source of the message.

Such a service could be based upon safe storage of all current public keys and also all public keys used in the past. However, this might demand a great deal of storage space at the signature arbitration centre. A more economical solution is for all signatures received by users to be resubmitted to the arbitration service for re-signature (if the original sender's signature is deemed to be valid) with the secret key of the centre. All that is then necessary is to preserve the public keys used at various times by the arbitration service. Later arbitration on the validity of a signature is then based upon the public key of the arbitration service.

The whole integrity of this system of signature generation depends upon the secrecy of the secret key held by each participant. If such a key is disclosed, then the integrity of any associated signatures is immediately suspect. Indeed one could conceive of a dishonest user who claimed falsely that his secret key had been stolen and who thus tried to repudiate contracts which he had in reality entered into. Clearly an individual who tried this more than once would rapidly lose credibility, but secure handling of secret keys by honest users needs underlining. The problem of deliberate fraud is really no worse than that met when paper documents are forged using some physical authenticating device such as a seal.

THE ROLE OF PUBLIC KEY CRYPTOSYSTEMS IN KEY MANAGEMENT

It has often been suggested that the invention of the public key cryptosystem has transformed the problem of management of encryption keys. We saw in an earlier session that where a key distribution centre is involved in making DES session keys available at communicating users, it may be necessary to establish DES master keys at each location in order to allow secure communication between users and the KDC. This is much more economical that having separate master keys for session key transmission between all user pairs on the network. However, in a large network it does mean that the KDC must store a large number of master keys and that all users must possess a key to communicate with the KDC. We shall now consider whether the introduction of a public key cryptosystem makes key management more convenient.

If it is intended that the encryption of data shall be carried out with a PKC, then it is only necessary to ensure that the public keys of the communicating users are available to their opposite numbers. This is done by distributing public keys under the signature of the KDC. The only key that needs wide permanent distribution is the public key of the KDC. If this is available, then each user can check the signature on any communications arriving from the KDC and thus can obtain an authentic copy of the public key of the user with whom he wishes to exchange communications. An outline of a possible protocol for this purpose is shown in figure 5.

If, on the other hand, it is considered that encryption under a PKC is too inefficient for handling general data, then it is possible to define a way in which DES session keys can be distributed securely under PKC encryption by the KDC. A suggested protocol for this purpose is illustrated in figure 6. As in the protocol of figure 5, the only widely disseminated information is the public key of the KDC. Signature is not involved in this protocol; the session keys are conveyed under encryption with the secret key of each user. The number of steps in this protocol is greater than that of figure 5, but it

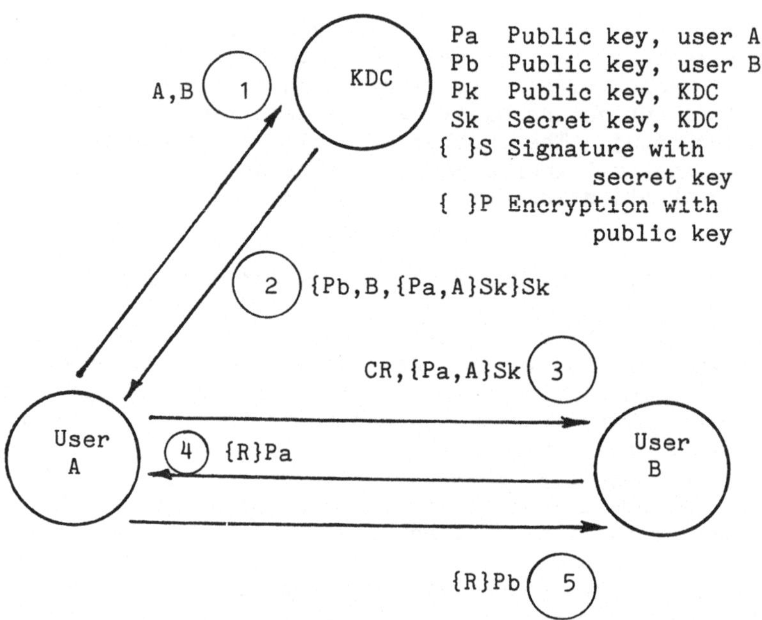

Figure 5. Public key distribution with KDC

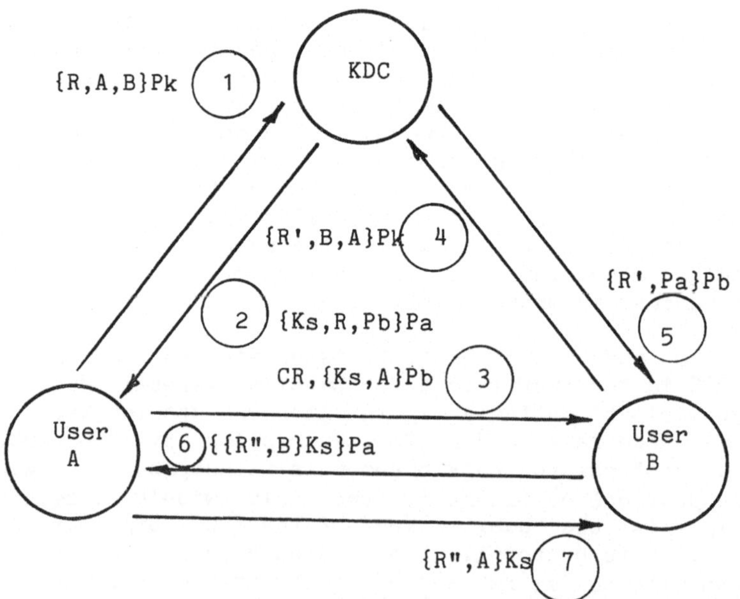

Figure 6. DES key distribution with public key protection

does not seem possible to reduce these and yet preserve the integrity of the key management. The protocol is discussed at greater length in Price and Davies (13); for the present it must be regarded as experimental and tentative.

CONCLUSIONS

Public key cryptosystems have caused a minor revolution in attitudes to data encryption. It is evident that they have potentially a very important role in generating digital signatures. They also have a useful function in making key management easier. Whether they are yet at a stage where their security can be regarded as proven is questionable. They provide a challenge to theoretical mathematicians and it is quite certain that their strength will continue to be intensively investigated. In a few years' time it may be possible to be more confident in asserting their viability as practicable cryptosystems.

REFERENCES

1. Diffie, W. & Hellman, M.E. "New directions in cryptography." Trans. IEEE Inf. Theory, IT-22, 6, November 1976, pp. 644-654.

2. Merkle, R.C. "Secure communications over insecure channels." Comm. ACM, 21, 4, April 1978, pp. 294-299.

3. Merkle, R.C. & Hellman, M.E. "Hiding information and signatures in trap-door knapsacks." Trans. IEEE Inf. Theory, IT-24, September 1978, pp. 525-530.

4. Rivest, R.L., Shamir, A. & Adleman, L. "A method of obtaining digital signatures and public-key cryptosystems." Comm. ACM., 21, 2, February 1978, pp. 120-126.

5. Schroeppel, R. & Shamir, A. "A T*(S**2) = O(2**n) time/space tradeoff for certain NP-complete problems." Proc. 20th IEEE Symposium on Foundations of Computer Science, October 1979, pp. 328-336.

6. Shamir, A. & Zippel, R.E. "On the security of the Merkle-Hellman cryptographic scheme." Trans. IEEE Inf. Theory, IT-26, 3, May 1980, pp. 339-340.

7. Shamir, A. "A fast signature system." MIT, Laboratory for Computer Science, Report No. MIT/LCS/TM-107, July 1978.

8. Davies, D.W., Price, W.L. & Parkin, G.I. "An evaluation of public key cryptosystems." NPL Report CTU1, March 1979. (Also published in Information Privacy, 2, 4, July 1980, pp. 138-154.)

9. Simmons, G.J. & Norris, M.J. "Preliminary comments on the MIT public-key cryptosystem." Cryptologia, 1, 4, 1977, pp. 406-414.

10. Smid, M.E. "A key notarization system for computer networks." National Bureau of Standards, Special Publication 500-54, October 1979.

11. Lipton, S.M. & Matyas, S.M. "Making the digital signature legal - and safeguarded." Data Communications, February 1978, pp. 41-52.

12. Davies, D.W. & Price, W.L. "The application of digital signatures based on public key cryptosystems." Proc. 4th ICCC, Atlanta, October 1980, pp. 525-530.

13. Price, W.L. & Davies, D.W. "Issues in the design of a key distribution centre." NPL Report DNACS 43/81, April 1981.

SYNCHRONIZATION OF CONCURRENT PROCESSES WITHOUT GLOBALITY ASSUMPTIONS

P.E. Lauer

Computing Laboratory,
University of Newcastle upon Tyne,
Newcastle upon Tyne, NE1 7RU, U.K.

The mathematically precise definition of a simple notation for designing and implementing programs involving synchronization of parallel processes introduced by C.A.R. Hoare in (2) and (3) is expressed in the COSY formalism. The latter formalism not only generalizes the semantic trace model used by Hoare but also reduces the complexity of the formalism for representing concurrency and simplifies the proofs of correctness carried out in the model. A number of versions of Hoare's examples are developed and absence of partial or total system deadlock is demonstrated.

INTRODUCTION

The advent of LSI and VLSI technology and microprogramming techniques has greatly increased the number of choices a digital system designer has for decomposing a system into ultimate subsystems which grow considerably more powerful and complex in function as technology progresses. Furthermore, designers have been learning how to combine such subsystems in new ways giving rise to systems which perform the same general functions as earlier systems but with a much greater degree of parallelism and distribution.

We will be concerened with the general problem of the definition and analysis of synchronization in concurrent systems without the assumption of a global nature for such entities as the system clock, the system state, the control, the observer, etc. Systems without such globality assumptions will be called distributed.

K. G. Beauchamp (ed.), New Advances in Distributed Computer Systems, 341–365.

If a combination of subsystems is to cooperate coherently to perform a particular system function synchronization is necessary to ensure proper joint behaviours of subsystems with respect to any parts of the system they share. If the resulting system is distributed and capable of parallel behaviours in subsystems, then the required synchronization must be specified without recourse to either a central clock, central control or global state. Many of the conventional formal tools the system designer has at his disposal, such as Automata Theory, are only suitable for adequately expressing sequential systems and they do this under the assumption of a global system state. Hence, designers have had to support such formalisms with additional formal and informal notions when applying them to the specification and analysis of concurrent and distributed systems.

A number of researchers have been seeking for an appropriate extension to the notions of automata theory and other formalisms to permit the inclusion of formal specification and analysis of synchronic properties of systems, such as concurrency, distribution, deadlock and starvation. Among the well developed approaches put forward we mention Petri net theory (1), the model of synchronized parallel processes of Hoare (2) (3), the calculus of communicating systems of Milner (4), and the concurrent system formalism COSY (7) (9), all of which seem to be converging to an analogous level of abstraction in dealing with the synchronic aspects of systems. Even though there are superficially certain striking similarities in all the approaches mentioned a more careful comparison reveals considerable divergence between some of the approaches with regard to the semantics of concurrency and distribution. Space will not allow a comparison of all these approaches here but we can roughly divide them into two groups depending on the choice made in modelling concurrency.

Petri Nets COSY	do not reduce concurrency to non-deterministic interleaving
Synchronized Parallel Processes (Hoare) Communicating Systems (Milner)	reduce concurrency to non-deterministic interleaving (at least for the purpose of conceptualisation)

Our research has indicated that the reduction of concurrency to non-deterministic interleaving is adequate for the complete treatment of such problems as proving the absence of partial or total system deadlock, even though the reduction makes the resulting formalism more complex than we feel is required. However, we are convinced that such a reduction makes it even more difficult or even impossible to deal with other system behaviour such as the so called starvation of some subsystem, or to formalize the notion of priority in the presence of concurrency and distribution.

The present paper will introduce the COSY formalism in direct comparison with the model of synchronized parallel processes of Hoare (2) to give these introductory statements more substance.

WHAT IS COSY IN GENERAL?

COSY (the name is derived from Concurrent System) is a formalism intended to simplify the study of synchronic aspects of concurrent and distributed systems where possible by abstracting away from all aspects of systems except those which have to do with synchronization. In other words, if we consider systems as consisting of activities or events we will abstract from any interpretation these events may have when we apply the system and merely consider how occurrences of such uninterpreted atomic events are related to occurrences of other such uninterpreted events in the system. Thus an occurrence of one event may presuppose a corresponding occurrence of another event, or an occurrence of one event may preclude the concurrent occurrence of another event, or several events may occur concurrently (by which we mean "not necessarily in some sequence" rather than "simultaneously", the latter of which appears as an ill-defined concept in our relativistic view of time). Of course it is possible to formally relate interpretations to such uninterpreted systems in order to demonstrate that a system models the behaviour of some actual or possible real world system. We shall take up this point again in a later section.

In the process of developing COSY, we gradually came to formulate what we now consider to be some inportant requirements that should be possessed by a design methodology for concurrent and distributed systems. Firstly we considered that it was important to encourage the designer to rid himself, as much as possible, of those preconceptions he might have developed as a result of his experience in sequential programming, where it is natural to think of systems as sequential centralised and synchronous. Indeed, we felt that it would not suit our purpose to base our approach on any kind of modification of a sequential programming language, since in such cases, the designer is obliged to make his understanding of his problem conform to the available programming language constructs. This is undesirable in the sense that the solution of his problem becomes confused with the implementation of the solution, that the structure of the solution may be hidden behind a mass of implementation details which are not in themselves an essential part of it. Another language might serve just as well. This is particularly important if one wishes, as we do, to concentrate on the specification of synchronization properties of systems rather than the problem-oriented functions these systems perform. To base the development (rather than the implementation) of a solution to a systems organisation problem on some specific programming language construct (such as semaphores or monitors) would be rather like basing the business of solving a problem in linear algebra (as opposed to implementing the solution) on FORTRAN as opposed to the appropriate mathematics. In view of this, we chose to specify synchronization properties in terms of the general notion of permitted behaviours in a linguistic or black box manner. A system will be associated with a grammatical type object whose "language" models a set of permitted or required behaviours.

Our system design methodology should therefore be built around a notation which should abstract from non-synchronization semantics of conventional programming constructs and specific synchronization mechanisms.

However, it should be possible to relate the notation both to any other well defined programming notation and to any major theory of concurrent and distributed systems.

The path notation, due to Campbell and Habermann (15), which was designed so that one could state the proper coordination of concurrent processes as the permissible order of execution of operations on shared system objects as part of the object definition, seemed a good starting point for the development of the notation. However, we felt that the original suggestions about how concurrency should be expressed in paths was inappropriate and not a natural or convenient extension of the notion of finite state automaton or regular expression. Furthermore, we felt that the notion of type involving paths was not abstract enough but involved implementation detail which tended to unnecessarily increase the complexity of the problem of system analysis. More discussion of these points can be found in (13) and (8). Hence we developed what we consider a more natural way of expressing concurrency with paths which are called generalized paths (5). Here we have, as required, an abstract notation of a linguistic kind. Indeed a path expression is essentially a collection of regular grammars, each represented as a regular expression. Just as a single regular expression determines a set of strings, each of which may be considered as a labelled total order modelling a sequence of executions of the operation which label it, so may a collection of regular expressions determine a set of vectors of strings, where each vector may be considered as a labelled partial order, modelling a non sequential behaviour of operation executions.

The path notation thus satisfies our requirements for a notation for abstractly describing concurrent systems. It also satisfies our requirement that our notation should be easily relatable to other major theories of concurrent systems. The relationship between regular expressions and state machines is well known. There is a relationship between path expressions and Petri-nets which generalises this relationship. In fact every path expression defines a safe net which decomposes into a set of state machines corresponding to the regular expressions from which the path expression is composed. The path expression and corresponding net may each be shown to define exactly the same set of asynchronous behaviours (5).

A number of precise results about this relationship have been obtained and the interested reader is referred to (16) (17) and (6). We only mention that any safe net can be expressed as a behaviourally equivalent basic COSY program and hence we can think of COSY as involving a programming notation alternative to, for example, graphic representations of nets in two dimensional planes. This close correspondence between linguistic features of COSY and those of net theory is deliberate and extends to higher-level and more abstract and general representations of COSY programs and nets. For example, our macro generators, the so called replicators discussed in (9) (7) were developed from some suggestions of H. Genrich, a major contributor to general net theory, and similar generators have been introduced into the net formalism (see (1) p. 523 multiplication of subnets).

An additional requirement is that the methodology should contain facilities for the verification of systems designed in its notation. The semantics in terms of nets, or equivalently, the semantics which regards a path expression as generating a language of vectors, provide a mathematical milieu for the formal definition of systems properties and for the analysis of a path expression to see whether the system it defines possesses such properties. The choice of regular expressions, rather than some more general structure with the power of Turing machines means, in addition, that all such properties are in principle decidable.

So far, then, we have a notation for describing concurrent systems in an abstract manner, clearly related to Net Theory and possessed of a clean behavioural semantics which provides a firm mathematical foundation for verifying behavioural properties of systems. The further evolution of the COSY methodology was conditioned by other desiderata that we considered essential in a software design environment.

The first is a matter of convenience for the programmer and a facility for generalisation; it is required that regular expressions of arbitrary size and complexity should be capable of being defined in a succinct and illuminating manner particularly when the regular expression in question has the kind of regularity of structure which might be expressed using iteration. For this, the replicator notation, which contains facilities for the iterative definition of regular expressions involving indexed operation names, was introduced (7) and (9).

It also seemed that the notation should contain facilities for the expression of hierarchy and modularity, in the sense that it contains features which allow the programmer to introduce levels of abstractness into his design and to practice information hiding and which support other techniques of structured programming. Furthermore it should facilitate the design of distributed systems consisting of subsystems which are capable of proper concurrent behaviour without the presupposition of a centralised or global system state or single clock. For these purposes, the notation has been equipped with a class-like construct, for which we use the term system, (7) and (9).

Finally, we need to be able to systematically relate COSY system specifications to implementations in any programming notation. The system construction also contains means for associating specifications with implementations, (7) and (9).

To allow direct comparison with Hoare's treatment (2) of some of the aspects of systems mentioned above, the rest of the present paper will follow the organization of (2) and deal with the same examples used there. Space will only allow us to quote Hoare's main concepts and notation in the appendix of the present paper and we refer the reader to the cited paper for further details.

BASIC CONCEPTS AND NOTATIONS FOR SEQUENTIAL SYSTEMS

Concepts of the COSY model

1. The ultimate components of a system are events each of which is capable of giving rise to a number (possibly zero) of occurrences of that event during any period of (discrete) behaviour of the system.

For example, a vending machine may involve the events:

(E1) insertion of a 5 penny coin
(E2) insertion of a 10 penny coin
(E3) withdrawal of a large packet of biscuits
(E4) withdrawal of a small packet of biscuits
(E5) return of a 5 penny coin

and any number of others.

2. The *basis* of a system is the set of all events considered to constitute the system.

For example, if we denote the explicitly mentioned events above by their labels, then the basis of the vending machine is

(B1) {E1,E2,E3,E4,E5,...}

3. A (finite) *trace* of a sequential system is any (finite) possibly empty sequence of event occurrences which constitutes the history of an actual or potential sequential behaviour of the system from one initial event occurrence up to some final event occurrence.

For example, the following are traces of the vending machine:

(T1) E2.E4.E5 the sequence consisting of an occurrence of E2 followed by an occurrence of E4 followed by an occurrence of E5

(T2) ε the empty sequence of occurrences of events

where the dot "." denotes the juxtaposition of the event occurrences in the order as written left to right.

4. A *sequential system* is completely determined by the set of all its traces. From the definition of trace it follows that for any system

(i) ε is a trace of the system, and
(ii) If s.t is a trace of the system, then so is s .

Above, (i) indicates that the empty behaviour is a behaviour of any system and it represents the behaviour of the system before any of its events have occurred; and (ii) indicates that any initial segment of a behaviour of a system is itself a behaviour of the system.

In (2) Hoare states that he intends the trace (T1) to correspond to what he calls a *successful initial transaction* of the vending machine (see appendix). Hence, we might determine the machine by listing the set of all traces corresponding to some successful initial transaction of the machine. Then one could obtain the remaining traces by (ii) above as the set of all prefixes (initial segments) of multiples of successful initial transactions. In fact this is what we have often done in the COSY model where we usually work with systems each of which behaves cyclically when it is operating successfully. In the case of a cyclic sequential system we identify the notion of successful initial history (trace) of the system with the notion of a history (trace) which returns the system to the point which preceded the first occurrence of any of that system's constituent events, or briefly, in the sequential and centralized case, what would be called its initial state.

For example, the vending machine VM could be determined by listing the set of successful initial traces SIT(VM) we intend:

SIT(VM)={E1.E1.E3,E1.E4,E2.E4.E5,E2.E3}.

If $\{t_1,\ldots,t_n\}$ is a set of traces, then by $\{t_1,\ldots,t_n\}^*$ we mean the set of all traces of length zero or more obtained by juxtaposing traces t_i and t_j from the set; and if the set of all prefixes of a set of traces T is defined as

$$\mathrm{Pref}(T) = \{y \mid y.\,z=x \text{ for some } z,x\in T\}$$

then the set of all traces Trace(VM) of the system VM is defined as

Trace(VM) = Pref(SIT(VM)*) = SIT(VM)*.Pref(SIT(VM))

where "." between two sets of traces denotes the set of all possible juxtapositions of elements from the two sets in the order left to right as written.

In other words

Trace(VM) = Pref({E1.E1.E3,E1.E4,E2.E4.E5,E2.E3}*)
= {E1.E1.E3,E1.E4,E2.E4.E5,E2.E3}*.{ε,E1,E2,E1.E1,E2.E4}.

NOTATION FOR SEQUENTIAL COSY SYSTEMS

1. Events belonging to some system are denoted by event symbols in our notation. For example, to express the vending machine system in COSY we might let the event symbols "5p", "10p", "large", "small" and "5pchange" denote events E1 through E5, respectively.

2. The alphabet of an expression P_S denoting a (sequential) system S, denoted by Alpha (P_S), is the set of event symbols occurring in P corresponding one-to-one to the events in the basis of S. For example,

Alpha(P_{VM}) = {5p,10p,large,small,5pchange}.

3. A sequential system is expressed by writing

<u>path</u> (sequence)*<u>end</u>

where "sequence" is a regular expression formed from the elements of the alphabet of the system by the use of the connectives ";","," and "*" denoting sequentialization, non-deterministic choice and iteration zero or more times, respectively. More formally a sequential system is expressed by a path expression which is an expression derived from the non-terminal "path" in the little grammar below:

```
path       =   path(sequence)*end
sequence   =   orelement/orelement;sequence
orelement  =   element/element,orelement
element    =   event_symbol/element*/(sequence)
```

4. As indicated before, the <u>set of traces</u> of the sequential system expressed by a path expression P is defined in two steps:

4.1 First we define its corresponding set of successful initial traces SIT(P) by

4.1.1 SIT(<u>path</u>(sequence)*<u>end</u>) = SIT(sequence)

The set of successful initial traces of a cyclic path is the set of successful initial traces of its component sequence.

4.1.2 $\text{SIT}(\text{orelement}_1;\ldots;\text{orelement}_n) = \text{SIT}(\text{orelement}_1)\ldots\text{SIT}(\text{orelement}_n)$

The set of successful initial traces of a sequence is the set of all possible juxtapositions of successful initial traces of its constituent orelements taken in the order left to right as written.

4.1.3 $\text{SIT}(\text{element}_1,\ldots,\text{element}_n) = \text{SIT}(\text{element}_1)\cup\ldots\cup\text{SIT}(\text{element}_n)$

The set of successful initial traces of an orelement is the set theoretic union of the sets of successful initial traces of its constituent elements.

4.1.4 SIT(element*) = SIT(element)*

The set of all successful initial traces of a starred element is the set of traces obtained by juxtaposing members of the set of all successful initial traces of the element any number (possibly zero) of times.

4.1.5 SIT(event_symbol)={event_symbol}

The set of all successful initial traces of an event symbol is just the singleton set whose only element is the event symbol.

4.1.6. SIT((sequence)) = SIT(sequence)

Redundant parentheses do not affect the set of successful initial traces.

4.2 Then we define the set of traces of P by:

Trace (P) = Pref(SIT(P)*) = SIT(P)*.Pref(SIT(P))

The set of traces of P is the set of all prefixes of multiples of successful initial traces of P.

Examples

(P1) path (a;b)* end

Trace (P1) = Pref(SIT(P1)*) = Pref ({a.b}*) = {a.b}*.{ε,a} = {ε,a,a.b,a.b.a,...}
(PVM) path((5p;(5p;large),small),(10p;(small;5pchange),large))* end
SIT(PVM) = {5p.5p.large,5p.small,10p.small.5pchange,10p.large}
and
Trace(PVM) = SIT(PVM)*.{ε,5p,10p,5p.5p,10p.small}

A comparison with the notation for synchronized parallel processes introduced by Hoare in (2) (see appendix), shows that at this level of abstraction the only significant divergence is the use of right recursion in the definition of processes in (2) instead of iteration denoted by the "*" in COSY. When all the right recursions in the definition of a synchronized process P specify the repetition of the same process P then the notation in (2) just reduces to the simple path expressions which form one of the basic components of the COSY formalism (see (5)-(15)). Although we have not yet developed mechanical rules for this reduction in general our experience with COSY makes us confident that we can also reduce all right recursion to equivalent paths involving only iteration.

CONCURRENT SYSTEM MODEL

Concepts of the COSY model

1. The ultimate components of a concurrent system $S = S_1 \ldots S_n$ consisting of sequential subsystems $S_1, \ldots, S_n$, are also events as in the case where S is a single sequential system.

2. The basis of a concurrent system $S = S_1 \ldots S_n$ is the set theoretic union of the bases of its constituent subsystems.

More formally

$Bas(S) = Bas(S_1) \cup \ldots \cup Bas(S_n)$

3. A vector of (finite) congreable traces of a concurrent system $S = S_1 \ldots S_n$ is a vector whose i-th component is a (finite) trace of the subsystem S_i for all $1 \leq i \leq n$, and in which the traces of all subsystems agree about the number and order of occurrences of events they share. In other words a vector of (finite) traces of S is any vector of (finite, possibly empty) sequences of event occurrences which constitutes the history of an actual or potential concurrent behaviour of the system from some (possibly concurrent) event occurrences up to some (possibly concurrent) event occurrences. The i-th subhistory of the vector of histories is a history of an actual or potential sequential behaviour of the i-th sequential subsystem from some event occurrence to some event occurrence. But in addition, if several subsystems involve the same event then their respective traces have to agree about the number and order of occurrences of the event in which they coincide. Hence we talk about a vector of congreable traces.

We formally characterize vectors of congreable traces by introducing the notions of projection, event vector, and juxtaposition of event vectors. Next we associate an event vector with each event in the basis Bas(S) of $S = S_1 \ldots S_n$, and call the set of event vectors corresponding to S the vector basis of S, VBas(S) for short. As in the case of "Bas(S)*" we mean by "VBas(S)*" the set of all vectors of traces of length zero or more obtained by juxtaposing event vectors from VBas(S). Finally, the set of vectors of congreable traces corresponding to S, denoted VTrace(S), is that part of the cartesian product of the sets of traces of the subsystems for which we can also find a corresponding element of the set of all vectors of traces.

4. A concurrent system is completely determined by the set of all its vectors of congreable traces. From the definition of vector of traces it follows again that

(i) $\underline{\varepsilon} = (\varepsilon, \ldots, \varepsilon)$ is a trace of the system, and
(ii) If $\underline{s}.\underline{t}$ is a trace of the system, then so is $\underline{s}$;
where: if $\underline{s} = (s_1, \ldots, s_n)$ and $\underline{t} = (t_1, \ldots, t_n)$ are traces then $\underline{s}.\underline{t} = (s_1, \ldots, s_n).(t_1, \ldots, t_n) = (s_1.t_1, \ldots, s_n.t_n)$.

Note that juxtaposition of vectors of traces is defined as component wise juxtaposition of corresponding traces.
$\underline{\varepsilon}$ stands for the vector of empty traces.

We will now be slightly more formal but we will use examples to illustrate the formalism. The <u>vector basis</u> of a system $S = S_1 \ldots S_n$ is defined to be

$$VBas(S) = \{\underline{a} \mid \exists a \in Bas(S) \ \underline{and} \ \underline{a} = (proj_1(a), \ldots, proj_n(a))\}$$

where for $1 \leq i \leq n$:

$$proj_i(a) = \begin{cases} a \text{ if } a \in Bas(S_i) \\ \varepsilon \text{ otherwise} \end{cases}$$

For example, assume we have a concurrent system VMC consisting of the vending machine VM and a customer CUST, where the basis and the traces of CUST are:

$$Bas(CUST) = \{E1, E2, E3, E4\} = Bas(VM) - \{E5\}$$

where '–' indicates set theoretic subtraction, and

$$Trace(CUST) = \{E1.E3, E2.E3, E4.E3\}^*.\{\varepsilon, E1, E2\}.$$

Then the basis of the combined system VMC is

$$Bas(VMC) = Bas(VM) \cup Bas(CUST) = Bas(VM)$$

and the vector basis of VMC is

$$VBas(VMC) = \{(E1,E1),(E2,E2),(E3,E3),(E4,E4),(E5,\varepsilon)\}.$$

Finally, implicitly characterized, the set of vectors of traces of VMC would be

$$VTrace(VMC) = (Trace(VM) \times Trace(CUST)) \cap VBas(VMC)^*$$

and explicitly it would yield

$$VTrace(VMC) = \{(E2.E3,E2.E3)\}^*.\{\underline{\varepsilon},(E1,E1),(E2,E2)\}$$

which in this case happens to be equivalent to

$$VTrace(VMC) = \{\underline{a} \mid \exists a \in Trace(VM) \cap Trace(CUST) \ \underline{and} \ \underline{a} = (a,a)\}.$$

Note that our vector traces tell us that the only successful initial (cyclic) congreable behaviour in the system is (E2.E3,E2.E3) which corresponds to an occurrence of the event "insertion of a 10 penny coin" followed by an occurrence of the event "withdrawal of a large packet of biscuits". Furthermore, the fact that (E1,E1) is not a prefix of any successful initial (cyclic) congreable behavior of the system indicates that the system will <u>deadlock</u> after the occurrence of the event "insertion of a 5 penny coin", as was Hoare's intent in constructing this example in (2). The deadlock occurs because after the insertion of a 5 penny coin the customer can only ask for a large packet of biscuits but the vending machine will only allow this after a previous insertion of a 10 penny coin or the successive insertion of two 5 penny coins.

Before we consider more examples and discuss our modelling of concurrent events we introduce the notation for expressing concurrent systems.

Notation for concurrent COSY systems

1. Given a set of sequential paths $P_1,\ldots,P_n$ where for $1 \le i \le n$, P_i denotes some sequential system S_i then a concurrent system $S = S_1\ldots S_n$ is expressed in COSY by writing a <u>concurrent</u> (or <u>generalised</u>) path <u>expression</u>

<u>system</u> $P_1\ldots P_n$ <u>endsystem</u>

2. The alphabet Alpha (P_S) of a concurrent path expression P_S corresponding to system S is the set theoretic union of the alphabets of its constituent P_i, $1 \le i \le n$:

$$\text{Alpha}(P_S) = \text{Alpha}(P_1) \cup \ldots \cup \text{Alpha}(P_n).$$

For example, we can express the system VMC by the concurrent path expression (PVMC)

<u>system</u>
(PVM) <u>path</u> (5p;(5p;large),small),(10p;(small;5pchange),large) <u>end</u>

(PCUST) <u>path</u> large,small,(10p;large),(5p;large) <u>end</u>

<u>endsystem</u>

Note that we have omitted the outermost parentheses and the "*" in writing paths but they are still assumed to be cyclic as before. SIT(PVM) and Trace (PVM) have been given earlier.

$$\text{Trace(PCUST)} = \{\text{large,small,10p.large,5p.large}\}^*.\{\varepsilon,\text{5p,10p}\}$$

Analogously to the construction for concurrent systems in the previous section we obtain the vectors of traces of P_S from vectors of event symbols. Hence,

$$\text{VTrace}(P_S) = (\text{Trace}(P_1)\times\ldots\times\text{Trace}(P_n)) \cap \text{VAlpha}(P_S)^* \quad \text{and,}$$

$$\text{VTrace(PVMC)} = \{(\text{10p.large,10p.large})\}^*.\{\underline{\varepsilon},(\text{5p,5p}),(\text{10p,10p})\}$$

We will from now on ignore the difference between events of the system being expressed and event symbols denoting them, since we will be concerned with properties of systems, such as absence of (total) deadlock and absence of partial deadlock, which can be studies solely in terms of the uniterpreted event symbols and their interconnection by the connectives ";", "," and "*" in sequential paths and by coincidence of event symbols in concurrent paths. The difference between events and event symbols must be made when one studies whether a system specification in COSY accurately models some real system. We have dealt with the latter topic in references (9)(12) and (13).

In our researches we have come to call a system which is incapable of total system deadlock a deadlock-free system and one which is incapable even of partial system deadlock an adequate system. We now define these notions in our formal model.

(DF) P is deadlock-free if and only if
$\forall \underline{x} \in VTrace(P) \exists a \in Alpha(P) : \underline{x}.\underline{a} \in VTrace(P)$

(A) P is adequate if and only if
$\forall \underline{x} \in VTrace(P) \forall a \in Alpha(P) \exists \underline{y} \in VAlpha(P)^* : \underline{x}.\underline{y}.\underline{a} \in VTrace(P)$

where $\underline{a}$ is the vector of events corresponding to a.

(DF) says that P is deadlock-free exactly when for every vector of (congreable) histories of P there always exists at least one event of P such that its occurrence could coincidentally extend that vector of histories in all components corresponding to subsystems involving that event.

(A) says that P is adequate exactly when for every vector of (congreable) histories of P and any event of P, there exists an extension of the vector of histories after which the occurrence of the event could coincidentally extend all components corresponding to subsystems involving that event.

According to these definitions PVMC is neither deadlock-free nor adequate since the vector of traces "(5p,5p)" cannot be coincidentally extended by the occurrence of any event of PVMC according to its structure.

The following two modifications of Hoare's vending machine example will provide us with an opportunity to discuss how one might use our model to demonstrate that a system is deadlock-free or adequate.

```
(PVMC1)
System
(PVM)path(5p;(5p;large),small),(10p;(small;5pchange),large)end
(C1)path 5p;small end
(C2)path 10p;large end
endsystem
```

VTrace(PVMC1) = {(5p.small, ε),(10p.large, ε, 10.large)}*.
{ε,(5p.5p, ε),(10p, ε, 10p)}

Examination of the set of prefixes Pref(VSIT(PVMC1)), that is the second set above, shows that every prefix can be extended by an occurrence of at least one event. Hence, PVMC1 is deadlock-free. However, it is not adequate since examination of VTrace(PVMC1) indicates that certain possible behaviours of the vending machine on its own, such as "5p.5p.large" or "10p.small.5pchange" can never occur in the system involving the two customers with their restricted behaviour. Hence, for no x ∈ VTrace(PVMC1) does there exist a continuation y ∈ VAlpha(PVMC1)* such that, for example x.y 5pchange ∈ VTrace (PVMC1). Hence PVMC1 is not adequate.

```
(PVMC2)
system
(PVM1) path (5p;small),(10p;large) end
(C1) path 5p;small end
(C2) path 10p;large end
endsystem
```

VTrace(PVMC2) = VTrace(PVMC1)

but PVMC2 is not only deadlock-free but adequate since the definition of adequacy is satisfied as the reader can verify by an exhaustive consideration of the set of vectors of traces.

But so far our combined system consisting of vending machine and customers is totally sequential, though the order in which customers will succeed with successful initial transactions is by non-deterministic choice. To obtain an example which would allow us to discuss concurrency modelling and to explain one of our theorems which simplifies the proof of adequacy, we specify a vending machine which may be used by two customers concurrently, that is, a machine that has distinct slots for 5p and 10p coins and two distinct points for extraction of small and large packets of biscuits.

```
(PVMC3)
system
(P1) path 5p;small;plunk end     }
(P2) path 10p;large;plonk end    }  vending machine
(P3) path 5p;small end           }
(P4) path 10p;large end          }  customers
endsystem
```

In PVMC3 "plunk" and "plonk" denote the sounds made by a small and large packet of biscuits dropping out of the machine, respectively. This kind of machine is called the noisy machine by Hoare in (2), and involves events of the machine which do not need cooperation of customers to occur.

The event vectors of VAlpha(PVMC3) are:

$\underline{5p} = (5p, \varepsilon, 5p, \varepsilon)$ $\qquad$ $\underline{10p} = (\varepsilon, 10p, \varepsilon, 10p)$
$\underline{small} = (small, \varepsilon, small, \varepsilon)$ $\qquad$ $\underline{large} = (\varepsilon, large, \varepsilon, large)$
$\underline{plunk} = (plunk, \varepsilon, \varepsilon, \varepsilon)$ $\qquad$ $\underline{plonk} = (\varepsilon, plonk, \varepsilon, \varepsilon)$.

Considering any two event vectors on a single line above we see that each such pair has at most one non-empty event as corresponding components. This means that the events corresponding to any such pair of event vectors may be executed concurrently relative to each other and formally this is modelled by the fact that the juxtaposition of the event vectors commutes. In general, we define for distinct events a and b and any vector of traces $\underline{x}$ of P: a and b are concurrent relative to each other after history $\underline{x}$ if $\underline{x}.\underline{a} \in$ VTrace(P) and $\underline{x}.\underline{b} \in$ VTrace(P) and $\underline{a}.\underline{b} = \underline{b}.\underline{a}$.

For example, in PVMC3, 5p and 10p are concurrent relative to each other initially. Since $\underline{\varepsilon}.\underline{5p} \in$ VTrace(PVMC3) and $\underline{\varepsilon}.\underline{10p} \in$ VTrace(PVMC3) and
$\underline{5p}.\underline{10p} = (5p, \varepsilon, 5p, \varepsilon).(\varepsilon, 10p, \varepsilon, 10p) = (5p, 10p, 5p, 10p) = (\varepsilon, 10p, \varepsilon, 10p).$
$(5p, \varepsilon, 5p, \varepsilon) = \underline{10p}.\underline{5p}$. That is, an occurrence of 5p and a concurrent occurrence of 10p can begin a concurrent behaviour of PVMC3. These concurrent occurrences need not be interleaved, for example, as after arbitration relative to a single clock.

The system must be capable of functioning correctly even without the assumption that concurrent occurrences of 5p and 10p must be "observable". Consideration of PVMC3 indicates that the events 5p and 10p occur in no single path and hence there is no sequential subsystem to which both belong. Their occurrences are to be thought of as independent and should not in general be reduced to arbitrary interleaving. We will return to this point later.

If we reconsider PVMC3 we see that none of the event symbols occur more than once in a single sequential path and that the comma "," is nowhere used. Concurrent paths of this form are called GE_o-paths (see (5) and (9)), and in our formal theory of COSY systems there exists a theorem which yields a more effective way of deciding whether a GE_o-path is adequate or not, than by considering all histories of possible behaviours of the path, as we have done in our analysis of the examples so far. In fact, the theorem says that it is sufficient and necessary to find a single history of the whole system P such that each of its components is an element of the set of successful initial traces of the corresponding subsystem P_i. Formally, we say P is adequate if and only if

$$\exists \underline{x} \in \mathrm{VTrace}(P)\ \forall i \in \{1,\ldots,n\} : [\underline{x}]_i \in \mathrm{SIT}(P_i),$$

where $[\underline{x}]_i$ denotes the i-th component of the vector $\underline{x}$.

Reconsidering PVMC3 it is easy to verify that

$\underline{vmc}$ = (5p. small.plunk,10p.large.plonk,5p.small,10p.large)

is a vector of traces of PVMC3 by noting that each component is a trace of the corresponding path P_i, $1 \leq i \leq 4$, and by exhibiting $\underline{vmc}$ as a juxtaposition of vectors of events of PVMC3; for example

$\underline{vmc}$ = $\underline{5p}$.$\underline{small}$.$\underline{plunk}$.$\underline{10p}$.$\underline{large}$.$\underline{plonk}$.

The fact that $\underline{vmc}$ satisfies the theorem is established by showing that

$[\underline{vmc}]_1$ = 5p.small.plunk $\in$ SIT(P1)
$[\underline{vmc}]_2$ = 10p.large.plonk $\in$ SIT(P2)
$[\underline{vmc}]_3$ = 5p.small $\in$ SIT(P3)
$[\underline{vmc}]_4$ = 10p.large $\in$ SIT(P4)

We have obtained a number of theorems of this nature for considerably larger classes of programs, and we have applied these results to the verification of numerous operating system strategies for sharing resources in concurrent and distributed computer systems (see references (5)-(14), and (16), (17)).

COMPARISONS

We have now introduced the reader to the COSY formalism and indicated how one can specify concurrent systems which can be verified with regard to such global behavioural properties as absence of deadlock and adequacy. Along the way we have also briefly pointed out similarities and differences to Hoare's notation and model for synchronized parallel processes (2)(3).

This final section of the paper will add a few more remarks of a comparative nature but a thorough and precise comparison of COSY with Hoare's work in (2) and (3) is beyond the scope of the present paper.

Briefly recall what we have already noted about the two approaches. Hoare argues for an interleaving model of concurrency and we argue for a non-interleaving model of concurrency. As far as the notations are concerned, Hoare's notation (2) can be considered to be equivalent in expressive power to concurrent path expressions as introduced in (5) if we ignore the abort process. Finally, we make a stronger distinction between the system being described or specified and the notation used to specify the system. The first is an actually or potentially existing and dynamic system, not necessarily having a linguistic nature, and the latter always being a linguistic entity. We fell that there is need to maintain a distinction even at the level of informality of the papers under discussion if the possibility of misinterpretation of statements concerning systems involving concurrency and distribution is to be avoided.

We will take up the above points in some more detail now.

Interleaving or non-interleaving

In general, we can give several reasons for having a formal non-interleaving model of concurrency in addition to an interleaving model, even though most workers in this problem area still feel that the latter is sufficient and convenient.

1. In order to demonstrate that an interleaving model of concurrency can fully model all concurrent systems that can be modelled by a non-interleaving model of concurrency, it is neccessary to give a formal notion of a non-interleaving model and to prove the two models equivalent in some sense.

2. There is strong evidence that the two approaches are not equivalent when one has to express behaviours of concurrent systems which are maximal in the sense, for example, that everything that could have happened has happened. Such maximal behaviours are used to prove properties of a compound system such as absence of starvation of some subsystem due to, for example, unfair scheduling or malicious cooperation on the part of other subsystems. The latter kind of starvation is possible for the philosophers in Hoare's paper (2) though no deadlock can occur.

3. Even for those problems where such an equivalence can be demonstrated it is still arguable that the non-interleaving model yields a simpler mathematics than the interleaving model.

To make these points clearer consider the very simple system of concurrent paths below:

(PS1) system path a end path b end endsystem
(PS2) system path a end path b end path a,b end endsystem

VAlpha (PS1) = $\{(a,\varepsilon),(\varepsilon,b)\}$

and $\underline{a}.\underline{b} = \underline{b}.\underline{a}$. means that a is concurrent to b,

but

VAlpha(PS2) = $\{(a,\varepsilon,a),(\varepsilon,b,b)\}$ and

$\underline{a}.\underline{b} = (a,\varepsilon,a).(\varepsilon,b,b) = (a,b,a.b) \neq (a,b,b.a) = (\varepsilon,b,b).(a,\varepsilon,a) = \underline{b}.\underline{a}$

which means that a is not concurrent with b.

If one uses an interleaving model of concurrency the two systems are the same unless one introduces further complexity into the system by dividing concurrent events into two events each, one for the beginning of an occurrence of the event and one for the ending of an occurrence of the event. We shall explain further. For certain purposes we have used an interleaving definition of concurrent composition of paths in the period 1975-1977 (see (5), (6),(8),(11),(17)) which is identical with Hoare's definition of parallel composition of processes P and Q, denoted by "(P||Q)", in the appendix. We give our form of the definition for direct comparison with the vector model of the present paper. The interleaving semantics of the concurrent path $P = P_1 \ldots P_n$ is:

(IS) $Trace(P) = \{x \in Alpha(P)^* \mid \forall i \in \{1,\ldots,n\}\ proj_i(x) \in Trace(P_i)\}$

where $proj_i$ is defined for single events as before and furthermore for $a_1 . \ldots . a_n \in Alpha(P)^*$ we define

$proj_i(a_1 . \ldots . a_n) = proj_i(a_1) . \ldots . proj_i(a_n)$ for all $1 \leq i \leq n$.

If one divided every operation of PS1 and PS2 into begin and end events we would obtain:

(PS1') system path a_begin;a_end end path b_begin;b_end end endsystem

and

(PS2') system path a_begin;a_end end
path b_begin;b_end end
path (a_begin;a_end),(b_begin;b_end) end
endsystem

then one can see that according to PS1'

a_begin.b_begin.a_end.b_end

is a possible (concurrent) behaviour of PS1', whereas according to PS2' it is not since its third path forbids any overlapping of the events a and b. However, we feel that this splitting of events enforced by the interleaving semantics is an unnecessary over complication.

Another overcomplication of the semantical tools enforced by the interleaving model is the fact that one has to work with equivalence classes of traces where one would be working with a single vector in the vector of traces approach we are advocating. Hence, the vector vmc of the previous section represents the entire equivalence class of traces which differ only with respect to permutation of the interleaving of concurrent events.

Finally, the reader will have noticed that we have enclosed the word "finite" in parentheses in many of our definitions because Hoare did so in the corresponding definitions in (2). We can drop the word finite from all our definitions and all true statements made about the model will be true as before. However, the interleaving model does not generalize as easily if at all. Suppose we wanted to say formally what constitutes a maximal behaviour of PS1, that is, a behaviour in which everything that could have happened has happened. In the interleaving model a trace is represented by a sequence of event occurrences (or symbol occurrences) and a trace is maximal if it is not the prefix of any trace longer than it. Hence, the following traces of PS1 are maximal traces:

aaa...	infinite number of occurrences of a
baaa...	occurrence of b followed by an infinite number of occurrences of a
bbb...	infinite number of occurrences of b

but none of these corresponds to a maximal behaviour of PS1 which consists for example of an infinite number of a's followed by an infinite number of b's which cannot be represented as a trace.

If however we go to the non-interleaving model of vectors of traces then the following vectors of traces corresponding to the above examples of maximal traces are not vectors of maximal traces

(aaa..., ϵ)
(aaa...,b)
(ϵ,bbb...) ...

For example the first is a vector of prefixes of the second and hence is not maximal. But (aaa...,bbb...), which cannot be represented as an infinite sequence of a's followed by an infinite sequence of b's in the interleaving model, is a vector of maximal traces and is precisely the unique formal expression denoting the maximal behaviour of PS1.

CONCLUSION

In this short paper we have tried to show how the mathematically precise definition of a simple notation for designing and implementing programs involving synchronization of parallel processes introduced by C.A.R. Hoare in (2) can be expressed in the COSY formalism. We have also indicated how the COSY formalism generalizes the semantic trace model used by Hoare and what might be some advantages of the generalized model. Finally, we used the COSY notation to develop a number of versions of Hoare's example and used the COSY semantic formalism to verify that partial or total system deadlock is impossible.

ACKNOWLEDGEMENTS

I would like to acknowledge my indebtedness to Tony Hoare with regard to my understanding of concurrency, in particular for his repeated creation of definitive concurrent programming language concepts. His ideas in (2) and (3) were, as always, highly stimulating and challenging to me and the present paper represents an initial attempt to formulate the similarities and differences of the two approaches. Acknowledgement is due to John Cotronis who has given much of his time to discuss these topics with me and with whom I am producing a detailed comparison of the work going on at Oxford and Newcastle upon Tyne. The work reported in this paper was supported by a grant from the Science Research Council of Great Britain. My indebtedness to the thorough and fruitful researches of Eike Best and Mike Shields should be evident from the appearence of their names in the Bibliography. The author would like to thank Mrs. Joan Armstrong for her patience and efficiency in preparing this typescript.

REFERENCES

1. Net theory and Applications: Proceedings of the Advanced Course on General Net Theory of Processes and Systems, Hamburg 1979 (Ed. W. Brauer) <u>Lecture Notes in Computer Science 84</u>, Springer Verlag 1980.

2. Hoare, C.A.R.: Synchronisation of Parallel Processes. In: <u>Advanced Techniques for Microprocessor Systems</u>. (Ed. F.K. Hanna), Peter Peregrinus Ltd. 1980.

3. Hoare, C.A.R.: Communicating Sequential Processes. In: On the construction of programs. (Eds. McKeag and MacNaghten) Cambridge University Press 1980.

4. Milner, R.: A Calculus of Communicating Systems. Lecture Notes in Computer Science 92, Springer Verlag 1980.

5. Lauer, P.E., Campbell, R.H.: Formal semantics for a class of high level primitives for coordinating concurrent processes. Acta Informatica 5, 247-332 (1975).

6. Lauer, P.E., Shields, M.W., Best, E.: The design and certification of asynchronous systems of processes. Proc. of EEC Advanced Course on Abstract Software Specification, Lyngby, Jan. 22 - Feb. 2, 1979. Lecture Notes in Computer Science, No. 86, Springer Verlag, 1979, pp. 451-503.

7. Lauer, P.E., Torrigiani, P.R., Shields, M.W.: COSY: a system specification language based on paths and processes. Acta Informatica, Vol. 12, pp. 109-158, 1979.

8. Lauer, P.E., Shields, M.W.: Abstract specification of resource accessing disciplines: adequacy, starvation, priority and interrupts. SIGPLAN Notices, Vol. 13, No. 12, Dec. 1978.

9. Lauer, P.E., Shields, M.W.: COSY An environment for development and analysis of concurrent and distributed systems. In: Software Engineering Environments (Ed. H. Hünke) North Holland 1981.

10. Shields, M.W.: Adequate Path Expressions. Proc. Symp. on the Semantics of Concurrent Computation, Evian-les-Bains, July 2-4, 1979. Lecture Notes in Computer Science Vol. 70, Springer Verlag 1979.

11. Shields, M.W., Lauer, P.E.: On the abstract specification and formal analysis of synchronization properties of concurrent systems. Proc. of Int. Conf. on Mathematical Studies of Information Processing. Aug. 23-26, Kyoto, 1978. Lecture Notes in Computer Science 75, Springer Verlag 1979, pp. 1-32.

12. Shields, M.W., Lauer, P.E.: A formal semantics for concurrent systems. Proc. 6th Int. Colloqu. for Automata, Languages and Programming, July 16-21, 1979 Graz, Lecture Notes in Computer Science, 71, Springer Verlag, 1979, pp. 569-584.

13. Shields, M.W., Lauer, P.E.: Verifying concurrent system specification in COSY. Proc. 8th Symposium on Mathematical Foundations of Computer Science, Aug. 31 – Sept. 6, 1980, Poland. Lecture Notes in Computer Science, No. 88, Springer Verlag 1980, pp. 576–586.

14. Lauer, P.E., Torrigiani, P.R., Devillers, R.: A COSY Banker: Specification of highly parallel and distributed resource management. Proc. 4th International Symposium on Programming, Paris, April 22–24, 1980. Lecture Notes in Computer Science 83, Springer Verlag 1980, pp. 223–239.

15. Campbell, R.H., Habermann, A.N.: The specification of process synchronization by path expressions. Lecture Notes in Computer Science V. 16, Springer Verlag, pp. 89–102.

16. Best, E.: Adequacy of Path Programs In: Net Theory and Applications: Proceedings of the Advanced Course on General Net Theory of Processes and Systems. Hamburg, 1979. (Ed. Prof. Wilfried Brauer). Lecture Notes in Computer Science 84, Springer Verlag 1980.

17. Lauer, P.E., Shields, M.W., Best, E.: Formal Theory of the Basic COSY Notation. The Computing Laboratory, University of Newcastle upon Tyne, Tech. Rep. Series No. 143, November 1979.

APPENDIX: VERBATIM QUOTES OF DEFINITIONS FROM REFERENCE (2)

Concepts

1. The ultimate constituent of our model is a symbol, which may be intuitively understood as denoting a class of event in which a process can participate.

a) "5p" denotes insertion of a coin into the slot of a vending machine VM.

b) "large" denotes withdrawal from VM of a large packet of biscuits.

2. The alphabet of a process is the set of all symbols denoting events in which that process can participate.

c) {5p, 10p, large, small, 5pchange} is the alphabet of the vending machine VM.

3. A trace is a finite sequence of symbols recording the actual or potential behaviour of a process from its beginning up to some moment in time.

d) <10p, small, 5pchange> is a trace of a successful initial transaction of VM.

e) <> (the empty sequence) is a trace of its behaviour before its first use.

4. A <u>process</u> P is defined by the set of all traces of its possible behaviour. From the definition of a trace, it follows that for any process P,

   < > is in P (i.e. P is non-empty)

   If st (the concatenation of s with t) is in P then so is s by itself. (i.e. P is <u>prefix-closed</u>).

NOTATIONS

1. The process ABORT is one that does nothing ABORT = {< >}

2. If c is a symbol and P is a process, the process (c → P) first does "c" and then behaves like the process P.

   $(c \to P) = \{\langle\ \rangle\} \cup \{\langle c \rangle s \mid s \text{ is in } P\}$

   where <c> is the sequence consisting solely of the symbol c. By convention the arrow associates on the right, so that

   $c \to d \to P = c \to (d \to P)$.

3. The process P ▯ Q behaves either like the process P or like the process Q; the choice will be determined by the environment in which it is placed.

   P ▯ Q = P∪Q (normal set union)

   By convention → binds more tightly than ▯ , so that

   c → P ▯ d → Q = (c → P) ▯ (d → Q).

4. The alphabet of a process P will be denoted by $\alpha(P)$. Usually we will assume that the alphabet of a process is given by the set of all symbols occurring in its traces.

5. We shall use recursive definitions to specify the behaviour of long-lasting processes. These recursions are to be understood in the same sense as the recursive equations of (say) a context-free grammar expressed in BNF.

EXAMPLES

f) $P = (a \rightarrow b \rightarrow P)$
$\quad = \{\langle\,\rangle, \langle a\rangle, \langle a, b\rangle, \langle a,b,a\rangle, \ldots\}$

g)

```
VM = (5p → (5p → (large → VM □ 5p → ABORT)
                □ small → VM
                 )
      □ 10p → (small → (5pchange → VM)
                 □ large → VM
       )       )
```

On its first step VM accepts either 5p or 10p. In the first case, its following step is either the acceptance of a second 5p (preparatory to withdrawal of a large packet of biscuits) or the immediate withdrawal of a small packet. The second case should be self-explanatory. In all cases, after a successful transaction, the subsequent behaviour of VM is to offer a similar service to an arbitrary long sequence of later customers. But if any customer is so unwise to put three consecutive 5p coins into the slot, the machine will break (ABORT), and never do anything else again.

6. The process $(P||Q)$ is the process resulting from the operation of P and Q in parallel.

$$\begin{aligned}(P||Q) = \{s \mid\, & s \text{ is in } (\alpha(P) \cup \alpha(Q))^* \\ \& \;\; & s \restriction \alpha(P) \text{ is in } P \\ \& \;\; & s \restriction \alpha(Q) \text{ is in } Q\}\end{aligned}$$

where $\quad s \restriction X$ (s restricted to X) is obtained from s by simply omitting all symbols outside X.

and $\quad X^*$ is the set of finite sequences of symbols from X

PARALLEL PROCESSING ARCHITECTURE AND PARALLEL ALGORITHMS

Dennis Parkinson

DAP Support Unit, Queen Mary College, (University of London) Mile End Road, London E1 4NS, England.

INTRODUCTION

Parkinson's law states that the demand for a resource rises to meet the capacity available to satisfy demand. Although the law was invented before widespread introduction of computers, there is probably no field of endeavour for which it is more appropriate. Any scientist modelling three dimensional fluid flows can, simply be halving the mesh size increase the computing requirements at least eightfold (both in computational needs and in data storage requirements).

There are many important problems which fall into the broad classification of fluid flow in 3 dimensional complex geometries, whose accurate solution is of great importance. One typical problem is the computation of the flow of oil/gas/water mixtures through the porous rocks which make an oil resevoir, the accurate modelling of oil resevoirs is an important factor in maximising the benefits from resources. Other important areas include safety calculations for the design of nuclear reactors (Three Mile Island), long term meteorological calculations for climatic predictions (will increasing CO_2 concentrations change our mean climate?), flow of air around complex shapes (will shock waves tear the tiles off the space shuttle?). The choice of topics is almost endless and hence there is a real demand for better computers which will allow more accurate calculations to be made.

K. G. Beauchamp (ed.), New Advances in Distributed Computer Systems, 367- 377.

In some cases there are alternatives to computers eg. building models. However, these alternatives are becoming very costly and inflexible. A modern wind tunnel consumes 60 MW power. The costs of this energy are becoming significant, but a computer to replace the wind tunnel would need to be 100 - 1000 times as powerful as today's most powerful machines.

The apparently insatiable demand for more and more scientific/technical computers is causing major changes in the way that computers are built and the majority of these new computers claim to be parallel. Unfortunately the words are often highly confusing and one of my major objectives is to attempt to clarify some of these words.

All entrants for the "World's Most Powerful Computer" competition claim to be parallel and so firstly, we must try to find out what 'parallel' means. We may paraphrase George Orwell and say "all computers are parallel but some are more parallel than others". Parallel is used to imply that in some part of the computer a number of operations are going on simultaneously. Unfortunately, even in the humble 8 - bit microprocessor there is parallel access to memory (ie all the bits of a word are fetched simultaneously). Another type of parallelism that has been with us almost since the first computer is the operation of peripheral units simultaneously with the operation of a central processing unit, and such operations are obviously parallel. The desire to claim parallelism is a manifestation of the widely held belief that more and more facilities of simultaneous operations must be built into future computers and that therefore one must establish one's position on the parallel bandwagon as soon as possible.

What then are these new types of computer architecture which claim to be parallel? There are two main types - the Vector processors such as CRAY-1 and CYBER 203/205, and processor arrays such as, DAP and DATAFLOW architectures. To illustrate the fundamental difference between these types it is helpful to use an analogy between methods of movement between two floors in a building. Vector processors are similar to moving staircases or escalators where processor arays are similar to lifts.

VECTOR PROCESSORS

Vector processors may be compared to escalators. An escalator accomplishes the task of moving people between two levels by breaking the operation into a number of smaller steps and moving people continuously between these steps. If a large number of people want to go up the escalator then they enter at the bottom and move to the top. There is a fixed delay before the first person gets to the top but the following people arrive in rapid succession. The people arrive in a strictly sequential order, but the device can claim to be parallel with the degree of parallelism equal to the number of steps. Vector or pipeline processors adopt the same strategy primarily to improve arithmetic performance. The task of, say, multiplication is broken down into a number of sub-operations and special hardware is provided to stream all pairs of operands through these sub-tasks. The special hardware is usually called a pipeline unit and is specialisation of a principle that has been common in the computer industry for the last 20 years.

In the case of say, the CRAY computer, 8 special registers have been added to the central processing unit. Each of these registers is capable of handling 64 variables. The instruction set has been augmented so that there are instructions which will multiply all elements of one vector register by the corresponding elements of another vector register and store the result in a third register. When this single instruction is used there is a delay before the first result becomes available. When the pipeline is producing results at its full rate, results appear at the rate 80 million per second. It is therefore often claimed that the machine is working at 80 MEGAFLOPS (millions of floating point operations per second). It should be realised that this is a peak rate and the figure ignores the initial delay in obtaining the first result. One way of increasing the performance of our escalator system is to provide a set of escalators in the hope that by using two together we can obtain twice the throughput of a single escalator. In computer terms this is the provision of multiple pipeline units, and for example the CRAY machine has 3 pipelined arithmetic units. However, these units are such that one unit is restricted to multiplication, another to addition and a third to the calculation of reciprocals, so provided one has a special job in which one wished to do all these operations simultaneously, then even higher performance can be obtained using a technique known as "chaining".

The CYBER series of Vector processors are an improvement on the pioneering STAR processor made by CDC and have more general purpose pipelined units that the CRAY processor. The disadvantage of a general purpose unit is that it takes longer to initialise the operation, so it is necessary for it to work on longer vecotrs before its use becomes efficient.

Different versions of the CYBER series have different numbers of vector pipeline units which work on data fetched from the memory of the machine and return the results back to the memory - unlike the CRAY where the vector operations are restricted to the data held in the special registers. The important characteristic of the Vector processors is their performance when working on very small vectors. Vectors of length one are called scalars and user experience suggests that the most important property of a vector processor is its scalar processing capacity. The reason for this is that vector pipelines are only efficient when they work on all members of a set of data, and any calculation which includes a lot of conditional, or data dependent operations, will require a large number of initialization operations, greatly decreasing the performance of the system. There is therefore a growing number of experts who specialise in modifying existing algorithms to make them suitable for running on Vector processors. The correct name for their activities is vectorization, although it is often losely called parallelization.

PROCESSOR ARRAYS

Processor arrays arise from the simple concept of trying to increase power by multiplying the number of power units. By processor arrays we mean computer systems made from a set of processors and which collaborate in order to accomplish a single computational task.

An aside must be made to discuss "array processors". On the market there are a number of objects which call themselves array processors which unfortunately are not processor arrays. The most common of these so-called array processors is the AP 120B of Floating Point Inc. The AP 120B is more closely related to the Vector processor class computer and is a special purpose pipelined floating point arithmetic unit made primarily to augment the performance of mini-computers. The ideal problems for this class of machine are those in which one wishes to perform many operations on relatively small amounts of data. Such calculations include the Fast Fourier Transform algorithm, for which "array processors" were really designed. As these calculations are calculations on arrays of data the manufacturers have christened their machines array processors but they do not consist of multiple processors, and so they are not processor arrays.

An analogy which we can use for processor arrays is that of a lift. In a lift a number of people are moved simultaneously, between one level of the building and another. They all leave simultaneously and arrive simultaneously, and providing the capacity of the lift is sufficient, the time taken does not depend on the number of passengers. Obviously this is a kind of parallelism but different from that of a vector processor. There are two types of processor arrays. The ILLIAC IV is the architypal processor array and is a member of the Single Instruction Multiple Datastream (S.I.M.D.) class of systems.

SIMD systems consist of a master processor and a number of identical slave processors each with a private memory. The master processor transmits the same instruction stream to each and every slave processor, so that if an instruction is an addition each processor adds two numbers together simultaneously. The intention in the ILLIAC IV was to have 256 slave processors and so produce 256 times the power of a single processor. Although only one quarter of the ILLIAC IV was built (ie. a 64 processor system) a theory of parallel processing has developed which assumes that all SIMD machines will be broadly similar to the ILLIAC IV.

The second type of processor array is called Multiple Instruction Multiple Datastream (M.I.M.D) machine, which is a more general type of system in which the individual elements, are allowed to perform different operations, so that one processor may be doing multiplication whilst another an addition. Many observers believe that the future must lie in the direction of the M.I.M.D. system, but unfortunately, as yet, there are no commercial systems with more than a handful of processors. The majority of research on Dataflow architectures is being done in universities primarily in the USA, UK and France and it is still at an exploratory stage.

The most widely studied multiple processor architecture is the S.I.M.D. system and there are now available commercially a number of designs. In the UK, ICL have have developed two of the 4096 processor DAP system and in France CISMA is making a PROPAL machine 512 - 2048 processor system. These systems are analogous to the ILLIAC IV design in that they consist of a large number of processors under the control of a master processor from which they all recieve the same instruction team. However, unlike the ILLIAC IV the processors in the array are very elementary, and indeed are little more than elementary one bit logical processors. Hence normal arithmetic on such machines has to be created by low level software and this is relatively slow. The power of such machines comes not from the fact that the elements are fast but from the fact that there are many of them. The use of logical building units means that in general the systems are powerful for logical operations as well as for arithmetic operations but do not satisfy many of the axioms assumed by theoreticians of parallel computing.

PARALLEL ALGORITHMS

The choice of optimum algorithm for a computer is generally affected most strongly by the execution time of the various options. As the timing characteristics of parallel computers are different from those of serial computers, the optimum algorithms are often different and also the optimum algorithm for a vector processor differs from that of a processor array. To illustrate the reasons for this, it is instructive to consider an elementary calculation which wished to repeat some operation N times. On an idealised serial computer the total time will be given by $T_s = Nt_s$ where t_s is the time required to do the operation once.

The ideal processor array will be able to do all N cases simultaneously so the time taken would be $T_p = t_p$ where t_p is the single operation time for the array element. Any real system will only have a finite number of processor (say M_p) and so the time taken will be

$$T_p = [N/M_p]\ t_p$$

where $|x|$ smallest integer is equal to or greater than x. The time taken on a vector processor would be

$$T_v = Nt_v + [N/M_v]\ S_v$$

where S_v is the time to start a vector operation and M is the number of vector operations done for each start. As our purpose is to demonstrate the qualitative differences between the architectures it is important to remove the effects of the relative speeds and we hence examine the dimensionless quantity $R_i = T_i / t_i$ $(i = s,v,p)$

Then for a serial computer

$$R_s = N$$

a processor array

$$R_p = [N/M_p]$$

a vector processor

$$R_v = N + [N/M_v]\ q_{vi}\ (q_v = S_v/t_v).$$

From these simple equations we can immediately note that the vector processor is a type of hybrid system with performance characteristics derived from both serial computers and processor arrays.

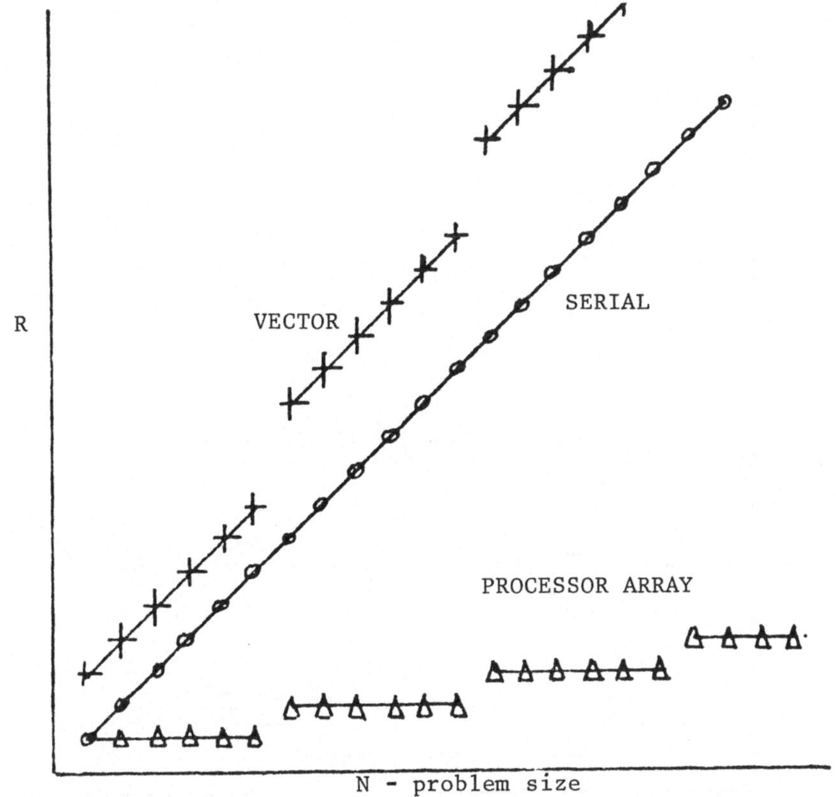

Figure 1. Time dependence of problems with system type.

In figure 1 we have plotted schematically the variation of R with N for three computer types. The example machines in figure 1 are not based on any real systems. For some real systems we find that

CRAY 1 $M_V = 64$, $q_V \sim 8$

CYBER 205 $M_V = \infty$, $q_V \sim 40$

DAP $M_P = 4096$

The range M_V and M_P is so great that the linear scales in Figure 1 do not allow the real Vector and Parallel systems to be easily represented. Figure 2 is obtained by drawing Figure 1 with logarithmic scales and using real machine data. Inspection of Figure 2 shows that the curves for vector processors differ from those for serial processors only for small values of N. The curve for the DAP has two very distinct segments with a linking section.

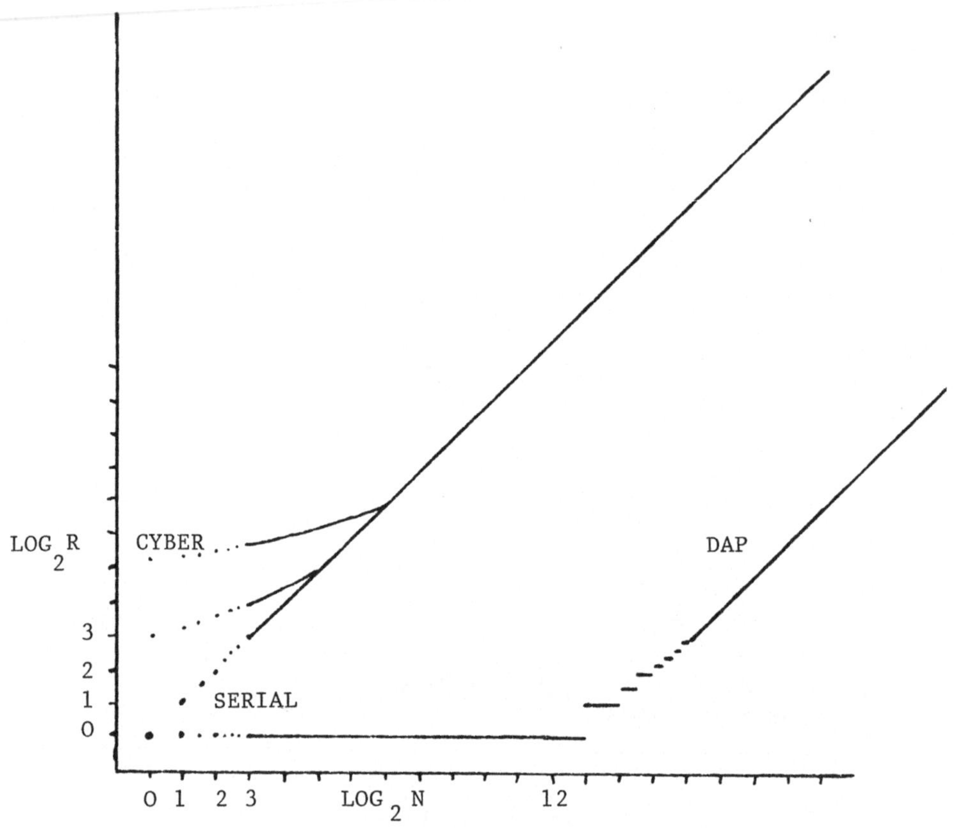

Figure 2. Time dependence of problems for real processor systems.

The horizontal section of the DAP curve is a truly parallel section with performance being independent of the size of the problem. For problems which fit into this range the classical operation count is invalid and many of our ideas as to what operations are optimal must be re-evaluated. The left hand branch of the curve is parallel to the serial curve and simply expresses the obvious statement that if problems require more operations than the computer perform simultaneously, then it is necessary to step through them serially in batches with size equal to capacity of the system. There is little new numerical or algorithmic theory arising from this serial-like branch of the curve so I shall concentrate on the parallel branch. It should be emphasised the following discussion has little relevance to pseudo-parallel systems such as vector processors.

THE SOLUTION OF SETS OF LINEAR EQUATIONS

As a simple example of the fundamental differences between parallel and sequential algorithms we can study the problem of solving N equations in N unknowns with one right hand side. For simplicity of presentation we ignore the pivoting operations recommended by numerical analysis. We shall assume that we have a processor array with N*(N+1) processors. The most commonly used method for solving linear equations must be the LU decomposition algorithm which solves the equations $A\,\tilde{x} = \tilde{b}$ by first decomposing A into the product form A=LU where L is a Lower Triangle matrix. The equations are then solved by first solving

$L\,\tilde{y} = \tilde{b}$ to evaluate $\tilde{y}$

then x is obtained from $U\tilde{x} = \tilde{b}$

As L and U are triangular forms the solution of the equations $L\tilde{y} = \tilde{b}$ and $U\tilde{x} = \tilde{b}$ is simple. The operation for the algorithm are well known. For our purpose it is only necessary to examine the leading terms in the operation counts and columns marked serial in Table 1 show operation counts for the various phases of the algorithms.

TABLE 1
Leading Terms in LU Decomposition Algorithm for Solving $A\,\tilde{x} = \tilde{b}$

PHASE	MULTIPLICATIONS
A = LU	N3/3
Solve $L\,\tilde{y} = \tilde{b}$	$N^2/2$
Solve $U\,\tilde{x} = \tilde{y}$	$N^2/2$
To solve for a single RHS	$N^3/3$
To solve for a new RHS	N^2

At this level of approximation the number of additions is identical to the number of multiplications and so their inclusion will not modify the conclusions. Time for execution of an algorithm on a serial machine is proportional to the operational count, so Table 1 demonstrates that there are benefits to be derived in obtaining the LU decomposition on serial machines. The benefits are greatest in solving sets of equations with different right hand sides. Another well known algorithm for solving sets of linear equations is the GAUSS-JORDAN algorithm. The GAUSS-JORDAN algorithm diagonalises the matrix by eliminating terms above and below the main diagonal simultaneously. The algorithm can be modified to produce the inverse of matrix A if required. Table 2 shows the operation counts for the various phases.

TABLE 2
Leading Terms in Solution of $A\underset{\sim}{x} = \underset{\sim}{b}$ by GAUSS-JORDAN Algorithm

PHASE	MULTIPLICATION
Solve $A\underset{\sim}{x} = \underset{\sim}{b}$	$N^3/3$
Compute A^{-1}	N^3
Solve for a new RHS using $x = A^{-1}b$	N^2

Comparing Table 1 and Table 2 we see that the GAUSS-JORDAN algorithm requiores 1.5 times as many operations as the LU decomposition algorithm and so is a slower algorithm for serial computers.

If we have a processor array with N^2 processors we can use the high parallelism to speed up the LU decomposition and the GAUSS-JORDAN algorithm. Both algorithms require that multiples of a row of a matrix are subtracted from other rows. And the GAUSS-JORDAN algorithm uses more of the processors than the LU algorithm. The total multiplications for solving $A\underset{\sim}{x} = \underset{\sim}{b}$, computiong A^{-1}, or LU is the same (N) so the three phases all take the same time. Solving $Ly = b$ on N^2 computers is not nearly such a convenient parallel operation and N steps (each with a multiplication and an addition are needed). Table 3 summarizes the parallel operation counts for all tasks covered in Tables 1 and 2.

TABLE 3
Parallel Operation Counts for Solving $A\underset{\sim}{x} = \underset{\sim}{b}$ on N^2 Computers

TASK	MULTIPLICATIONS	ADDITIONS
A - LU	N	N
Solve Ly = b	N	N
Solve Lx = y	N	N
Solve for a single RHS	2N	2N
Solve for a new RHS	2N	2N
GAUSS-JORDAN		
Solve $A\underset{\sim}{x} = b$	N	N
Compute A^{-1}	N	N
Solve for a new RHS given A^{-1}	1	$\mathrm{Log}_2 N$

Table 3 displays the qualitative differences between parallel computers and serial-like computers. The Table shows that given sufficient parallelism the equations $A\underset{\sim}{x} = \underset{\sim}{b}$ can be solved 'de novo' in less operations (ie. faster) than can the equations $LU\underset{\sim}{x} = b$, hence triangular forms seem to have no advantage over normal forms for parallel computers. This example has been covered in depth as it illustrates two important principles of parallel computing.

(1) The optimum algorithm for parallel computing is not always found by translating the optimum serial algorithm.

(2) It is important to study the problem whose solution is really needed (ie $A\underset{\sim}{x} = \underset{\sim}{b}$) rather than a derived problem (Ly = b)

SUMMARY

Parallel computing is a growing field of computer study, unfortunately the terminology used is not uniform and is used in contradictory and confusing fashions. 'Parallel' architectures hold great promise for both faster and cheaper computation but their proper exploitation will require a re-examination of our algorithmic methods and programming languages. This re-examination may be as fundamental as that which took place when hand driven calculators were replaced by digital computers.

PARALLEL NON-LINEAR ALGORITHMS

M. Feilmeier and W. Rönsch

Institut für Rechentechnik
Technische Universität Braunschweig, Pockelsstrasse 14
D-3300 Braunschweig, West-Germany

0. INTRODUCTION

Since 1950 the development of modern Numerical Analysis methods has been influenced primarily by the architecture of the currently available digital computers. Whereas the original concept was that of the von Neumann computer (uniprocessor system), we can see today the emergence of true multiprocessor systems - affected above all by the rapid progress of computer hardware. There is now an evergrowing interest in numerical algorithms which efficiently solve the respective problems on such machines. The numerical algorithms developed to exploit parallelism can be grouped together under the heading of 'Parallel Numerical Analysis'.

After introducing some fundamental concepts for analyzing parallel numerical algorithms (chapter 1), we shall illustrate the different starting points for parallelization using some examples (chapter 2). In chapter 3 the problem of stability of parallel algorithms, which has up to now not been the subject of much attention, is briefly discussed. Chapter 4 is dedicated to parallel non-linear algorithms for the solution of ordinary differential equations, single non-linear equations, non-linear systems of equations and optimization problems.

1. THE ANALYSIS OF PARALLEL ALGORITHMS

Algorithms can be estimated and analyzed using different criteria. In considering the methods of parallelization, the degree of time complexity T(n) of an algorithm is an important aspect. T(n) denotes the number of time steps necessary to compute the solution

K. G. Beauchamp (ed.), New Advances in Distributed Computer Systems, 379–393.

of a problem having n parameters. The time complexity essentially determines the limitations to the growth in the scale of the problem which can still be attempted by fast computers. (see Hoßfeld [13]).

Table 1

Time complexity $T(n)$	Number of parameters n					
	20	50	100	200	500	1000
$10^3 \cdot n$	0.02 sec	0.05 sec	0.1 sec	0.2 sec	0.5 sec	1 sec
$10^3 \cdot n \cdot \log n$	0.09 sec	0.3 sec	0.6 sec	1.5 sec	4.5 sec	10 sec
$10^2 \cdot n^2$	0.04 sec	0.25 sec	1 sec	4 sec	25 sec	2 min
$10 \cdot n^3$	0.08 sec	1.25 sec	10 sec	1.3 min	21 min	2.7 h
2^n	1 sec	35 a	$4 \cdot 10^{14}$ c	-	-	-
3^n	58 min	$2 \cdot 10^8$ c	-	-	-	-

(a = year, c = century, 1 time step = 10^{-6} sec)

Table 1 shows that, despite the rapid technological progress of the last few years, there still remain problems which exceed the capacity and performance of currently available computers. It is therefore an important task in 'Parallel Numerical Analysis' to look for algorithms which reduce time complexity and are at the same time numerically stable. (As an aside the progress in scientific computation since 1950 is due in equal measure to both faster computers and improved algorithms.)

Let $T_p(n)$ denote the number of time steps for a parallel algorithm on a parallel computer with $p > 1$ processors and $T_1(n)$ denote the number of time steps for the 'best' serial algorithm. Then

$$S_p(n) = \frac{T_1(n)}{T_p(n)} \leq p \quad \text{is termed the speed-up ratio and}$$

$$E_p(n) = \frac{S_p(n)}{p} \leq 1 \quad \text{is termed the efficiency of the}$$

parallel algorithm.

These notions are of more importance in the classification of parallel algorithms, but do not reveal very much about the actual time used to perform an algorithm (see Parkinson [20]).

2. METHODS FOR PARALLELIZATION OF NUMERICAL ALGORITHMS

There are essentially three different methods of parallelization:

(i) The starting point for constructing a parallel algorithm is the problem itself.

(ii) A known serial algorithm (before implementation) serves as the basis for parallelization.

(iii) An already implemented (serial) algorithm, i.e. the program code, is analyzed for potential parallelization.

It seems obvious that method (i) will provide the best result in exploiting the parallelism inherent in the problem. Nevertheless, the other methods have been developed and used successfully.

In the following, each of the three methods are illustrated by examples:

(i) Consider the triangulation of a matrix A:

$$A = L \cdot U \quad .$$

The computation of the elements of L and U reduces to a set of non-linear recurrence problems, which can only be parallelized by using certain tricks (see Feilmeier [7]). On the other hand, Evans/Hatzopoulos [4], Evans [5] analyzed a different decomposition, the so-called WZ-decomposition:

$$A = W \cdot Z \quad .$$

(The non-zero elements of the decomposition matrices W and Z have the shape of the letters W and Z respectively within these matrices). This decomposition leads in a quite natural manner to a parallel algorithm.

A second example is Swarztrauber's algorithm for the solution of tridiagonal linear systems. Swarztrauber [26] solved the problem by abandoning parallelization via a pivoting strategy, which seems hopeless, using instead Cramer's rule. Although his algorithm does not use a pivoting strategy, its numerical behaviour is nonetheless satisfactory. In contrast, the numerical stability of the other parallel algorithms known up to now can only be guaranteed for diagonal-dominant matrices.

(ii) This class of parallelization methods is the most extensive and best developed. There are many serial algorithms which contain vectorial elements, i.e. operations between vectors or matrices and vectors. Since the parallelization of these operations is straightforward, the terms 'Vectorization' and 'Parallelization' are often used synonymously. Nevertheless, vectorization is a special case of parallelization.

Many iterative methods - in particular iterative methods in Linear Algebra - are formulated in terms of matrices and vectors. But there are important iterative methods, for example Successive Over Relaxation (SOR), which require the application of certain tricks before parallelization becomes possible. Consider the problem of solving the tridiagonal linear system

$$A\cdot x = r \tag{2.1}$$

with

$$A\cdot x = \begin{bmatrix} a_1 & b_1 & & 0 \\ c_2 & a_2 & b_2 & \\ & \ddots & \ddots & \ddots \\ 0 & & c_n & a_n \end{bmatrix} \cdot \begin{bmatrix} x_1 \\ \\ \\ x_n \end{bmatrix} = \begin{bmatrix} r_1 \\ \\ \\ r_n \end{bmatrix} =: r \tag{2.2}$$

The SOR method

$$\begin{aligned} a_1x_1^{(k+1)} &= (1-\omega)a_1x_1^{(k)} & & - \omega b_1x_2^{(k)} + \omega r_1 \\ a_ix_i^{(k+1)} &= (1-\omega)a_ix_i^{(k)} - \omega c_ix_{i-1}^{(k+1)} & & - \omega b_ix_{i+1}^{(k)} + \omega r_i \\ a_nx_n^{(k+1)} &= (1-\omega)a_nx_n^{(k)} - \omega c_nx_{n-1}^{(k+1)} & & + \omega r_n \end{aligned} \tag{2.3}$$

$$i = 2,\ldots,n-1$$

seems to have no potential for parallelization. The clue to the solution of the problem is a permutation ("red-black ordering", "checkerboard-ordering", e.g. Young [31], Lambiotte/Voigt [14], Buzbee/Golub/Howell [3]):

$$PAP^Ty := \bar{A}y = z \iff$$

$$\underbrace{\left[\begin{array}{cccc|cccc} a_1 & & & 0 & b_1 & & & 0 \\ & a_3 & & & c_3 & b_3 & & \\ & & \ddots & & & \ddots & \ddots & \\ 0 & & & a_{2^k-1} & 0 & & c_{2^k-1} & b_{2^k-1} \\ \hline c_2 & b_2 & & 0 & a_2 & & & 0 \\ & c_4 & b_4 & & & a_4 & & \\ & & \ddots & \ddots & & & \ddots & \\ 0 & & & c_{2^k} & 0 & & & a_{2^k} \end{array}\right]}_{\bar{A}} \cdot \underbrace{\left[\begin{array}{c} x_1 \\ x_3 \\ \vdots \\ x_{2^k-1} \\ \hline x_2 \\ x_4 \\ \vdots \\ x_{2^k} \end{array}\right]}_{y} = \underbrace{\left[\begin{array}{c} r_1 \\ r_3 \\ \vdots \\ r_{2^k-1} \\ \hline r_2 \\ r_4 \\ \vdots \\ r_{2^k} \end{array}\right]}_{z} \tag{2.4}$$

(without loss of generality: $n := 2^k$)

Application of the SOR method to (2.4) yields

$$
\begin{aligned}
&a_1 \quad y_1^{(k+1)} = (1-\omega)a_1 \quad y_1^{(k)} \qquad\qquad\qquad - \omega b_1 \quad y^{(k)}_{\frac{n}{2}+1} + \omega z_1 \\
&a_{2i-1} y_i^{(k+1)} = (1-\omega)a_{2i-1} y_i^{(k)} - \omega c_{2i-1} y^{(k)}_{\frac{n}{2}+i-1} - \omega b_{2i-1} y^{(k)}_{\frac{n}{2}+i} + \omega z_i \\
&\qquad i = 2,\ldots,\tfrac{n}{2}
\end{aligned}
\tag{2.5}
$$

$$
\begin{aligned}
&a_{2i} \quad y^{(k+1)}_{\frac{n}{2}+i} = (1-\omega)a_{2i} \quad y^{(k)}_{\frac{n}{2}+i} - \omega c_{2i} \quad y_i^{(k+1)} - \omega b_{2i} y_{i+1}^{(k+1)} + \omega z_{\frac{n}{2}+i} \\
&a_n \quad y_n^{(k+1)} = (1-\omega)a_n \quad y_n^{(k)} - \omega c_n \quad y^{(k+1)}_{\frac{n}{2}} \qquad\qquad + \omega z_n \\
&\qquad i = 1,\ldots,\tfrac{n}{2}-1
\end{aligned}
\tag{2.6}
$$

Supposing $y_1^{(k)},\ldots,y_n^{(k)}$ are known, $y_1^{(k+1)},\ldots,y_{n/2}^{(k+1)}$ can be computed in parallel according to (2.5). (2.6) yields (in parallel!) $y_{n/2+1}^{(k+1)},\ldots,y_n^{(k+1)}$: It is possible to formulate the SOR method in terms of vectors of length n/2.

Parallelization by permutation also forms the basis of the "odd-even reduction" ("cyclic reduction", e.g. Lambiotte/Voigt [14], Madsen/Rodrigue [17]). The idea is to eliminate certain coefficients in the equations by elementary row transformations. By permutation we obtain a tridiagonal linear system of equations of order n/2.

For the sake of simplicity the method is now illustrated for n=8. Let R(k) denote the k-th row of Ax=r. The elementary row transformation

$$
R(2i) \to R(2i) - (c_{2i}/a_{2i-1}) * R(2i-1) - (b_{2i}/a_{2i+1}) * R(2i+1) \tag{2.7}
$$

eliminates references to odd numbered variables in the even numbered equations. We get a tridiagonal system of equations for x_2, x_4, x_6, x_8:

$$
\begin{aligned}
&-(c_{2i-1}c_{2i}/a_{2i-1})x_{2i-2} + (a_{2i} - b_{2i-1}c_{2i}/a_{2i-1} \\
&\qquad - b_{2i}c_{2i+1}/a_{2i+1})x_{2i} - (b_{2i}b_{2i+1}/a_{2i+1})x_{2i+2} \\
&\qquad = r_{2i} - (c_{2i}/a_{2i-1})r_{2i-1} - (b_{2i}/a_{2i+1})r_{2i+1} \quad ; \; i=1,2,3,4
\end{aligned}
\tag{2.8}
$$

$(c_1 := b_8 := 0)$

The repeated application of this reduction process leads to a single equation for x_8. We obtain the other unknowns by back substitution.

The general case is discussed in the papers cited above.

(iii) The parallelization methods introduced up to now cannot be applied to already implemented algorithms (i.e. programs) without changing existing software. It is therefore useful to look for parallelizations of programmed serial algorithms which may be performed automatically by computer. This is very important for the reuse of expensive serial software investment on parallel computers.

If one ignores the mathematical content, parallelism in ALGOL and FORTRAN programs are essential hidden in DO loops. This is the reason why compilers for vector computers (CYBER 203/205, Cray-1) transform serial DO loops into DO loops executable in parallel (see Lamport [15], 'Auto-vectorizer').

Another approach stems from the inspection of a serial compiler's output. This output may contain 'large' arithmetic expressions - formed at least to some extent, from a large number of small subexpressions - whereby these arithmetic expressions may be transformed into their corresponding parallel algorithms. This procedure is, however, riddled with problems owing to the numerical instability of the algorithms.

3. THE PROBLEM OF NUMERICAL STABILITY OF PARALLELIZATION METHODS

It certainly cannot be expected that the proven numerical stability of a serial algorithm will remain invariant after parallelization. This fundamental problem is now discussed using arithmetic expressions as example.

In the last few years, many algorithms for the parallelization of arithmetic expressions have been developed (see Segerer [24]). All these algorithms are based on the principle of tree hight reduction which can be illustrated as follows:

$$E = ((a_1 \cdot a_2 + a_3) \cdot a_4 + a_5) \cdot a_6 + a_7$$

$$\tilde{E} = (a_1 \cdot a_2 \cdot a_4 \cdot a_6 + a_3 \cdot a_4 \cdot a_6) + (a_5 \cdot a_6 + a_7)$$

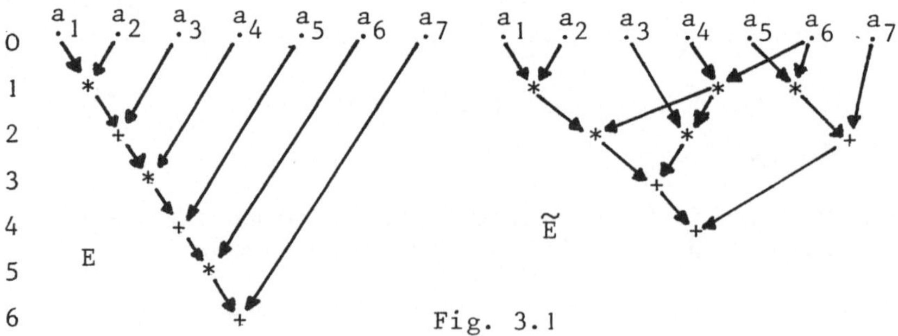

Fig. 3.1

E and $\widetilde{E}$ are equivalent, i.e. they can be transformed into one another by applying the associative, commutative and distributive laws.

We shall term an algorithm 'numerically stable' if the errors caused by inaccurate input data, assuming the arithmetic operations are performed exactly, have the same order of magnitude as the errors caused by rounding errors at every execution step, provided that the input data has already been given in machine number form. There are two methods of analyzing the numerical stability of an algorithm:

- backward analysis (see Wilkinson [27])
- forward analysis (see Larson/Sameh [16], Stummel [25]).

While backward analysis interprets the perturbed result of an arithmetic expression as the result of an exact evaluation with slightly changed input data, forward analysis is based on the introduction of so-called magnification factors. These factors can be derived from linearization:

$$x \to x(1+\varepsilon_x) \quad , \qquad y \to y(1+\varepsilon_y)$$

$$\varepsilon_{x\cdot y} = \varepsilon_x + \varepsilon_y \quad , \quad \varepsilon_{x/y} = \varepsilon_x - \varepsilon_y$$

$$\varepsilon_{x\pm y} = \frac{x}{x\pm y}\,\varepsilon_x \pm \frac{y}{x\pm y}\,\varepsilon_y \qquad \text{if } x\pm y \neq 0 \ .$$

Forward analysis has, in comparison with backward analysis, four definite advantages:

(1) Forward analysis can be applied to every arithmetic expression regardless of its structure.

(2) It allows an excellent measurement of all rounding errors which occur during the evaluation of the expression.

(3) It makes it possible to locate the occurrence of an instability exactly and therefore enables a relationship to be established between the structure of the expression and the instability.

(4) Forward analysis can be used to perform an automatic rounding error analysis.

An analysis of the Winograd algorithm without divisions (see Winograd [29]) using the backward analysis technique proves numerical stability of the algorithm (see Feilmeier/Segerer [6]). An analogous result was obtained by Brent in his algorithm for expressions without divisions (see Brent [2]).

There are examples which demonstrate that backward analysis may fail for an arithmetic expression with divisions (see Feilmeier/Segerer [6]). In these cases, forward analysis assists in proving the <u>instability</u> of the Winograd algorithm for expressions with divisions. The same <u>instability</u> result can be obtained for many other parallel algorithms developed for expressions with divisions.

The reason for instability does not lie in the presence of divisions, as the following theorem shows:

Theorem 3.1: Let E be an arbitrary arithmetic expression with distinct variables. Then the relative, accumulated rounding error R is bounded by the product

$$R \leq R_1 \cdot R_2 \quad , \qquad (3.1)$$

where R_1 denotes the relative error due to the most inaccurate input data which can occur and R_2 (≥ 1) denotes the number of plus and minus operations in E.

Proof: (see Feilmeier/Gomm/Rönsch/Töpfer [8]). □

Remark: The estimation (3.1) is based on the linearization method and is therefore correct only to the first order. The factor R_2 can considerably reduced if all the sums in the expression E are evaluated by recursive doubling; on the other hand, recursive doubling by products will yield no improvement in R_2.

It turns out that the reason for the instability of expressions with divisions is that divisions and variables, which occur more than once in the parallelized expression, appear together. This fact has to be heeded in developing a parallelization method for arithmetic expressions. This means that maximal complexity and numerical stability are in conflict with each other; both cannot be realized simultaneously. The price for numerical stability must be paid for by abandoning the maximum possible complexity.

A parallelization method for arithmetic expressions with divisions, in which each variable occurs only once, can be found in the DFG Report 'Parallele Numerik' (see Feilmeier/Gomm/Rönsch/Töpfer [8]). There it is proved that this method even transforms a certain class of arithmetic expressions having variables appearing more than once in a stable manner. Experimental results support the theoretical predictions. Since this parallelization method does not change the number of the binary arithmetic operations, it would appear well-suited to vector computers such as the Cray-1. For a detailed discussion of all these results see the above-mentioned DFG Report.

It can be shown through examples that the brutal force method using double precision numbers for the evaluation of arithmetic expressions transformed by unstable parallelization methods will not render the above considerations irrelevant, as the reason for instability is not the number of digits in the mantissa, but rather its finiteness. Thus evaluations in double precision mode will not provide a satisfactory solution to instability problems.

4. PARALLEL NON-LINEAR ALGORITHMS

We shall consider three kinds of parallel non-linear algorithms

- Numerical solution of ordinary differential equations (ODE)
- Determination of zeros for functions of one or more variables
- Optimization of parameters

These fields are important in practice and play an indispensable role as a tool in non-linear problems, e.g. partial differential equations (P.D.E.). Nevertheless, the efficiency of such parallel methods may be limited.

ODEs: Block Methods

Consider the equation (for simplicity we restrict our description to $\mathbb{R}^1$):

$$y' = f(x,y) \; ; \quad y(x_o) = y_o \; . \tag{4.1}$$

The set of all new function values which are evaluated during each application of the iteration formula is called a block. For each k-point block, k new function values are evaluated simultaneously. We shall discuss one-step block methods for SIMD-type computers (i.e. methods using only the last point of a block for the next block).

Let y_n be the approximate solution of the given initial value problem at x_n, the initial point of the actual block. The points of a block are assumed to be equidistant: $x_{n+r} = x_n + r \cdot h$; $r = 1,\ldots,k$.

Rosser [22] has developed serial formulae, e.g. of order 4 for a 2-point block. The corresponding parallel formulae are (see Worland [30]):

$$\begin{aligned}
y_{n+r,0} &= y_n + r \cdot h \cdot y'_n && r = 1,2 \\
y_{n+1,s+1} &= y_n + \frac{h}{12}(5y'_n + 8y'_{n+1,s} - y'_{n+2,s}) && \Big\} \; s = 0,1,2 \\
y_{n+2,s+1} &= y_n + \frac{h}{3}(\, y'_n + 4y'_{n+1,s} + y'_{n+2,s}) &&
\end{aligned} \tag{4.2}$$

On comparing the amount of time spent in function evaluations using both methods, we obtain a ratio of 5:3, if the simultaneous evaluation of $y'_{n+1,s}$ and $y'_{n+2,s}$ in the parallel case is taken into account.

By increasing the number of points in a block, it is possible to increase the order of the integration method: each point x_{n+r} of a block is attached to one of the processors.

ODEs: Multirate Methods (see Gomm [11])

The above-mentioned block methods are implicit methods. However, many practical problems stem from the real-time simulation of dynamic systems. In this case, the real-time constraints usually allow only one functional evaluation at each integration step. The integration step size h can be so small that its computation time becomes larger than that of the real-time step. To be more specific, this situation frequently occurs with "stiff" systems of equations. Here, those eigenvalues of the system which have large negative real parts have only a marginal influence on the total solution. However, they impose - as is well known - a small step size h. The numerical solution of the non-stiff components, computed with this small step size h, are often unnecessarily exact for many applications. If it is possible to divide the system of differential equations into two subsystems, a "stiff and a "non-stiff" subsystem

$$y' = f(t,y) = \begin{bmatrix} s(x,z,t) \\ g(x,z,t) \end{bmatrix} = \begin{bmatrix} x' \\ z' \end{bmatrix} \; ; \; y_o = \begin{bmatrix} x_o \\ z_o \end{bmatrix} , \tag{4.3}$$

then the so-called "multi-rate" method can be used efficiently. This method consists of integrating the stiff subsystem with a shorter step size than the non-stiff subsystem.

$$\begin{aligned} x_{n\cdot k+i+1} &= M(x_{n\cdot k+i}, t_{n\cdot k+i}, h \;, \hat{s}), \; i = 0,\ldots,k-1 \\ z_{(n+1)k} &= M(z_{n\cdot k} \;, t_{n\cdot k} \;, k\cdot h, g) \end{aligned} \tag{4.4}$$

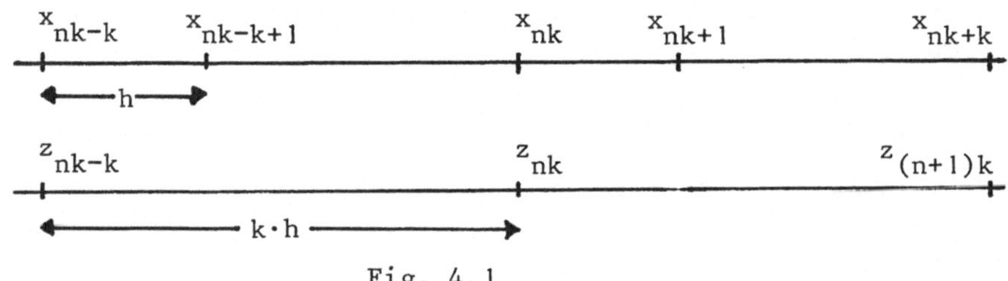

Fig. 4.1

(h is the integration step size for the stiff subsystem, k·h the integration step size for the non-stiff subsystem, $k \in \mathbb{N}$ the multirate factor, M a well-known integration method and $\hat{s} = s(x,\hat{z},t)$, where $\hat{z}$ is an approximation of z).

A particular approximation $\hat{z}$ of the z-values is necessary in evaluating the stiff subsystem s at the points $t_{n\cdot k+i+1}$; $i = 0,\ldots,k-1$. At these points, the values of z are not computed by an integration procedure but rather by using extrapolation polynomials from the values z_{nk-ik} and $g(x_{nk-ik}, z_{nk-ik}, t_{nk-ik})$, $i = 0, 1, \ldots, k-1$.

Under real-time constraints, multi-rate methods require very painstaking and difficult programming on a serial computer (see Gear [9]). However, even if the difficult problem of programming has been achieved, the serial computer may still prove to be too slow. This fact, together with economic considerations, provides a reason for the parallel evaluation of the multi-rate methods on MIMD computers: each of the different time stages - one time stage corresponds to one step size - is associated with one or more processors.

A successful use of parallel multi-rate methods is only possible after making a complete analysis of the error and stability properties of the multi-rate methods. Gomm [10],[11] has developed a method of constructing the stability polynomials of multi-rate versions of linear multi-step methods. He has made an extensive stability analysis of multi-rate Adams-Bashforth methods. Gomm's analysis shows that stiff systems with a weak coupling between the stiff and the non-stiff subsystem can be solved using multi-rate methods almost as well as with conventional ones, whilst taking less computation time.

Single Non-Linear Equations

The simplest serial search method is the bisection method, which clearly can be performed in parallel: for each iteration step, the function is simultaneously evaluated at r equidistant points. The new interval points are chosen on the basis of the signs of the function values. The bisection method needs $\log_2 \frac{1}{\varepsilon}$ function evaluations to enclose the zero in an interval of length ε, whereas the parallel method only requires $\log_{r+1} \frac{1}{\varepsilon}$ evaluations. The speed-up ratio is therefore

$$S = \frac{\log_2 \frac{1}{\varepsilon}}{\log_{r+1} \frac{1}{\varepsilon}} = \log_2 (r+1) \quad .$$

With functions having a high degree of smoothing, methods can be constructed which will converge yet faster (see Miranker [18]).

Locally iterative methods have been discussed by different authors. The limitations of these methods are formulated in a theorem of Winograd [28]: the order of convergence of every locally iterative method using r processors to evaluate simultaneously function values and derivatives up to order d is bounded by $1 + (d+1)r$. Since this theorem includes the serial case ($r = 1$), the speed-up ratio, which can obtained directly, is

$$S = \frac{\log_{d+2} \log \frac{1}{\varepsilon}}{\log_{1+(d+1)r} \log \frac{1}{\varepsilon}} = a \cdot \log_{d+2} (1+(d+1)r)$$

and this value is far removed from the best case S = r. In addition, it is worth mentionig that most parallel algorithms for finding the zero of a function of one variable are designed for SIMD computers. This situation is different with

Non-Linear Systems of Equations

For multi-dimensional functions, it would appear advantageous to use MIMD computers whenever available.

Most synchronous methods can be analyzed by means of a fixed point theorem. This makes the mathematical reasoning much easier. Nevertheless, the time needed by each processor to do its task in such an algorithm may vary markedly. The best solution may therefore involve using the MIMD computer in an asynchronous manner.

Asynchronous iteration is the asynchronous version of şucsessive approximation. This iteration consists of a sequence x^j, j = 1,2,... of vectors of $\mathbb{R}^k$. The sequence is defined by

$$x_i^j := \begin{cases} x_i^{j-1} & \text{for } i \notin J_i \\ f_i(x_1^{s_1(j)}, \ldots, x_k^{s_k(j)}) & \text{for } i \in J_i \end{cases} ,$$

with a startingvector x^o. $J = \{J_j | j = 1,2,\ldots\}$ is a sequence of non-empty subsets of $\{1,2,\ldots,k\}$ and $S = \{(s_1(j),\ldots,s_k(j)) | j=1,2,.$ a sequence of elements of $\mathbb{N}^k$.

J and S must satisfy the following conditions for all i = 1,2,...,k:

- $s_i(j) \leq j-1$ for all j = 1,...,k
- $s_i(j) \to \infty$ holds for $j \to \infty$
- Each i occurs an infinite number of times in the sets J_j.

A statement about the convergence of the asynchronous iteratio has been made by Baudet [1]. In addition, further assertions can be made concerning the speed of convergence. The number R of single iterations (a single iteration is one iteration of a single component), which are necessary to reduce the error by the factor 10, can be estimated from an upper bound

$$R \leq - \frac{2 \sum_{i=1}^{k} \left\lfloor \frac{n_{max}}{n_i} \right\rfloor - 1}{\log \rho(A)} ,$$

where $\rho(A)$ is the spectral radius of the Lipschitzmatrix A and n_i is the number of time steps during which the i-th component of the function f_i can be evaluated. (see Feilmeier/Gomm/Rönsch/Töpfer [8] and Miranker [19]).

Linear Programming

In approaching the general solution to the linear optimization problem, we first of all refer to Gründler [12], who has implemented the well-known simplex algorithm on a Siemens SMS 201. The effective savings in computation time gained by this algorithm are not given. It seems that the speed-up ratio will probably grow linearly with the number p of modules. Some proposals for the true parallelization of simplex, duoplex and multiplex algorithms can be found in Prause [21]. In addition, this paper tackles the minimization of unimodal functions without constraints and the general non-linear optimization problem without constraints. Prause summarizes many different papers, reports on some calculations and gives additional hints for possible solutions.

REFERENCES

1. Baudet, G.M., *Asynchronous Iterative Methods for Multiprocessors*, J. ACM 25,(1978) pp. 226-244
2. Brent, R.P., *The Parallel Evaluation of Arithmetic Expressions in Logarithmic Time*, in: Traub, J.F. (ed.), Complexity of Sequential and Parallel Numerical Algorithms, Academic Press, 1973, pp. 83-102
3. Buzbee, B.L./Golub, G.H./Howell, J.A., *Vectorization of the Cray-1 of Some Methods for Solving Elliptic Difference equations*, in: Kuck, D.J./Lawrie, D.H./Sameh, A.H. (eds.), High Speed Computer an Algorithm Organization, Academic Press, 1977, pp. 255-271
4. Evans, D.J./Hatzopoulos, M., *A Parallel Linear System Solver*, Intern. J. Comp. Math. 7,3(1979) pp. 227-238
5. Evans, D.J., *Parallel Numerical Algorithms for Linear Systems*, in: Evans, D.J. (ed.), Parallel Processing Systems, Cambridge University Press, 1980 (to appear)
6. Feilmeier, M./Segerer, G., *Numerical Stability in Parallel Evaluation of Arithmetic Expressions*, in: Feilmeier, M. (ed.), Parallel Computers - Parallel Mathematics, North-Holland Publ. Co., 1977, pp. 107-112
7. Feilmeier, M., *Parallel Numerical Algorithms*, in: Evans, D.J. (ed.), Parallel Processing Systems, Cambridge University Press, 1980 (to appear)
8. Feilmeier, M./Gomm, W./Rönsch, W./Töpfer, K., *Parallele Numerik*, DFG-Bericht, Institut für Rechentechnik, Technische Universität Braunschweig, März 1981
9. Gear, C.W., *Conflicts between Realtime and Software*,

in: Rice, J.R. (ed.), Mathematical Software III, Academic Press, 1977

10. Gomm, W., *Stabilitätsuntersuchungen von expilziten Multirate-Methoden zur numerischen Lösung von Anfangswertproblemen bei gewöhnlichen Differentialgleichungen*, Dissertation, Technische Universität Braunschweig, 1980
11. Gomm, W., *Stability Analysis of Explicit Multirate Methods*, Math. Comp. Simul. 23,1(1981) pp. 34-50
12. Gründler, D., *Die Lösungsmöglichkeiten von Operations-Resarch-Problemen mit einer Parallelrechnerstruktur und Implementierung des Simplex-Algorithmus als Beispiel*, Diplom-Arbeit, Technische Universität München, 1979
13. Hoßfeld, F., *Parallelverarbeitung - Konzepte und Perspektiven*, Angew. Informatik 22,12(1980) pp. 485-492
14. Lambiotte, J.J./Voigt, R.G., *The Solution of Tridiagonal Linear Systems on the CDC STAR-100 Computer*, ACM Trans. Math. Softw. 1,(1975) pp. 308-329
15. Lamport, L., *The Parallel Execution of DO Loops*, C. ACM 17,2 (1974) pp. 83-93
16. Larson, J./Sameh, A., *Efficient Calculation of the Effects of Rounding Errors*, ACM Trans. Math. Softw. 4,(1978) pp. 228-236
17. Madsen, N.K./Rodrigue, G.H., *Odd-Even Reduction of Pentadiagonal Matrices*, in: Feilmeier, M. (ed.), Parallel Computers - Parallel Mathematics, North-Holland Publ. Co., 1977, pp. 103-106
18. Miranker, W.L., *Parallel Search Methods for Solving Equations*, in: Feilmeier, M. (ed.), Parallel Computers - Parallel Mathematics, North-Holland Publ. Co., 1977, pp. 9-15 and in: Math. Comp. Simul. 20,2(1978) pp. 93-101
19. Miranker, W.L., *Hierarchical Relaxation*, Computing 23,3(1979) pp. 267-285
20. Parkinson, D., *Practical Parallel Processors and their Uses*, in: Evans, D.J. (ed.), Parallel Processing Systems, Cambridge University Press, 1980 (to appear)
21. Prause, D., *Parallelisierung von Optimierungsproblemen*, Diplom-Arbeit am Institut für Rechentechnik, Technische Universität Braunschweig, Jan. 1980
22. Rosser, J.B., *A Runge-Kutta for all Seasons*, SIAM Rev. 9, 3(1967) pp. 417-452
23. Schendel, U., *Einführung in die parallele Numerik*, R. Oldenbourg Verlag, 1981
24. Segerer, G., *Numerische Qualität paralleler Algorithmen*, in: Parallele Datenverarbeitung und parallele Algorithmen, Technische Universität Berlin, Brennpunkt Kybernetik, 1979, pp. 51-67
25. Stummel, F., *Perturbation Theory for Evaluation Algorithms of Arithmetic Expressions*, 1979 (to appear in Math. Comp.)
26. Swarztrauber, P.N., *A Parallel Algorithm for Solving General Tridiagonal Equations*, Math. Comp. 33,145(1979) pp. 185-199
27. Wilkinson, J.H., *Rounding Errors in Algebraic Processes*, Prentice Hall, 1963

28. Winograd, S., *Parallel Iterative Methods*, in: Miller, R.E./ Thatcher, J.W. (eds.), Complexity of Computer Computations, Plenum Press, 1972, pp. 53-60
29. Winograd, S., *On the Parallel Evaluation of Certain Arithmetic Expressions*, J. ACM 22,4(1975) pp. 477-492
30. Worland, P.B., *Parallel Methods for the Numerical Solution of Ordinary Differential Equations*, IEEE Trans. Comp. C-25, 10(1976) pp. 1045-1048
31. Young, D., *Iterative Solution of Large Linear Systems*, Academic Press, 1971

MICROCOMPUTER NETWORKS FOR PROCESS CONTROL

S. M. Prince

Department of Computing, Imperial College, London.

There are many advantages in the use of a distributed network of microcomputers for process control applications, including fault tolerance, local intelligence, and incremental expansion. This paper presents a communication system architecture for a Distributed Computer Control System (DCCS). It is based on interconnected subnetworks, allowing both rings and serial highways to be used for the subnetworks. Section 2 describes the communication requirements of a DCCS. Section 3 presents the topology, and section 4 the architecture of the communication system. Protocols are dicussed in section 5.

1. INTRODUCTION

The current trend in process control systems is towards the use of networks of microcomputers. The low cost of microcomputers makes it possible to use one for each function or device. A Distributed Computer Control System (DCCS) has several advantages over a centralised system. The effects of faults can be localised so they will not affect the rest of the system. Redundancy can be provided without having to duplicate the entire system. Local intelligence reduces communication needs and allows a stand-alone capability in case of failure of the communication system. The use of Local Area Network (LAN) technology considerably reduces the cabling required, and makes it easier to extend and reconfigure the system.

These advantages are offset by the need for communications software and hardware, which form a major part of the DCCS. The loose coupling between stations results in relatively long

K. G. Beauchamp (ed.), New Advances in Distributed Computer Systems, 395–404.

and variable delays, with the possibility of errors.

This paper presents a communication system architecture for a DCCS. The next section discusses the characteristics of a DCCS, which differ in important ways from other LAN applications such as resource sharing and office automation. A network topology is then described based on interconnected subnetworks. The Architecture presented in section 4 is a simplification of the ISO seven layer reference model. Section 5 discusses suitable protocols for the application.

To avoid confusion the term 'process' is applied only to the physical process being controlled. The term 'task' is used to refer to a software or hardware module performing a particular function. It corresponds to the term 'process' as used in computing science. A 'station' consists of one or more processors supporting both application tasks and communication functions.

2. CHARACTERISTICS OF A DCCS.

Typical applications for a DCCS include (2-4):

Chemical processes
Steel production
Paper industry
Food industry
Oil refining
Power generation and distribution
Mine automation
Laboratory automation
Building environment monitoring

There is considerable variation in requirements between different applications, and even within applications, but the following are common to all DCCS applications (5):

2.1. Communication Delays.

Communications concerned with direct control of the plant, such as between a controller and an actuator, are very time critical. Communications delays usually have to be small, with a predictable maximum delay. Communication at this level is usually between stations that are close together. Higher level functions, such as monitoring and supervisory control are less time critical. Delays can be longer and more variable, but a well defined maximum delay is still required.

In order to provide short, predictable response times the

network should be operated well below full load. Line utilisation and throughput are not important since private high bandwidth lines will be used.

2.2. Safety: Integrity and Availability.

Error checking is particularly important in process control because a corrupt message could cause a serious accident, possibly with loss of life, if accepted as correct.

For safety reasons a very high high level of availability is required. Failures should be localised, and should be detected and reported quickly. There should be facilities for bypassing faults, and for fast, remote diagnosis of faults. This is necessary because stations may be inaccessible.

The hardware itself must also meet the safety standards of any application it is used for. For example it may have to operate in the potentially explosive atmosphere of a mine or a chemical plant.

2.3. Error Rates.

A DCCS may have to operate in a noisy or hostile environment. Line error rates are likely to be higher than for other applications. Noise may be caused by power switching, or by the presence of large magnetic or electric fields. In some cases it may be necessary to use fibre-optics or infra-red because of their immunity to such interference.

2.4. Stations.

The trend is towards very simple stations each performing a single function, such as interfacing a sensor or actuator. There may be large numbers of such stations in a major installation, so they should be inexpensive.

2.5. System Configuration.

Control systems are usually static over months or years and do not need to create tasks dynamically or support task migration. Most of the communication traffic is due to regular polling so traffic patterns are predictable. However alarm situations can give rise to a sudden increase in traffic, and the network must not fail at such times. This is another reason for operating the network well below full capacity (see 2.1).

2.6. Types of Traffic.

Most of the traffic in a DCCS takes the form of short, high

priority messages, usually with responses. These include commands and responses, synchronisation signals, service requests, and alarms.

There are also logging messages and file transfers (including program transfers). Logging messages do not require short response times. The information is statistical so that loss of some logging messages is acceptable. File transfers are also of low priority and would be handled by system tasks. These tasks would have to transfer the file as a series of short blocks, reassembling it at the destination.

3. NETWORK TOPOLOGY.

LAN technology offers either ring or serial highway based topologies. Neither has any clear advantage for process control. Serial highways are potentially more reliable because the highway is passive. They also have the advantage of being fairly easy to tap into. However the length of a highway is limited unless repeaters are used, and as it gets longer propagation delays become significant. Rings are susceptible to failure of the ring interface which can disrupt all communication, but reliability can be improved by the use of bypass switches, a parallel ring, or a braided ring. Size is not limited since each ring interface regenerates the signal, but as the number of stations is increased the total delay round the ring is increased.

Since rings and highways are shared media some form of access control mechanism is required. These fall into two categories: polled and contention. Contention based systems such as Ethernet (6) and the register insertion ring of DLCN (7) are fairly simple to implement. Access control is fully distributed so there are no special stations. The statistical nature of contention techniques makes them unpopular for time critical communications, but they can be used for supervisory and management communication. At low loads contention systems can give very good performance with minimal overheads (8).

Access control by polling each station in turn has the advantage of well defined timing. An upper limit can always be placed on the poll cycle time, assuming each station tranmits only one frame per poll. This means polling is the preferred method for time critical functions. Polled rings are simple to implement because of the natural ordering of the stations. Rings can be polled using either a circulating token or fixed size slots (9). Polled highways are much more complex and incur a large overhead for polling messages. The IEC has developed a standard for a polled serial highway for process control called

PROWAY (10), which will probably use token passing (hub polling). The main problem with polling techniques is that they are partially or wholly centralised. One station has to be designated to initiate and control the polling operation. In some cases this station is necessary only for start-up and to monitor the highway for failures. In others it is vital to the operation of the highway. Ensuring there is always one and only one such station can be a complex problem.

Point to point lines, although not normally associated with local area networks, have been used for process control where communication is not too time critical. An example is TOPSYNET (11), a mesh network carrying supervisory messages. Although point to point lines offer lower delays because they are not shared, they usually have to be used with store and forward routing. A message may have to travel via several intermediate stations, each hop increasing the end to end delay. However point to point lines can be useful where two stations need a direct, low delay connection.

Local area networks are limited to around 100 to 200 stations by electrical factors, delays, and the capacity of the shared medium. This limitation can be overcome by interconnecting multiple rings or serial highways through store and forward stations called bridges (12). An advantage of this is that the subnetworks are independent so faults in one do not affect others.

This topology corresponds to the structure of process control systems and forms the basis of the proposed DCCS network architecture. Each unit of the process is fairly self contained, controlled by a cluster of stations. These stations would be connected to a single subnet across which short, bounded delays are required. The cluster communicates with other groups only at a supervisory level, so that inter-subnet communication can be slower. To allow flexibility in the configuration of the communication system the subnets can be of any type, ring or serial highway, contention or polled, as suits the particular application. The choice will be influenced as much by cost, availability, and political factors as by suitability. Point to point lines can be used to interconnect bridge stations where a low delay path is required. Future expansion is not restricted by the initial choice of topology, and new technologies can be incorporated as available.

4. COMMUNICATION SYSTEM ARCHITECTURE.

The architecture is based on the ISO Open Systems Interconnection reference model, simplified to five layers:

application, transport, network, link, and physical (13,14). The functions within these layers are kept as simple as possible because of the need to achieve low delays while using microprocessors.

4.1. The Physical Layer.

This layer consists of the communication medium and any medium dependent functions, such as modulation and demodulation. It presents the link layer with transmission of bit streams between stations on a common medium.

4.2. The Link Layer.

This layer is concerned with intra-subnet communication. It handles access control and error detection. It provides a standard highway interface to the Network Layer which is independant of the actual type of highway used. Error recovery may be provided, but is not necessary it this level. Most messages are likely to be between stations in the same subnet, so that end to end error control would duplicate link level error control. Where messages travel outside the subnet they are generally supervisory in nature and less time critical. Omitting link level error recovery considerably reduces the complexity of a station. If a link is particularly error prone forward error correction can be used. Note that error detection (using cyclic redundancy checks or similar mechanisms) must always be provided. If there is no error recovery corrupt frames are just ignored.

4.3. The Network Layer.

This layer handles inter-subnet routing, providing the Transport Layer with an unreliable datagram service between any two stations in the network. Each packet carries a destination address of the form:

<subnet><station>.

The Network Layer routes the packet to the destination subnet and it is then delivered directly to the destination station.

For simplicity static routing tables should be used. These can contain alternative routes for initial recovery from faults. A Network Control Centre (NCC) can update routing tables as faults are detected and corrected, and when the network is modified.

4.4. The Transport Layer.

This layer provides end to end protocols for error control. It provides the Application Layer with message delivery at the required level of integrity. It may provide several levels of priority and integrity and possibly more than one protocol. This level interfaces to the application through the operating system, which provides the communication primitives as part of the system interface. Possible protocols are discussed in section 5.

4.5. The Application Layer.

The Application Layer consists of a set of independent modules (15,16). Each consists of one or more tasks and has a number of ports through which it can communicate with other modules. The module is the unit of design and compilation. Most of the modules in a DCCS are standard types, such as a three-term controller or a sensor interface. A module may be instantiated many times at different stations. Each module instance resides at one station and cannot be split across stations. However there can be more than one module at each station.

Ports are separated into two types: entryports and exitports. This reflects the asymmetrical nature of application communication. Messages are sent from exitports to entryports, and responses may be returned. The exitport is always the master. Responses can be returned only in response to messages. The linkage between exitports and entryports is not restricted to one to one. Multiple exitports can be linked to one entryport (e.g. for a server) or an exitport can be linked to multiple entry ports (multi-destination messages).

The application software is viewed in two ways: the programmer's view and the system configuration view. The programmer is concerned only with writing modules to a specification. He does not need to know how the module is to be instantiated or which other modules each instance will be linked to. Therefore he uses local names for the module's ports, and the port specifications completely define the module interface.

The system configuration view is concerned with specifying modules to be written, instantiating modules, allocating them to stations, and setting up the linkage between them. Most of the linkage is static and is set up only when the system is created or modified. This is not under the control of the modules. Some ports will be linked dynamically while the system is running, as a result of system calls by the modules. This will be the case for servers. Dynamic linkage is also useful where one module communicates with too many others to maintain separate ports for all of them.

5. TRANSPORT LAYER PROTOCOLS.

The Transport Layer is the main protocol layer of the architecture. It should provide protocols oriented to the applications needs, yet should be simple and efficient. It can take advantage of the master/slave asymmetry of application communication to simplify the protocols used. Error recovery can be under the control of the master (exitport) end.

Name to address mapping is not a problem in a static, externally linked system. Each exitport normally gets the address of the corresponding entryport when linkage is set up at start-up. Dynamic linkage requires the address to be supplied to the task using the port. In most cases the address will be the source address from a received message.

The Transport layer can use either datagram (DG) or virtual circuit (VC) techniques. VC protocols are more efficient for reliable communication. This efficiency is achieved by maintaining permanent status information at each port. This information has to be set up by a three way control message exchange before the ports can communicate. For static links this overhead occurs only during start-up and reconfiguration, so it does not affect normal running. However for dynamic links the overhead may be significant, especially if only a few user messages are sent before closing the VC again.

Datagram techniques have the advantage of not requiring permanent status information. It is very easy to change linkage because there is no set-up step. This makes datagrams particularly useful for dynamic linkage, and speeds up reconfiguration. However datagrams can provide full reliability only by using extra control messages for each application message transfer. In most systems a large part of the traffic is concerned with reading data from input devices, such as sensors. For this it does not matter if an out of date read request is received. A simplified DG protocol can handle this case with the same number of messages as a VC. Logging messages and other statistical reports can be sent without any error recovery at all.

6. CONCLUSION.

The requirements that must be met by a communication system for a distributed computer control system differ in certain areas from other LAN applications. In particular bounded response times and high reliability are important, and short response times are usually required between local stations. A network architecture that meets these requirements has been presented. It is based on broadcast subnetworks interconnected by store and forward bridge stations. Subnetworks can use any LAN technology, ring or serial highway. Point to point lines can also be used to provide low delay connections. The architecture, based on the ISO reference model, is kept simple to allow implementation on microprocessor stations while keeping communication delays to a minimum.

REFERENCES.

(1) Syrbe, M.: 'Basic principles of advanced process control system structures and a realisation with distributed microcomputers'. IFAC 1978, pp. 393-401.

(2) 'Distributed computer control systems conference'. IEE Conf. Publ. 153, 1977.

(3) 'Centralised control systems'. IEE Conf. Publ. 161, 1978.

(4) Harrison, T.J. (Ed.): 'IFAC workshop on distributed computer control systems'. Pergammon Press, 1979.

(5) Prince, S.M. and Sloman, M.S.: 'Communication requirements of a distributed computer control system'. IEE proc., vol 128, pt. E, No 1, January 1981, pp. 21-34.

(6) Metcalfe, R.M. and Boggs, D.R.: 'Ethernet: distributed packet switching for local computer networks'. CACM, 1976, 19, pp.395-404.

(7) Reames, C.C. and Liu, M.T.: 'Design and simulation of the Distributed Loop Computer Network (DLCN)'. Compcon conf. on Distributed Processing, Sept 6-9, 1977, pp.3-25.

(8) Shoch, J.F. and Hupp, J.A.: 'Performance of an Ethernet local network: a preliminary report'. Proc. of the Local Area Network symposium, Boston, May 1979, pp. 113-125.

(9) Loomis, D.C.: 'Ring communication protocols'. Tech. Rep. #26, Dept. of Information and Computer Science, University of California, January 1973.

(10) IEC/65A (Secretariat) 18 /WG6: 'Process data highway (PROWAY) for distributed process control systems'.

(11) Bishop, P.G.: 'TOPSYNET'. CERL internal report.

(12) Clark, D.D., Pogran, K.T., and Reed, D.P.: 'An introduction to local area networks'. Proc IEEE, 1978, 66, pp. 1497-1517.

(13) ISO TC 97/SC 16 N537 Revised: 'Open system interconnection - Basic Reference Model'. Nov. 1980.

(14) Sloman, M.S. and Prince, S.M.: 'Local network architecture for process control'. IFIP WG6.4 Int. workshop on local area computer networks, 1980.

(15) Lister, A., Magee, J., Sloman, M., and Kramer, J.: 'Distributed Process Control Systems: programming and configuration'. Imperial College research report no. 80/12, May 1980.

(16) Kramer, J., Magee, J., and Sloman, M.: 'A software architecture for distributed computer control systems'. IFAC symposium on theory and applications of digital control, 5-7 January 1981, New Delhi, India.

LIST OF PARTICIPANTS

MR. G.S. ARROZ, Instituto Superior Tecnico, Complexo I do Inic, A V. Rovisco Pais, 1000 LISBOA, Portugal.

MR. R. BANNERJEE, University of Cambridge, Computer Laboratory, Corn Exchange Street, CAMBRIDGE CB2 3QG, U.K.

MR. R. BATES, Nils Lavritssønsv 20, OSLO 8, Norway.

DR. K.G. BEAUCHAMP, Computer Services Department, University of Lancaster, Bailrigg, LANCASTER LA1 4YW, U.K.

DR. D. BIRAN, Ministry of Communications, State of Israel, P O B 39250, TEL-AVIV 61390, Israel.

DR. H. BOTTENBURG, NATO Integrated Communications, System Management Agency, 8 Rue de Geneve, 1140 BRUXELLES, Belgium.

IR. J.M. VAN DEN BURG, Headquarters PTT, Directorate for Commercial Telecommunications, P O Box 30,000, 2500 GA THE HAGUE, Netherlands.

MR. J.W. BURREN, Science Research Council, Rutherford Laboratory, Chilton, Didcot, OXON. OX11 0QW, U.K.

MR. J.A. CARDOSO, Faculty of Science and Technology, University of Coimbra, 3000 COIMBRA, Portugal.

PROF. A.G. CERVEIRA, Universidade Nova de Lisboa, Seminario dos Olivais, 1899 LISBOA CODEX, Portugal.

DR. T. CHEUNG, Department of Computer Science, University of Ottawa, ONTARIO K1N 9B4, Canada.

DR. O. CIFTCIOGLU, Technical University of Istanbul, Electrical Engineering Faculty, Nukleer Guc. Kursusu, Teknik Universite, ISTANBUL, Turkey.

MR. N.J.P. COOPER, Department SES 1.1.1, British Telecom Research Labs., Martlesham Heath, IPSWICH, U.K.

DR. I. CNOP, Vrije Universiteit Brussel, WE Pleinlaan 2, B 1050 BRUSSELS, Belgium.

PROF. A. DANTHINE, Systems et Automatique, Institut d'Electricite, Montefiore B28, Universite de Liege, au Sant Tilman, B4000 LIEGE, Belgium.

MR. A. DRAKE, British Telecoms, Product Development Unit, PD2.4.1 Room 463, Proctor House, LONDON, U.K.

PROF. DR. M. FEILMEIER, Institut für Rechentechnik, Technische Universität Braunschweig, Pockelsstrasse 14, 3300 BRAUNSCHWEIG, W. Germany.

DR. J.E. GARDNER, Medical Physics Dept., College Hospital, 11-20 Capper Street, LONDON WC1E 6JA, U.K.

MR. G.P. GIRAUDBIT, SITA, 112 Ave Charles de Gaulle, 92522 NEVILLY S/SEINE, France.

MR. T. GJERTSEN, Norwegian Defence Research Establishment, Division of Electronics, P O Box 25, N-2007 KJELLER, Norway.

MR. H. GOLD, Prilaz JNA 17, 41000 ZAGREB, Yugoslavia.

MR. M. GROSS, Department Landscape, Architecture and Regional Planning, University of Massachusetts, Hills North, AMHERST MA 01003, U.S.A.

DR. M. HENNESSY, Department of Computer Science, University of Edinburgh, Kings Building, Mayfield Road, EDINBURGH EH9 3JZ, U.K.

DR. A.J. HINCHLEY, London Network Team, U L C C, 20 Guilford Street, LONDON WC1N 1DZ, U.K.

MR. O. HOLM, Nodeca Attn Capt. O BAE, Box 7020H, OSLO 3, Norway.

DR. A. HOPPER, University of Cambridge, Computer Laboratory, Corn Exchange Street, CAMBRIDGE, CB2 3QG, U.K.

DR. J. HOWLETT, 20b Bradmore Road, OXFORD, OX2 6QP, U.K.

DR. G. HUFF, IBM Corporation, 6905 Rainwater Road, Raleigh, N. CAROLINA 27609, U.S.A.

MR. C. HUITEMA, Project Pilote NADIR, c/o I N R I A, Roquencourt BP105, 78153 Le Chesnay, CEDEX, France.

DR. P.I.B. KING, Computing Laboratory, University of Newcastle-upon-Tyne, NEWCASTLE-UPON-TYNE, NE1 7RU, U.K.

ASST. PROF. DR. A. KIPER, Department of Computer Science, Middle East Technical University, ANKARA, Turkey.

MR. V. KUMAR, CSG Mathematics Faculty, The University, SOUTHAMPTON, SO9 5NH, U.K.

PROF. F.F. KUO, Department of Electrical Engineering, University of Hawaii, HONOLULU HI 968 22, U.S.A.

MR. J.E. LAMBERT, EAPS2, University of Sussex, BRIGHTON BN1 9QT, U.K.

MR. N. LANYON, Marcia Lanyon Limited, 34 Pembroke Gardens, LONDON W8 6HU, U.K.

DR. P. LAUER, Computing Laboratory, The University, Newcastle-upon-Tyne, NE1 7RU, U.K.

MR. I. LIE, Norwegian Defence Research Establishment, Division of Electronics, P O Box 25, N-2007 KJELLER, Norway.

MR. O.G. LOQUES, Department of Computing, Imperial College, LONDON SW7 2BZ, U.K.
MR. W.A. McCRUM, Govt. Canada Dept. of Communications, Room 1644, 300 Slater Street, Ottawa, ONTARIO K1A OC8, Canada.
DR. J. MAJITHIA, Computer Communications Network, Group CPH 2373G, Waterloo, ONTARIO N2L 3G1, Canada.
MR. A. MATTASOGLIO, C I L E A, Via R Sanzio 4, I - 20090 Segrate, MILAN, Italy.
MR. M. MELI, 1 Via G. Gozzi, 20129 MILAN, Italy.
MR. S. MIEGE, Service Informatique, Ecole Superieure d' Electricite, BP 20, 35510 CESSON-SEVIGRE, France.
MR. C.G. MILLER, British Telecom, Rm 1210, 151 Gower Street, LONDON SC1E 6BA, U.K.
MS. A.P. NOCK, 20 Pasquier Road, LONDON E17 6HB, U.K.
MR. H.J. NORTON, British Telecom (TE/SES) 3.3.2), Room 1215, 151 Gower Street, LONDON WC1E 6BA, U.K.
PROF. D. PARKINSON, ICL, 322 Euston Road, LONDON NW1 3BD, U.K.
DR. W.L. PRICE, National Physical Laboratory, Teddington, MIDDLESEX TW11 OLW, U.K.
MR. S.M. PRINCE, Post Graduate Research, Department of Computing, Imperial College, 180 Queensgate, LONDON SW7 2BZ, U.K.
PROF. J.B. RIERA, E T S I Telecommunication, Cindad Universitaria, MADRID 3, Spain.
MR. W. ROENSCH, Institüt für Rechentechnik, Technishe Universitat, Pockelstrasse 14, 3300 BRAUNSCHWEIG, W. Germany.
DR. L. ROESSING, Forschungsinstitute für Funk u Mathematik, Königstrasse 2, D5307 Wachtberg-Werthoven, W. Germany.
DR. R.A. ROSNER, Computer Board, Joint Network Team, Science Research Council, Rutherford Laboratory, Chilton, Didcot, OXON OX11 OQX, U.K.
DR. B. SARIKAYA, School of Computer Science, McGill University, 805 Sherbrooke Street W., MONTREAL, PQ H2A 2K6, Canada.
DIPL.ING. W.H.P. SCHMIDT, Shape Technical Centre, P O Box 174, 2501 CD THE HAGUE, Netherlands.
MR. P.L. SKAN, Room B22, Digital Processes Group, UMIST, Dept. of Electrical Engineering and Electronics, P O Box 88 Sackville Street, MANCHESTER M60 1QD, U.K.
MR. D.A. STEEDMAN, Dept. 3D23, Bell-Northern Research, P O Box 3511 Station 'C', Ottawa, ONTARIO, Canada.
PROF. J. VINAS, E T S I Telecommunication, Cindad Universitaria, MADRED 3, Spain.
DR. R.W. WILLIAMS, Dept. of Electrical Engineering, Votey Building, University of Vermont, Burlington, VERMONT, 05405, U.S.A.
GP. CPT. L. WING, I S D, B F P O 28.
DR. Y.S. WU, U.S. Office of Naval Research, 223 Old Marylebone Road, LONDON NW1 5TH, U.K.

INDEX